Theory and Practice of Mathematics

Theory and Practice of Mathematics

Editor: Victor Nason

NY Research Press
New York

Published by NY Research Press
118-35 Queens Blvd., Suite 400,
Forest Hills, NY 11375, USA
www.nyresearchpress.com

Theory and Practice of Mathematics
Edited by Victor Nason

International Standard Book Number: 978-1-63238-574-1 (Hardback)

Cataloging-in-Publication Data

Theory and practice of mathematics / edited by Victor Nason.
p. cm.
Includes bibliographical references and index.
ISBN 978-1-63238-574-1
1. Mathematics. I. Nason, Victor.
QA7 .T44 2018
510--dc23

Contents

Preface

The main aim of this book is to educate learners and enhance their research focus by presenting diverse topics covering this vast field. This is an advanced book which compiles significant studies by distinguished experts in the area of analysis. This book addresses successive solutions to the challenges arising in the area of application, along with it; the book provides scope for future developments.

Mathematics is the study of numbers, measurements, quantities and their correlation with each other. Arithmetic, algebra, geometry and analysis are the main branches of mathematics. Logic, uncertainty, probability, set theory, etc. are modern concerns that have become central to mathematics. This discipline is applied to diverse fields such as architecture, economics, etc. Some of the diverse topics covered in this book on mathematics address the varied branches that fall under this category. Students, researchers, experts and all associated with mathematics will benefit alike from this book.

It was a great honour to edit this book, though there were challenges, as it involved a lot of communication and networking between me and the editorial team. However, the end result was this all-inclusive book covering diverse themes in the field.

Finally, it is important to acknowledge the efforts of the contributors for their excellent chapters, through which a wide variety of issues have been addressed. I would also like to thank my colleagues for their valuable feedback during the making of this book.

Editor

On Diff(M)-Pseudo-Differential Operators and the Geometry of Non Linear Grassmannians

Jean-Pierre Magnot

Academic Editor: Lei Ni

Lyc'ee Jeanne d'Arc, Avenue de Grande Bretagne, F-63000 Clermont-Ferrand, France;
jean-pierr.magnot@ac-clermont.fr

Abstract: We consider two principal bundles of embeddings with total space $Emb(M, N)$, with structure groups $Diff(M)$ and $Diff_+(M)$, where $Diff_+(M)$ is the groups of orientation preserving diffeomorphisms. The aim of this paper is to describe the structure group of the tangent bundle of the two base manifolds:

$$B(M, N) = Emb(M, N)/Diff(M) \text{ and } B_+(M, N) = Emb(M, N)/Diff_+(M)$$

from the various properties described, an adequate group seems to be a group of Fourier integral operators, which is carefully studied. It is the main goal of this paper to analyze this group, which is a central extension of a group of diffeomorphisms by a group of pseudo-differential operators which is slightly different from the one developped in the mathematical litterature e.g. by H. Omori and by T. Ratiu. We show that these groups are regular, and develop the necessary properties for applications to the geometry of $B(M, N)$. A case of particular interest is $M = S^1$, where connected components of $B_+(S^1, N)$ are deeply linked with homotopy classes of oriented knots. In this example, the structure group of the tangent space $TB_+(S^1, N)$ is a subgroup of some group GL_{res}, following the classical notations of (Pressley, A., 1988). These constructions suggest some approaches in the spirit of one of our previous works on Chern-Weil theory that could lead to knot invariants through a theory of Chern-Weil forms.

Keywords: Fourier-Integral operators; non-liner Grassmannian; Chern-Weil forms; infinite dimensional frame bundle

---

1. Introduction

Given M and N two Riemannian manifolds without boundary, with M compact, the space of smooth embeddings $Emb(M, N)$ is currently known as a principal bundle with structure group $Diff(M)$, where $Diff(M)$ naturally acts by composition of maps. The base:

$$B(M, N) = Emb(M, N)/Diff(M)$$

is known as a Fréchet manifold, and there exists some local trivializations of this bundle. We focus here on the base manifold, which seems to carry a richer structure than $Emb(M, N)$ itself.

This paper gives the detailed description of the structure group of the tangent bundle of connected components of $TB(M, N)$. This structure group can be slightly different when changing of connected component of $B(M, N)$. It is viewed as an extension of the group of automorphisms $Aut(E)$ of a vector bundle E by some group of pseudo-differential operators. We show that this group is a regular Lie group (in the sense that it carries an exponential map), and that it is also a group of

Fourier integral operators, which explains the notations FIO_{Diff} and FCl_{Diff} ("Cl" for "classical"). All these groups are constructed along a short exact sequence of the type:

$$0 \to PDO \to FIO \to Diff \to 0$$

where PDO is a group of pseudo-differential operators, FIO is a group of Fourier integral operators, and $Diff$ is a group of diffeomorphisms; this sequence plays a central role in the proofs. The theorems described are general enough to be applied to many groups of diffeomorphisms: volume preserving diffeomorphisms, symplectic diffeomorphisms, hamiltonian diffeomorphisms, and to groups of pseudo-differential operators: classical or non-classical, bounded or unbounded, compact and so on, but we concentrate our efforts on $Diff(M)$ and $Diff_+(M)$, the group of orientation preserving diffeomorphisms. The constructions are made for operators acting on smooth sections of trivial or non trivial bundles. For a non trivial bundle E, the group of automorphisms of the bundle plays a central role in the description, because easy arguments suggest that there is no adequate embedding of the group of diffeomorphisms of the base manifold into the group of automorphisms of the bundle. Specializing to $M = S^1$, given a (real) vector bundle E over S^1, the groups $FIO_{Diff}(S^1, E)$ and in particular $FCl^{0,*}_{Diff_+}(S^1, E)$ is of particular interest, where $FCl^{0,*}_{Diff_+}(S^1, E)$ is defined through the short exact sequence:

$$0 \to Cl^{0,*}(S^1, E) \to FCl^{0,*}_{Diff_+}(S^1, E) \to Diff_+(S^1) \to 0$$

where $Cl^{0,*}(S^1, E)$ is the group of bounded classical pseudo-differential operators and $Diff_+(S^1)$ is the group of orientation-preserving diffeomorphisms. We have to notice that the necessary Fourier analysis on these operators naturally takes place in the complexification $E_{\mathbb{C}}$ of the vector bundle E, and that $E_{\mathbb{C}}$ as a complex vector bundle is trivial, but the real vector bundle E can be non trivial. Given any Riemannian connection on the bundle E, if ϵ is the sign of this connection (and this is a bounded pseudo-differential operators acting on smooth sections of E), it appears that $[FCl^{0,*}_{Diff_+}(S^1, E), \epsilon]$ is a set of smoothing operators. Thus, it is a subgroup of the group

$$Gl_{res} = \left\{ u \in Gl(L^2(S^1, E)) | [\epsilon, u] \text{ is Hilbert-Schmidt} \right\}$$

Even if the inclusion is not a bounded inclusion, this result extends the results given in [1] on the group: $Diff_+(S^1)$ (which inclusion map into Gl_{res} is not bounded too) and in [2] for the group $Cl^{0,*}(S^1, E)$. We get a non-trivial cocycle on the Lie algebra of $FCl^{0,*}_{Diff_+}(S^1, E)$ by the Schwinger cocycle, extending results obtained in [2,3] for a trivial complex bundle.

Coming back to $Emb(M, N)$, one could suggest that $Aut(E)$ is sufficient as a structure group, but we refer the reader to earlier works such as [4–6] to see how pseudo-differential operators can arise from Levi-Civita connections of Sobolev metrics when the adequate structure group for the L^2 metric is a group of multiplication operators. Moreover, especially for $M = S^1$, taking the quotient:

$$B_+(S^1, N) = Emb(S^1, N) / Diff_+(S^1)$$

we show that there is a sign operator $\epsilon(D)$, which is a pseudo-differential operator of order 0, and coming intrinsically from the geometry of $Emb(S^1, N)$, such that the recognized structure group of $TB_+(S^1, N)$ is $FCl^{0,*}_{Diff_+}(S^1, E) \subset Gl_{res}$. We finish with the starting point of this work, which was a suggestion of Claude Roger, saying that any well-defined Chern-Weil form of a connected component of $TB(S^1, N)$ can be understood as an invariant of a knot, whose homotopy class is exactly a connected component of $B(S^1, N)$. If one considers oriented knots, we get connected components of $B_+(S^1, N)$. The work begun has not been completely successful yet, but it is a pleasure to suggest some Chern-Weil froms that may lead to knot invariants, extending the approach of [6].

2. Preliminaries on Algebras and Groups of Operators

Now we fix M the source manifold, which is assumed to be Riemannian, compact, connected and without boundary, and the target manifold which is only assumed Riemannian. We note by $Vect(M)$ the space of vector fields on TM. Recall that the Lie algebra of the group of diffeomorphisms is $Vect(M)$, which is a Lie-subalgebra of the (Lie-)algebra of differential operators, which is itself a subalgebra of the algebra of classical pseudo-differential operators.

2.1. *Differential and Pseudodifferential Operators on a Manifold M*

Definition 1. Let $DO(M)$ be the graded algebra of operators, acting on $C^\infty(M,\mathbb{R})$, generated by:

- the multiplication operators: for $f \in C^\infty(M,\mathbb{R})$, we define the multiplication operator:

$$M_f : g \in C^\infty(M,\mathbb{R}) \mapsto f.g \text{ (by pointwise multiplication)}$$

- the vector fields on M: for a vector field $X \in Vect(M)$, we define the differentiation operator:

$$D_X : g \in C^\infty(M,\mathbb{R}) \mapsto D_X g \text{ (by differentiation, pointwise)}$$

Multiplication operators are operators of order 0, vector fields are operators of order 1. For $k \geq 0$, we note by $DO^k(M)$ the differential operators of order $\leq k$.

Differential operators are **local**, which means that:

$$\forall A \in DO(M), \forall f \in C_c^\infty(M,\mathbb{R}), supp(A(f)) \subset supp(f)$$

The inclusion $Vect(M) \subset DO(M)$ is an inclusion of Lie algebras. The algebra $DO(M)$, graded by the order, is a subalgebra of the algebra of classical pseudo-differential operators $Cl(M)$, which is an algebra that contains the square root of the Laplacian, and its inverse. This algebra contains trace-class operators on $L^2(M,\mathbb{R})$. Basic facts on pseudo-differential operators defined on a vector bundle $E \to M$ can be found in [7]. We assume known the definition of the algebra of pseudo-differential operators $PDO(M,E)$, classical pseudo-differential operators $Cl(M,E)$. When the vector bundle E is assumed trivial, *i.e.*, $E = M \times V$ or $E = M \times \mathbb{K}^p$ with $\mathbb{K} = \mathbb{R}$ or $\mathbb{C}$, we use the notation $Cl(M,V)$ or $Cl(M,\mathbb{K}^p)$ instead of $Cl(M,E)$. These operators are **pseudolocal**, which means that:

$$\forall A \in PDO(M,E), \forall f \in L^2(M,E), \text{ if } f \text{ is smooth on } K, \text{ then } A(f) \text{ is smooth on } K$$

Definition 2. A pseudo-differential operator A is **log-polyhomogeneous** if and only if its formal symbol reads (locally) as:

$$\sigma(A)(x,\xi) \sim_{|\xi| \to +\infty} \sum_{j=0}^{o} \sum_{k=-\infty}^{o'} \sigma_{j,k}(x,\xi)(log(|\xi|))^j$$

where $\sigma_{j,k}$ is a positively $k-$homogeneous symbol.

The set of log-polyhomogenous pseudo-differential operators is an algebra.

A global symbolic calculus has been defined independently by two authors in [8,9], where we can see how the geometry of the base manifold M furnishes an obstruction to generalize local formulas of composition and inversion of symbols. We do not recall these formulas here because they are not involved in our computations. More interesting for this article is to precise when the local formulas of composition of formal symbols extend globally on the base manifold.

We assume that M is equipped with charts such that the changes of coordinates are translations and that the vector bundle $E \to M$ is trivial. This is in particular true when $M = S^1 = \frac{\mathbb{R}}{2\pi\mathbb{Z}}$, or when $M = T^n = \prod_{i=1}^n S^1$. In the case of S^1, we use the smooth atlas $\mathcal{ATL}$ of S^1 defined as follows:

$$\begin{aligned} \mathcal{ATL} &= \{\varphi_0, \varphi_1\} \\ \varphi_n &: x \in]0; 2\pi[\mapsto e^{i(x+n\pi)} \subset S^1 \text{ for } n \in \{0;1\} \end{aligned}$$

Associated to this atlas, we fix a smooth partition of the unit $\{s_0; s_1\}$. An operator $A : C^\infty(S^1, \mathbb{C}) \to C^\infty(S^1, \mathbb{C})$ can be described in terms of 4 operators:

$$A_{m,n} : f \mapsto s_m \circ A \circ s_n \text{ for } (m,n) \in \{0,1\}$$

Such a formula is a straightforward application of a localization formula in the case of an atlas $\{\varphi_i\}_{i \in I}$ of a manifold M with associated family of partitions of the unit $\{s_i\}_{i \in I}$, see e.g., [7] for details.

Notations. We note by $PDO(M,E)$ (resp. $PDO^o(M,E)$, resp. $Cl(M,E)$) the space of pseudo-differential operators (resp. pseudo-differential operators of order o, resp. classical pseudo-differential operators) acting on smooth sections of E, and by $Cl^o(M,E) = PDO^o(S^1,E) \cap Cl(S^1,E)$ the space of classical pseudo-differential operators of order o.

If we set:

$$PDO^{-\infty}(M,E) = \bigcap_{o \in \mathbb{Z}} PDO^o(M,E)$$

we notice that it is a two-sided ideal of $PDO(M,E)$, and we define the quotient algebra:

$$\mathcal{F}PDO(M,E) = PDO(M,E)/PDO^{-\infty}(M,E)$$

$$\mathcal{F}Cl(M,E) = Cl(M,E)/PDO^{-\infty}(M,E)$$

$$\mathcal{F}Cl^o(M,E) = Cl^o(M,E)/PDO^{-\infty}(M,E)$$

called the algebras of formal pseudo-differential operators. $\mathcal{F}PDO(M,E)$ is isomorphic to the set of formal symbols [8], and the identification is a morphism of $\mathbb{C}$-algebras, for the multiplication on formal symbols defined before (see e.g., [7]). At the level of kernels of operators, a smoothing operator has a kernel $K_\infty \in C^\infty(M \times M, \mathbb{C})$, where as the kernel of a pseudo-differential operator is in general smooth only on the off-diagonal region $(M \times M) - \Delta(M)$, where $\Delta(M)$ denotes here, very exceptionnally in this paper, the diagonal set (and not a Laplacian operator). We finish by mentioning that the last property is equivalent to pseudo-locality.

2.2. Fourier Integral Operators

With the notations that we have set before, a scalar Fourier-integral operator of order o is an operator:

$$A : C^\infty(M, \mathbb{C}) \to C^\infty(M, \mathbb{C})$$

such that, $\forall (i,j) \in I^2$,

$$A_{k,j}(f) = \int_{supp(s_j)} e^{i\phi(x,\xi)} \sigma_{k,j}(x,\xi) (\widehat{s_j.f})(\xi) d\xi \tag{1}$$

where $\sigma_{k,j} \in C^\infty(supp(s_j) \times \mathbb{R}, \mathbb{C})$ satisfies:

$$\forall (\alpha, \beta) \in \mathbb{N}^2, \quad |D_x^\alpha D_\xi^\beta \sigma_{k,j}(x,\xi)| \leq C_{\alpha,\beta}(1 + |\xi|)^{o-\beta}$$

and where, on any domain U of a chart on M,

$$\phi(x,\xi) : T^*U - U \approx U \times \mathbb{R}^{dimM} - \{0\} \to \mathbb{R}$$

is a smooth map, positively homogeneous of degree 1 fiberwise and such that:

$$\det\left(\frac{\partial^2 \phi}{\partial x \partial_\xi}\right) \neq 0$$

Such a map is called **phase** function. (In these formulas, the maps are read on local charts but we preferred to only mention this aspect and not to give heavier formulas and notations.) An operator A is pseudo-differential operator if the operators $A_{k,l}$ in Formula (1) can be written as Fourier integral operators with $\phi(x,\xi) = x.\xi$. Notice that, in order to define an operator A, the choice of ϕ and $\sigma_{k,l}$ is not a priori unique for general Fourier integral operators. Let $E = S^1 \times \mathbb{C}^k$ be a trivial smooth vector bundle over S^1. An operator acting on $C^\infty(M,\mathbb{C}^n)$ is Fourier integral operator (resp. a pseudo-differential operator) if it can be viewed as a $(n \times n)$-matrix of Fourier integral operators with same phase function (resp. scalar pseudo-differential operators).

We define also the algebra of formal operators, which is the quotient space:

$$\mathcal{F}FIO = FIO/PDO^{-\infty}$$

which is possible because $PDO^{-\infty}$ is a closed two-sided ideal. When we consider classical Fourier integral operators, noted FCl, that is operators with classical symbols, we add to this topology the topology on formal symbols [10,11] which is an ILH topology (see e.g., [12] for state of the art). We want to quote that if the symbols $\sigma_{m,n}$ are symbols of order 0, then we get Fourier integral operators that are L^2−bounded. We note this set FIO^0. This set is a subset of FIO, and we have:

$$Cl^0 \subset PDO^0 \subset FIO^0 \subset FIO$$

The techniques used for pseudo-differential operators are also used on Fourier integral operators, especially Kernel analysis. Let us consider a local coordinate operator $A_{m,n}$ then, using the notation of of the Formula (1), the operator $A_{m,n}$ is described by a kernel:

$$K_{m,n}(x,y) = \int_\xi e^{i(\phi(x,\xi)) - y.\xi)} \sigma_{m,n}(x,\xi) d\xi$$

From this approach one derives the composition and inversion formulas that will not be used in this paper, see e.g., [13], but in the sequel we shall use the slightly restricted class of operators studied in [14–21] and also in [10,11] for formal operators.

2.3. *Topological Structures and Regular Lie Groups of Operators*

The topological structures can be derived both from symbols and from kernels, as we have quoted before, but principally because there is the exact sequence described below with slice. At the level of units of these sets, *i.e.*, of groups of invertible operators, the existence of the slice is also crucial. In the papers [10,11,14–22], the group of invertible Fourier integral operators receives first a structure of topological group, with in addition a differentiable structure, e.g., a Frölicher structure, which recognized as a structure of generalized Lie group, see e.g., [12].

We have to say that, with the actual state of knowledge, using [23], we can give a manifold structure (in the convenient setting described by Kriegl and Michor or in the category of Frölicher spaces following [24]) to the corresponding Lie groups. Let us recall the statement.

Theorem 3. *[23] Let G, H, K be convenient Lie groups or Frölicher Lie groups such that there is a short exact sequence of Lie groups:*

$$0 \to H \to G \to K \to 0$$

such that there is a local slice $K \to G$. Then:

$$G \text{ regular } \Leftrightarrow H \text{ and } K \text{ regular}$$

Remark 1. In [10,11,14–22], the group K considered is the group of 1-positively homogeneous symplectomorphisms $Diff_\omega(T^*M - M)$ where ω is the canonical symplectic form on the cotangent bundle. The local section considered enables to build up the phase function of a Fourier integral operator from such a symplectic diffeomorphism inside a neighborhood of Id_M. There is *a priori* no reason to restrict the constructions to classical pseudo-differential operators of order 0, and have groups of Fourier integral operators with symbols in wider classes. This remark appears important to us because the authors cited before restricted themselves to classical symbols.

2.4. $PDO(M, E)$, $Aut(E)$ and $Diff(M)$

We now get another group.

Theorem 4. *Let H be a regular Lie group of pseudo-differential operators acting on smooth sections of a trivial bundle $E \sim V \times M \to M$. The group $Diff(M)$ acts smoothly on $C^\infty(M, V)$, and is assumed to act smoothly on H by adjoint action. If H is stable under the $Diff(M)-$adjoint action, then there exists a corresponding regular Lie group G of Fourier integral operators through the exact sequence:*

$$0 \to H \to G \to Diff(M) \to 0$$

If H is a Frölicher Lie group, then G is a Frölicher Lie group. If H is a Fréchet Lie group, then G is a Fréchet Lie group.

Remark 2. The pseudo-differential operators can be classical, log-polyhomogeneous, or anything else. Applying the formulas of "changes of coordinates" (which can be understood as adjoint actions of diffeomorphisms) of e.g., [7], one easily gets the result.

Proof of Theorem 4. Let us first notice that the action:

$$(f, g) \in C^\infty(M, V) \times Diff(M) \mapsto f \circ g \in C^\infty(M, V)$$

can be read as, first a linear operator T_g with kernel:

$$K(x, y) = \delta(g(x), y) \quad (\text{Dirac } \delta - \text{function})$$

or equivalently, on an adequate system of trivializations [7],

$$T_g(f)(x) = \int e^{ig(x).\xi} \hat{f}(\xi) d\xi$$

This operator is not a pseudo-differential operator because it is not pseudolocal (unless $g = Id_M$), but since:

$$det\left(\partial_x \partial_\xi (g(x).\xi)\right) = det(D_x g)$$

we get that T_g is a Fourier-integral operator. Notice that another way to see it is the expression of its kernel.

Now, given $(A, g) \in H \times Diff(M)$, we define:

$$A_g = T_g \circ A$$

We get here a set G of operators which is set-theorically isomorphic to $H \times Diff(M)$. Since H is invariant under the adjoint action of the group $Diff(M)$, G is a group, and from the beginning of this proof, we get that G is a group, and that there is the short exact sequence announced:

$$0 \to H \to G \to Diff(M) \to 0$$

with a global slice:

$$g \in Diff(M) \mapsto T_g \in G$$

Since the adjoint action of Diff(M) is assumed smooth on H, we can endow G with the product Frölicher structure to get a regular Frölicher Lie group. Since $Diff(M)$ is a Fréchet Lie group, if H is a Fréchet Lie group, then G is a Fréchet Lie group. □

Remark 3. Some restricted classed of such operators are already considered in the literature under the name of $G-$pseudo-differential operators, see e.g., [25], but the groups considered are discrete (amenable) groups of diffeomorphisms. This gives a class of FIOs with linear phase, see [13].

Definition 5. Let M be a compact manifold and E be a (finite rank) trivial vector bundle over M. We define:

$$FIO_{Diff}(M, E) = \{A \in FIO(M, E) | \phi_A(x, \xi) = g(x).\xi ; g \in Diff(M)\}$$

The set of invertible operators $FIO^*_{Diff}(M, E)$ is obviously a group, that decomposes as:

$$0 \to PDO^*(M, E) \to FIO^*_{Diff}(M, E) \to Diff(M) \to 0$$

with global smooth section:

$$g \in Diff(M) \mapsto (f \in C^\infty(S^1, E) \mapsto f \circ g)$$

Hence, Theorem 4 applies trivially to the following context:

Proposition 6. *Let $FCl^{0,*}_{Diff}(M, E)$ be the set of operators $A \in FIO^*_{Diff}(M, E)$ such that A has a 0-order classical symbol. Then we get the exact sequence:*

$$0 \to Cl^{0,*}(M, E) \to FCl^{0;*}_{Diff}(M, E) \to Diff(M) \to 0$$

and $FCl^{0,*}_{Diff}(M, E)$ is a regular Frölicher Lie group, with Lie algebra isomorphic, as a vector space, to $Cl^0(M, E) \oplus Vect(M)$.

Notice that the triviality of the vector bundle E is here essential to make a $Diff(M)-$action on smooth section of $C^\infty(M, E)$. Let us assume now that E is not trivial. At the infinitesimal level, trying to extend straightway, one gets a first condition for the extension.

Lemma 7. *(see e.g., [26]) Let us fix a $0-$curvature connection ∇ on M. Then $X \in Vect(M) \mapsto \nabla_X \in DO^1(M, E)$ is a one-to-one Lie algebra morphism.*

We remark that the analogy with the setting of trivial bundles E stops here since the group $Diff(M)$ cannot be recovered in this group of operators. For example, when $M = S^1$, if E is non trivial, the (infinitesimally) flat connection ensures that the holonomy group $\mathcal{H}$ is discrete, but it cannot be trivial since the vector bundle E is not. On a non trivial bundle E, let us consider the group of bundle automorphism $Aut(E)$. The gauge group $DO^0(M,E)$ is naturally embedded in $Aut(E)$ and the bundle projection:

$$E \to M$$

induces a group projection

$$\pi : Aut(E) \to Diff(M)$$

Therefore ,we get a short exact sequence:

$$0 \to DO^{0,*}(M,E) \to Aut(E) \to Diff(M) \to 0$$

Following [27] there exists a local slice $U \subset Diff(M) \to Aut(E)$, where U is a C^0-open neighborhood on Id_M, which shows that $Aut(E)$ is a regular Fréchet Lie group. Therefore, the smallest group spanned by $PDO^*(M,E)$ and $Aut(E)$ is such that:

- the projection $E \to M$ induces a map $Aut(E) \to Diff(M)$ with kernel $DO^0(M,E) = Aut(E) \cap PDO(M,E)$
- $Ad_{Aut(E)}(PDO(M,E)) = PDO(M,E)$

therefore we can consider the space of operators on $C^\infty(M,E)$

$$FIO^*_{Diff}(M,E) = Aut(E) \circ PDO^*(M,E)$$

Lemma 8. *The map*

$$(B,A) \in Aut(E) \times PDO^*(M,E) \mapsto \pi(B) \in Diff(M)$$

induces a "phase map"

$$\tilde{\pi} : FIO^*_{Diff}(M,E) \to Diff(M)$$

Proof. Let $((B,A),(B',A')) \in Aut(E) \times PDO^*(M,E)$

$$\begin{aligned} B \circ A = B' \circ A' &\Leftrightarrow Id_E \circ A = B^{-1} \circ B' \circ A' \\ &\Leftrightarrow B^{-1} \circ B' = A \circ A'^{-1} \in PDO^*(M,E) \\ &\Rightarrow B^{-1} \circ B' \in DO^{0,*}(M,E) \\ &\Leftrightarrow \pi(B^{-1} \circ B') = Id_M \\ &\Leftrightarrow \pi(B) = \pi(B') \end{aligned}$$

□

The next lemma is obvious:

Lemma 9. $FIO^*_{Diff}(M,E)$ *is a group*

Lemma 10. $Ker(\tilde{\pi}) = PDO^*(M,E)$

Proof. Let $B \circ A \in FIO^*(M,E)$ such that:

$$\tilde{\pi}(B \circ A) = \pi(B) = Id_M$$

Then $B \in DO^{0,*}(M,E)$ and $B \circ A \in PDO^*(M,E)$ □

These results show the following theorem:

Theorem 11. *There is a short exact sequence of groups :*

$$0 \to PDO^*(M,E) \to FIO^*_{Diff}(M,E) \to Diff(M) \to 0$$

and, if $H \subset PDO^*(M,E)$ is a regular Fréchet or Frölicher Lie group of operators that contains the gauge group of E, if K is a regular Fréchet or Frölicher Lie subgroup of $Diff(M)$ such that there exists a local section $K \to Aut(E)$, the subgroup $G = K \circ H$ of $FIO^*_{Diff}(M,E)$ is a regular Fréchet Lie group from the short exact sequence:

$$0 \to H \to G \to K \to 0$$

2.5. Diffeomorphisms and kernel operators

Let $g \in Diff(M)$. Then a straightforward computation on local coordinates shows that the kernel of T_g is:

$$K_g = \delta(g(x), y)$$

where δ is the Dirac $\delta-$function. These operators also read locally as:

$$T_g(f) = \int_M e^{ig(x).\xi} \hat{f}(\xi) d\xi$$

on the same system of local trivializations used in [7], p.30-40.

2.6. Renormalized Traces on Diff(M)-Pseudodifferential Operators

Basics on renormalized traces are given in the appendix. Let us now investigate their extensions to the class of FIOs considered. Let us first explore the action of $Diff(M)$ and of $Aut(E)$ on $tr^Q(A)$.For this, we get:

Lemma 12. *Let $a \in \mathbb{Z}$. Let $A \in Cl^a(M,E)$ and let Q be a weight on E. Let B be an operator on $C^\infty(M,E)$ such that*

1. $Ad_B(Cl^a(M,E)) \subset Cl^a(M,E)$
2. *$Ad_B Q$ is a weight of the same order as Q*

Then

- $res(Ad_B A) = res(A)$
- $tr^{Ad_B Q}(Ad_B A) = tr^Q(A)$

The properties 1,2 are true in particular for operators $B \in Aut(E)$.

Proof.

Let Q be a weight on $C^\infty(M,E)$ and let $A \in Cl(M,E)$. Let $B \in Aut(E)$. Let $s \in \mathbb{R}_+^*$ then Ae^{-sQ} is trace class. By [7], we know that $Ad_B A$ (resp. $Ad_B Q$) is a classical pseudo-differential operator of the same order (resp. a weight of the same order). Then, since $e^{-\frac{s}{2}Q}$ is smoothing, $Ad_B(Ae^{-sQ})$, $BAe^{-\frac{s}{2}Q}$ and $e^{-\frac{s}{2}Q}B^{-1}$ are smoothing, and the following computations are fully justified:

$$\begin{aligned} tr\left(Ad_B(Ae^{-sQ})\right) &= tr\left(\left(BAe^{-\frac{s}{2}Q}\right)\left(e^{-\frac{s}{2}Q}B^{-1}\right)\right) \\ &= tr\left(\left(e^{-\frac{s}{2}Q}B^{-1}\right)\left(BAe^{-\frac{s}{2}Q}\right)\right) \\ &= tr\left(e^{-\frac{s}{2}Q}Ae^{-\frac{s}{2}Q}\right) \\ &= tr\left(Ae^{-sQ}\right) \end{aligned}$$

So that, we get the announced property. □

3. Splittings on the Set of S^1−Fourier Integral Operators

3.1. The Group $O(2)$ and the Diffeomorphism Group $Diff(S^1)$

Let us consider the $SO(2) = U(1)$-action on $S^1 = \mathbb{R}/\mathbb{Z}$ given by $(e^{2i\pi\theta}, x) \mapsto x + \theta$. This group acts on C^∞ by $(e^{2i\pi\theta}, f) \mapsto f(x+\theta)$ and we have:

$$\begin{aligned} f(x+\theta) &= \int e^{-i(x+\theta).\xi} \hat{f}(\xi) d\xi \\ &= \int e^{-i(x.\xi+\theta.\xi} \hat{f}(\xi) d\xi \end{aligned}$$

The term $e^{-i\theta.\xi}$ is oscillating in ξ and does not satisfies the estimates on the derivatives of symbols. So that, this operator is not a pseudo-differential operator but has obviously the form of a Fourier integral operator. The same is for the reflection $x \mapsto 1-x$ which corresponds to the conjugate transformation $z \mapsto \bar{z}$ when representing S^1 as the set of complex numbers z such that $|z| = 1$. This is a spacial case of the properties already stated for a general manifold M given $g \in Diff(S^1)$, g acts on C^∞ by right composition of the inverse, namely, for $f \in C^\infty$,

$$\begin{aligned} g.f(x) &= f \circ g(x) \\ &= \int e^{-ig(x).\xi} \hat{f}(\xi) d\xi \end{aligned}$$

which is also obviously a Fourier-integral operator, and the kernel of this operator is:

$$K_g(x,y) = \delta(y, g(x))$$

where δ is the Dirac δ-function. This is the construction already used in the proof of Theorem 4.

3.2. $\epsilon(D)$, Its Formal Symbol and the Splitting of $\mathcal{F}PDO$

The operator $D = -iD_x$ splits $C^\infty(S^1, \mathbb{C}^k)$ into three spaces:

- its kernel E_0, made of constant maps
- the vector space spanned by eigenvectors related to positive eigenvalues
- the vector space spanned by eigenvectors related to negative eigenvalues.

The following elementary result will be useful for the sequel, see [28] for the proof, and e.g., [3,6]:

Lemma 13.

(i) $\sigma(D) = \xi$
(ii) $\sigma(|D|) = |\xi|$ *where* $|D| = \left(\int_\Gamma \lambda^{1/2}(\Delta - \lambda Id)^{-1} d\lambda\right)$, *with* $\Delta = -D_x^2$.
(iii) $\sigma(D|D|^{-1}) = \frac{\xi}{|\xi|}$, *where* $D|D|^{-1} = |D|^{-1}D$ *is the sign of* D, *since* $|D|_{|E_0} = Id_{E_0}$
(iv) Let p_{E_+} *(resp.* p_{E_-}*) be the projection on* E_+ *(resp.* E_-*), then* $\sigma(p_{E_+}) = \frac{1}{2}(Id + \frac{\xi}{|\xi|})$ *and* $\sigma(p_{E_-}) = \frac{1}{2}(Id - \frac{\xi}{|\xi|})$

Let us now define two ideals of the algebra $\mathcal{F}PDO$, that we call $\mathcal{F}PDO_+$ and $\mathcal{F}PDO_-$, such that $\mathcal{F}PDO = \mathcal{F}PDO_+ \oplus \mathcal{F}PDO_-$. This decomposition is implicit in [29], section 4.4., p. 216, for classical pseudo-differential operators and we furnish the explicit description given in [28], extended to the whole algebra of (maybe non formal, non classical) pseudo-differential symbols here.

Definition 14. Let σ be a symbol (maybe non formal). Then, we define, for $\xi \in T^*S^1 - S^1$,

$$\sigma_+(\xi) = \begin{cases} \sigma(\xi) & \text{if } \xi > 0 \\ 0 & \text{if } \xi < 0 \end{cases} \text{ and } \sigma_-(\xi) = \begin{cases} 0 & \text{if } \xi > 0 \\ \sigma(\xi) & \text{if } \xi < 0 \end{cases}$$

At the level of formal symbols, we also define the projections: $p_+(\sigma) = \sigma_+$ and $p_-(\sigma) = \sigma_-$.

The maps $p_+ : \mathcal{F}PDO(S^1, \mathbb{C}^k) \to \mathcal{F}PDO(S^1, \mathbb{C}^k)$ and $p_- : \mathcal{F}PDO(S^1, \mathbb{C}^k) \to \mathcal{F}PDO(S^1, \mathbb{C}^k)$ are clearly algebra morphisms that leave the order invariant and are also projections (since multiplication on formal symbols is expressed in terms of pointwise multiplication of tensors).

Definition 15. We define $\mathcal{F}PDO_+(S^1, \mathbb{C}^k) = Im(p_+) = Ker(p_-)$ and $\mathcal{F}PDO_-(S^1, \mathbb{C}^k) = Im(p_-) = Ker(p_+)$.

Since p_+ is a projection, we have the splitting:

$$\mathcal{F}PDO(S^1, \mathbb{C}^k) = \mathcal{F}PDO_+(S^1, \mathbb{C}^k) \oplus \mathcal{F}PDO_-(S^1, \mathbb{C}^k)$$

Let us give another characterization of p_+ and p_-. Looking more precisely at the formal symbols of p_{E_+} and p_{E_-} computed in Lemma 13, we observe that:

$$\sigma(p_{E_+}) = \begin{cases} 1 & \text{if } \xi > 0 \\ 0 & \text{if } \xi < 0 \end{cases} \text{ and } \sigma(p_{E_-}) = \begin{cases} 0 & \text{if } \xi > 0 \\ 1 & \text{if } \xi < 0 \end{cases}$$

In particular, we have that $D_x^\alpha \sigma(p_{E_+})$, $D_\xi^\alpha \sigma(p_{E_+})$, $D_x^\alpha \sigma(p_{E_-})$, $D_\xi^\alpha \sigma(p_{E_-})$ vanish for $\alpha > 0$. From this, we have the following result:

Proposition 16. *Let $a \in \mathcal{F}PDO(S^1, \mathbb{C}^k)$. $p_+(a) = \sigma(p_{E_+}) \circ a = a \circ \sigma(p_{E_+})$ and $p_-(a) = \sigma(p_{E_-}) \circ a = a \circ \sigma(p_{E_-})$[28].*

3.3. The Case of Non Trivial (Real) Vector Bundle Over S^1

Let $\pi : E \to S^1$ be a non trivial real vector bundle over S^1 of rank k. Its bundle of frames is a $Gl(\mathbb{R}^k)-$ principal bundle, which means the following (see e.g., [30]):

Lemma 17. *Let $\varphi_1 :]a;b[\times \mathbb{R}^k \to E$ and $\varphi_2 :]a';b'[\times \mathbb{R}^k \to E$ be two local trivializations of E. Let $\mathcal{D} = \pi(\varphi_1(]a;b[\times \mathbb{R}^k) \cap \varphi_2(]a';b'[\times \mathbb{R}^k))$, let $\mathcal{D}_1 = \varphi_1^{-1}(\mathcal{D})$, and let $\mathcal{D}_2 = \varphi_2^{-1}(\mathcal{D})$. Then,*

$$\varphi_2^{-1} \circ \varphi_1 : \mathcal{D}_1 \times \mathbb{R}^k \to \mathcal{D}_2 \times \mathbb{R}^k$$

reads as:

$$\varphi_2^{-1} \circ \varphi_1 = \gamma \times M$$

where γ is a smooth diffeomorphism from $\mathcal{D}_1$ to $\mathcal{D}_2$, and where $M \in C^\infty(\mathcal{D}_1, Gl(\mathbb{R}^k))$.

Let us now turn to symbols of pseudo-differential operators acting on smooth sections of E. We first assume that we work with a system of local trivializations such that the diffeomorphisms γ are translations, and let us now look at the transformations of the symbols read on local trivializations. Under these assumptions, and with the notations of the previous lemma, a formal symbol σ_1 read on D_1 reads on D_2 as:

$$\sigma_2(\gamma(x), \xi) = M(x)\sigma_1(x, \xi)M(x)^{-1}$$

Proposition 18. *Let ∇ be a Riemannian covariant derivative on the bundle $E \to S^1$ and let $\frac{\nabla}{dt}$ be the associated first order differential operator, given by the covariant derivative evaluated at the unit vector field over S^1. We modify the operator $\frac{\nabla}{dt}$ into an injective operator $D = \frac{\nabla}{dt} + p_{ker\frac{\nabla}{dt}}$, where $p_{ker\frac{\nabla}{dt}}$ is the L^2 orthogonal projection on $ker\frac{\nabla}{dt} \subset C^\infty(S^1, E) \subset L^2(S^1, E)$, and we set:*

$$\epsilon(\nabla) = D \circ |D|^{-1}.$$

Then the formal symbol of $\epsilon(\nabla)$ is $\frac{i\xi}{|\xi|}$

Proof. Let us use the holonomy trivialization over an interval I. In this trivialization,

$$\frac{\nabla}{dt} = \frac{d}{dt}$$

and hence the formal symbol of $\frac{\nabla}{dt}$ reads as $i\xi$. Calculating exclusively on the algebra of formal operators on which composition and inversion governed by local formulas, we get $\sigma(|D|) = |\xi|$ and, by the same arguments as those of [28], we get the result. □

Proposition 19. *For each $A \in PDO(S^1, E)$, $[A, \epsilon(\nabla)] \in PDO^{-\infty}(S^1; E)$.*

Proof. We remark that, for any multiindex α such that $|\alpha| > 0$, $D_x^\alpha \sigma(\epsilon(\nabla)) = 0$ and $D_\xi^\alpha \sigma(\epsilon(\nabla)) = 0$. Hence, in $\mathcal{F}PDO(S^1, E)$,

$$\sigma([A, \epsilon(\nabla)]) = [\sigma(A), \sigma(\epsilon(\nabla))] = 0$$

so that $[A, \epsilon(\nabla)] \in PDO^{-\infty}(S^1, E)$. □

3.4. The Splitting Read on the Phase Function

The fiber bundle $T^*S^1 - S^1$ has two connected components and the phase function is positively homogeneous, so that we can make the same procedure as in the case of the symbols. However, we remark that we can split:

$$\phi = \phi_+ + \phi_-$$

where $\phi_+ = 0$ if $\xi < 0$ and $\phi_- = 0$ if $\xi > 0$. Unfortunately, ϕ_+ and ϕ_- are **not** phase functions of Fourier integral operators because there are some points where $\frac{\partial^2\phi_+}{\partial_x \partial_\xi} = 0$ or $\frac{\partial^2\phi_-}{\partial_x \partial_\xi} = 0$. However, we can have the following identities:

$$\begin{aligned}
\int_{\mathbb{R}} e^{i\phi(x,\xi)}\sigma(x,\xi)\hat{f}(\xi)d\xi &= \int_{\xi>0} e^{i\phi(x,\xi)}\sigma(x,\xi)\hat{f}(\xi)d\xi + \int_{\xi<0} e^{i\phi(x,\xi)}\sigma(x,\xi)\hat{f}(\xi)d\xi \\
&= \int_{\xi>0} e^{i\phi_+(x,\xi)}\sigma(x,\xi)\hat{f}(\xi)d\xi + \int_{\xi<0} e^{i\phi_-(x,\xi)}\sigma(x,\xi)\hat{f}(\xi)d\xi \\
&= \int_{\mathbb{R}} e^{i\phi_+(x,\xi)}\sigma_+(x,\xi)\hat{f}(\xi)d\xi + \int_{\mathbb{R}} e^{i\phi_-(x,\xi)}\sigma_-(x,\xi)\hat{f}(\xi)d\xi \\
&= \int_{\mathbb{R}} e^{i\phi(x,\xi)}\sigma_+(x,\xi)\hat{f}(\xi)d\xi + \int_{\mathbb{R}} e^{i\phi(x,\xi)}\sigma_-(x,\xi)\hat{f}(\xi)d\xi.
\end{aligned}$$

In the last line, we get the phase of a FIO.

3.5. The Schwinger Cocycle on $PDO(S^1, E)$ When E Is a Real Vector Bundle

The Scjwinger cocycle [31] is well-known in the theory of central extensions of algebras of pseudo-differential operators [3,6,32–34] are now analyzed from the viewpoint of operators acting on smooth sections of real vector bundles. Here, $\epsilon(\nabla)$ is not a sign operator, but an operator such that $\epsilon(\nabla)^2 = -Id$ up to a smoothing operator.

Theorem 20. *For any $A \in PDO(S^1, E)$, $[A, \epsilon(\nabla)] \in PDO^{-\infty}(S^1, E)$. Consequently,*

$$c_s^{\nabla} : A, B \in PDO(S^1, E) \mapsto \frac{1}{2}\mathrm{tr}\,(\epsilon(\nabla)[\epsilon(\nabla), A][\epsilon(\nabla), B])$$

is a well-defined $\mathbb{R}$-valued 2-cocycle on $PDO(S^1, E)$. Moreover, c_s^{∇} is non trivial on any Lie algebra $\mathcal{A}$ such that $C^{\infty}(S^1, \mathbb{R}) \subset \mathcal{A} \subset PDO(S^1, E)$.

Notice that $C^{\infty}(S^1, \mathbb{R})$ is understood as an algebra acting on $C^{\infty}(S^1, E)$ by scalar multiplication fiberwise. The proof follows the same arguments as in [3].

Proof. First, c_s^{∇} is the trace of operators acting on a real Hilbert space. so that, it is real valued. Since [1], see e.g., [6], if c_s^{∇} was trivial on Hoschild cohomology, there would have a 1-form $\nu : \mathcal{A} \to \mathbb{R}$ such that:

$$c_s^D = \nu([.,.])$$

and hence it would be true on $C^{\infty}(S^1, \mathbb{R})$ which is a commutative algebra. Hence, since $c_s^{\nabla} \neq 0$ on $C^{\infty}(S^1, \mathbb{R})$, it is non trivial on A. □

4. Sets of Fourier Integral Operators

4.1. The set $FIO(S^1, E)$

Here, for the definitions, $\varepsilon = \epsilon(D)$ or $\varepsilon = \varepsilon(\nabla)$, depending on the fact that E is a complex or a real vector bundle. Let us now define:

$$FIO_{res}(S^1; E) = \{A \in FIO(S^1, E) \text{ such that } [A; \epsilon] \in PDO^{-\infty}(S^1, E)\}$$

Proposition 21. *$FIO_{res}(S^1, E)$ is a set, stable under composition, with unit element.*

Proof. $FIO(S^1, E)$ is stable under composition [13]. Since $Cl^0(S^1, E)$ is contained in $FIO_{res}(S^1, E)$ by Theorem 20 so that $FIO_{res}(S^1, E)$ contains the identity map.

Let $A, B \in FIO_{res}(S^1, E)$,

$$[AB, \epsilon] \quad = \quad A[B; \epsilon] + [A; \epsilon]B$$

Since $[A, \epsilon]$ and $[B; \epsilon]$ are smoothing, we get that $[AB, \epsilon]$ is smoothing. □

We use the natural notations,

$$FIO^0_{res} = FIO^0 \cap FIO_{res}$$

We shall note by $FIO^*_{res}(S^1, E)$ the group of units of this set, and by $FIO^{0,*}_{res}(S^1, E)$ the group of units of the set $FIO^0_{res}(S^1, E)$.

Proposition 22. *$FIO^*_{res}(S^1, E) = FIO^*(S^1, E) \cap FIO_{res}(S^1, E)$ and $FIO^{0,*}_{res}(S^1, E) = FIO^{0,*}(S^1, E) \cap FIO_{res}(S^1, E)$.*

Proof. We already have trivially $FIO^*_{res}(S^1,E) \subset (S^1,E) \cap FIO_{res}(S^1,E)$. Let $A \in FIO^*(S^1,E) \cap FIO_{res}(S^1,E)$. We have to check that $A^{-1} \in FIO_{res}(S^1,E)$.

$$\begin{aligned} A[A^{-1},\varepsilon] &= [AA^{-1},\varepsilon] - [A,\varepsilon]A^{-1} \\ &= [Id,\varepsilon] - [A,\varepsilon]A^{-1} \\ &= -[A,\varepsilon]A^{-1} \\ &\in PDO^{-\infty}(S^1,E). \end{aligned}$$

So that,

$$\begin{aligned} [A^{-1},\varepsilon] &= A^{-1}A[A^{-1},\varepsilon] \\ &\in PDO^{-\infty}(S^1,E) \end{aligned}$$

The proof is the same for $0-$order operators. □

By the way, since $FIO^{0,*}(S^1,E)$ is a "'generalized Lie group"' in the sense of Omori, it is a Frölicher Lie group. By the trace property of Frölicher spaces, using the last proposition, $FIO^{0,*}_{res}(S^1,E)$ is a Frölicher Lie group [24]. Now, since we have that:

$$FIO^{0,*}_{res} \subset GL_{res},$$

the determinant bundle defined over GL_{res} can be pulled-back on $FIO^{0,*}_{res}$. The same way, it is shown in [2,3] that the Schwinger cocycle extends to the Lie algebra $PDO^0(S^1,E) + PDO^1(S^1,\mathbb{C}) \otimes Id_E$.

4.2. Yet Some Subgroups of $FIO^*_{res}(S^1,E)$

Let us first gather and reformulate many known results:

Lemma 23. *$Diff^+(S^1) \times C^\infty(S^1,\mathbb{C}^*) \subset FIO^{0,*}_{res}(S^1,\mathbb{C})$.*

Proof. First, we have that:

$$C^\infty(S^1,\mathbb{C}^*) \subset Cl^{0,*}(S^1,\mathbb{C})$$

so that,

$$C^\infty(S^1,\mathbb{C}^*) \subset FIO^{0,*}(S^1,\mathbb{C})$$

Let $g \in Diff^+(S^1)$. Following [1], the map $f \mapsto |g'|^{1/2}.(f \circ g)$ describes an operator in $U_{res} \subset GL_{res}$. Since the map $f \mapsto |g'|^{1/2}.f$ is a multiplication operator in $C^\infty(S^1,\mathbb{C}^*)$, we get that:

$$f \mapsto f \circ g = \int e^{ig(.).\xi}\hat{f}(\xi)d\xi \in GL_{res} \cap FIO^{0,*}(S^1,\mathbb{C})$$

□

Theorem 24. *Assume that E be a trivial vector bundle over S^1. Let $\tilde{\pi}$ be the projection $FIO^*_{Diff}(S^1,E) \to Diff(S^1)$. Then,*

$$\pi^{-1}(Diff_+(S^1)) \subset FIO_{res}(S^1,E)$$

This is a simple consequence of the previous results.

Theorem 25. *Assume that E is non trivial and let ϵ defined as before. Let $\tilde{\pi}$ be the projection $FIO^*_{Diff}(S^1,E) \to Diff(S^1)$. Then,*

$$FIO^*_{Diff_+}(S^1,E) = \pi^{-1}(Diff_+(S^1)) \subset FIO_{res}(S^1,E)$$

and there is a global smooth section (in the sense of Frölicher spaces, not necessarily in the sense of groups)

$$Diff_+(S^1) \to FIO_{res}(S^1, E)$$

of the short exact sequence:

$$0 \to PDO^*(S^1, E) \to FIO^*_{Diff}(S^1, E) \cap FIO_{res}(S^1, E) \to Diff_+(S^1) \to 0$$

Proof. Let $g \in Diff_+(S^1)$. We fix on E a connection ∇ and we set $n = rank(E)$. Since $Diff_+(S^1)$is the connected component of Id_{S^1} in $Diff(S^1)$, given η the unit vector field defined by orientation on S^1, we can choose a path

$$\gamma \in C^\infty([0,1], Diff_+(S^1)) \subset C^\infty([0,1] \times S^1, S^1)$$

such that:

$$\gamma(0) = Id_{S^1}, \gamma(1) = g$$

and

$$\forall x \in S^1, \forall t \in [0;1], (\frac{d\gamma}{dt}(t)(x), \eta(x))_{T_x S^1} > 0$$

This path is unique up to parametrization since we impose also the condition of minimal length. Let

$$H_x = Hol(\gamma(.)(x)) \in Gl(E_x, E_{g(x)})$$

be the induced parallel transport map. We get, for each $g \in Diff_+(S^1)$, a map H_g which is smooth by the properties of parallel transport, linear on the fibers, invertible, and which projects on S^1 to g. Thus, $H_g \in Aut(E)$, and it easy to see that it is a bijection on the collection of smooth trivializations of E. Now, turning to the map

$$g \mapsto H_g$$

is appears as a smooth map $Diff(S^1) \to Aut(E)$, but it cannot ba a group morphism when E is non trivial. We have, moreover, that

$$\forall g \in Diff^+(S^1), [\nabla, H_g] = 0$$

since $dim(S^1) = 1$ and H_g is a parallel transport map. So that, since ϵ is derived from $\frac{\nabla}{dt} = \nabla_\eta$, we get that: $H_g \in GL_{res}$. Now, an operator in $FIO^*_{Diff_+}(S^1, E)$ reads as:

$$H_g \circ A,$$

where $A \in PDO^*(S^1, E) \subset GL_{res}$. Then $H_g \circ A \in FIO_{res}$ □

Theorem 26. *The group*

$$FCl^{0,*}_{Diff_+}(S^1, E) = FIO^*_{Diff_+}(S^1, E) \cap FCL^0(S^1, E)$$

is a regular Frölicher Lie group.

Proof. We get the obvious exact sequence of Lie groups:

$$0 \to Cl^{*,0}(S^1, E) \to FCl^{*,0}_{Diff}(S^1, E) \to Diff_+(S^1) \to 0$$

Both $Cl^{*,0}(S^1, E)$ and $Diff_+(S^1)$ are regular, and $Aut(E) \subset Cl^{*,0}(S^1, E)$, so that the smooth section $Diff_+(S^1) \to Aut(E)$ described in the proof of the previous theorem gives the result by Theorem 11. □

Let us now describe a subgroup of $FIO^*_{Diff_+}(S^1, E)$.

Definition 27. Let $FIO^*_{b,Diff_+}(S^1, E)$ be the space of operators $A \in FIO^*_{Diff_+}(S^1, E)$ such that

1. $\pi(A)$ is a diffeomorphism of $S^1 = \mathbb{R}/\mathbb{Z}$ such that $\pi(A)(0) = 0$;
2. if u is a smooth section of E such that $u(0) = 0$, then $(Au)(0) = 0$.

These operators are called based operators, and the set of sections u of E such that $u(0) = 0$ is the space of based sections, noted $C^\infty_b(S^1; E)$. We note by $Diff_{b,+}(S^1)$ the infinite dimensional Lie group of diffeomorphisms g such that $g(0) = 0$.

We recall that $Diff_{b,+}(S^1)$ is a regular Lie subgroup of $Diff_+(S^1)$. Its Lie algebra is noted $\mathfrak{diff}_b(S^1)$. Given $c(t)$ a smooth curve in $FIO^*_{b,Diff_+}(S^1, E)$, starting at Id_E, $\frac{dc(t)}{dt}|_{t=0} = U + X$, where $X \in \mathfrak{diff}_b(S^1)$ and $U \in PDO(S^1, E)$ which stabilize $C^\infty_b(S^1; E)$.

Theorem 28.

- *Let $G \subset PDO^*(S^1, E)$ be a regular Lie group of based operators, that contains the space of based invertible multiplication operators, with Lie algebra $\mathfrak{g}$.*
- *Let $D \subset Diff_{b,+}(S^1)$ be a regular Lie subgroup of based diffeomorphisms, with regular Lie algebra $\mathfrak{d}$.*

There is a regular Lie group $FG_D \subset FIO^*_{b,Diff_+}(S^1, E)$ for which the following sequence is exact:

$$0 \to G \to FG_D \to D \to O$$

Proof. We consider first the regular Lie group of automorphisms $\pi^{-1}(D) \subset Aut(E)$. Then, with the same arguments, G and generate a group that we note FG_D, and adapting the computations of Lemma 8, we obtain the above exact sequence. Finally, by Theorem 3, FG_D is a regular Lie group. □

5. Manifolds of Embeddings

Notation: Let $E \to M$ be a smooth vector bundle over M with typical fiber x. For $k \in \mathbb{N}^*$, we denote by

- $E^{\times k}$ the product bundle, of basis M, with typical fiber $F^{\times k}$;
- $\Omega^k(E)$ the space of $k-forms$ on M with values in E, that is, the set of smooth maps $(TM)^{\times k} \to E$ that are fiberwise k-linear and skew-symmetric $(T_xM)^{\times k} \to E_x$ for any $x \in M$. If $E = M \times F$, we note $\Omega^k(M, F)$ the space of k-forms instead of $\Omega^k(E)$.

Let M be a compact manifold without boundary; let N be a Riemannian manifold, equipped with the metric $(.,.)$. Let $Emb(M, N)$ be the manifold of smooth embeddings $M \to N$. References for principal bundles of embeddings are [35,36].

5.1. $Emb(M, N)$ as a Principal Bundle

The group of diffeomorphisms of M, $Diff(M)$, acts smoothly and on the right on $Emb(M, N)$, by composition. Moreover,

$$B(M, N) = Emb(M, N)/Diff(M)$$

is a smooth manifold [23], and $\pi : Emb(M, N) \to B(M, N)$ is a principal bundle with structure group $Diff(M)$ (see [23]). Then, $g \in Emb(M, N)$ is in the $Diff(M)-$orbit of f if and only if $g(M) = f(M)$. Let us now precise the vertical tangent space and a normal vector space of the orbits of $Diff(M)$ on $Emb(M, N)$. $T_fEmb(M, N)$, the tangent space at f, is identified with the space of smooth sections of f^*TN, which is the pull-back of TN by f. VT_fP, the vertical tangent space at f is the space of smooth sections of $Tf(M)$. Let $\mathcal{N}_f$ be the normal space to $f(M)$ with respect to the metric $(.,.)$ on N.

For any $x \in M$, $T_{f(x)}N = T_{f(x)}f(M) \oplus \mathcal{N}f(M)$. Hence, denoting $f * \mathcal{N}_f$ the pull back of $\mathcal{N}_f$ by f, we have that:

$$C^\infty(f^*TN) = C^\infty(TM) \oplus f^*\mathcal{N}_f$$

Moreover, for any volume form dx on M, if:

$$< .,. >: X, Y \in C^\infty(f^*TN) \mapsto < X, Y >= \int_M (X(x), Y(x))dx$$

is a L^2-inner product on $C^\infty(f^*TN)$, this splitting is orthogonal for $< .,. >$. We get here a fundamental difference between the inclusion $Emb(M,N) \subset C^\infty(M,N)$, where the model space of the type $C^\infty(f^*TN)$, and $Emb(M,N)$ as a $Diff(M)-$ principal bundle: sections of the vertical tangent vector bundle read as order 1 differential operators, where as the operators acting on the normal vector bundle reads as $0-$order differential operators, just like the structure group of $TC^\infty(M,N)$. To be more precise, let $X \in C^\infty(f^*TN)$ and let $p : f^*TN \to Tf(M)$ be the orthogonal projection. The vector field $p(X) \in C^\infty(Tf(M))$ is seen as a differential operator acting on smooth functions $f(M) \sim M \to \mathbb{R}$, and the normal component $(Id - p)(X)$ is a smooth section on $\mathcal{N}_f$. In the sequel we shall note:

$$\mathcal{N} = \coprod_{f \in Emb(M,N)} \mathcal{N}_f$$

We turn now to local trivializations. Let $f \in C_b^\infty(M,N)$. We define the map $Exp_f : C_0^\infty(M, f^*TN) \to C_b^\infty(M,N)$ defined by $Exp_f(v) = exp_{f(.)}v(.)$ where exp is the exponential map on N. Then Exp_f is a smooth local diffeomorphism. Restricting Exp_f to a C^∞ - neighborhood $\tilde{U}_f$ of the 0-section of f^*TN, we define a diffeomorphism, setting:

$$(Exp_f)_{|\tilde{U}_f} : \tilde{U}_f \to V_f = Exp_f(\tilde{U}_f) \subset C_b^\infty(M,N)$$

Then, setting $U_f = I_f^{-1}\tilde{U}_f$, we can define a chart Ξ^f on V_f by:

$$\Xi^f(g) = (I_f^{-1} \circ (Exp_f)_{|\tilde{U}_f}^{-1})(g) \in U_f \subset C_b^\infty(M,E)$$

Given f,g in $C_b^\infty(M,N)$ such that $V_{f,g} = V_f \cap V_g \neq \emptyset$, we compute the changes of charts $\Xi^{f,g}$ from $U_{f,g}^f = \Xi^f V_{f,g}$ to $U_{f,g}^g = \Xi^g V_{f,g}$. Let $u \in U_{f,g}^f$, $v = (\Xi^f)^{-1}(u) \in V_{f,g}$.

$$\Xi^{f,g}(u) = \Xi^g \circ (\Xi^f)^{-1}(u) = (I_g^{-1} \circ (Exp_g)^{-1} \circ Exp_f \circ I_f)(u)$$

Since, $\forall x \in M$, the transition maps:

$$\Xi^{f,g}(u)(x) = (I_g^{-1} \circ (exp_{g(x)})^{-1} \circ exp_{f(x)} \circ I_f)(u(x))$$

are smooth, $(V_f, \Xi^f, U_f)_{f \in C_b^\infty(M,N)}$ is a smooth atlas on $C_b^\infty(M,N)$. Moreover, let $w \in C_0^\infty(M,E)$, setting $v = (\Xi^f)^{-1}(u)$, the evaluation of the differential at $x \in M$ reads :

$$D_u\Xi^{f,g}(w)(x) = (I_g^{-1} \circ D_{v(x)}(exp_{g(x)})^{-1} \circ D_{u(x)}(exp_{f(x)} \circ I_f))(w(x))$$

Hence, for $u \in C^\infty$, $D_u\Xi^{f,g}$ is a multiplication operator acting on smooth sections of E for any isomorphism I_f and I_g we can choose. Since I_f and I_g are fixed, the family $u \mapsto D_u\Xi^{f,g}$ is a smooth family of 0- order differential operators; this construction is described carefully in [37]. Now, let $f \in Emb(M,N)$ and let us consider the map:

$$\Phi^{U,f} : (f, v, X) \in TU \sim (1-p)TU \oplus pTU \mapsto \Xi^f(v).exp_{Diff(M)}(X) \in Emb(M,N)$$

This map gives a local (fiberwise) trivialization of the principal bundles $Emb(M,N) \to B(M,N)$ following [23,38,39], and we see that the changes of local trivializations have $Aut(\mathcal{N})$ as a structure group.

If M is oriented, we note by $Diff_+(M)$ the group of orientation preserving diffeomorphisms and we have the following trivial lemma:

Lemma 29.

$$\frac{Diff(M)}{Diff_+(M)} = \mathbb{Z}_2$$

Then, defining

$$B_+(M,N) = \frac{Emb(M,N)}{Diff^+(M)}$$

we get:

Proposition 30. *$B_+(M,N)$ is a 2-cover of $B(M,N)$*

Now, taking basepoints $x_0 \in M$ and $y_0 \in N$, we define the principal bundle of based embeddings

Proposition 31.
Let

$$Emb_b(M,N) = \{f \in Emb(M,N) | f(x_0) = y_0\}$$

Let

$$Diff_b(M) = \{g \in Diff(M) | g(x_0) = x_0\}$$

Let

$$Diff_{b,+}(M) = Diff_b(M) \cap Diff_+(M)$$

Let

$$B_b(M,N) = Emb_b(M,N)/Diff_b(M,N)$$

and

$$B_{b,+}(M,N) = Emb_b(M,N)/Diff_{b,+}(M,N)$$

Then $Emb_b(M,N)$ is a principal bundle with base $B_b(M,N)$ (resp. $B_{b,+}(M,N)$) and with structure group $Diff_b(M)$ (resp. $Diff_{b,+}(M)$)

Proof. It follows from the fact that $Emb_b(M,N) = ev_{x_0}^{-1}(y_0)$ in $Emb(M,N)$, and $Diff_b(M) = ev_{x_0}^{-1}(x_0)$ in $Diff(M)$. □

5.2. Almost Complex Structure on Based Oriented Knots

Here, we consider $B_{b,+}(S^1,N)$, which can be understood as a space of unparametrized oriented knots. Let $f \in Emb_b(S^1,N)$. Following the decompositions of the previous section, the tangent space

$$T_f Emb(S^1,N) = \{X \in f^*TN | X(0) = 0\}$$

decomposed into the sum:

$$\mathcal{N}_{b,f} \oplus Df(C_b^\infty(TS^1))$$

Let us consider the operator:

$$J = i\epsilon(D)$$

where $D = -i\frac{\nabla}{dt}$. We get that $J^2 = -Id$, so that J is an almost complex structure of $TB_+(S^1,N)$.

6. Chern-Weil Forms on Principal Bundle of Embeddings and Homotopy Invariants

6.1. Chern Forms in Infinite Dimensional Setting

Let P be a principal bundle, of basis M and with structure group G. Let $\mathfrak{g}$ be the Lie algebra of G. Recall that G acts on P, and also on $P \times \mathfrak{g}$ by the action $((p,v),g) \in (P \times \mathfrak{g}) \times G \mapsto (p.g, Ad_{g^{-1}}(v)) \in (P \times \mathfrak{g})$. Let $AdP = P \times_{Adg} = (P \times \mathfrak{g})/G$ be the adjoint bundle of P, of basis M and of typical fiber $\mathfrak{g}$, and let $Ad^k P = (AdP)^{\times k}$ be the product bundle, of basis M and of typical fiber $\mathfrak{g}^{\times k}$.

Definition 32. Let k in $\mathbb{N}^*$. We define $\mathfrak{Pol}^k(P)$, the set of smooth maps $Ad^k P \to \mathbb{C}$ that are k-linear and symmetric on each fiber, equivalently as the set of smooth maps $P \times \mathfrak{g}^k \to \mathbb{C}$ that are k-linear symmetric in the second variable and G-invariants with respect to the natural coadjoint action of G on $\mathfrak{g}^k$.

Let $\mathfrak{Pol}(P) = \bigoplus_{k \in \mathbb{N}^*} \mathfrak{Pol}(P)$.

Let $C(P)$ be the set of connections on P. For any $\theta \in C(P)$, we denote by $F(\theta)$ its curvature and ∇^θ (or ∇ when it carries no ambiguity) its covariant derivation. Given an algebra A, In this section, we study the maps, for $k \in \mathbb{N}^*$,

$$\begin{aligned} Ch \quad : \quad C(P) \times \mathfrak{Pol}^k(P) &\to \Omega^{2k}(M,\mathbb{C}) && (2)\\ (\theta, f) &\mapsto Alt(f(F(\theta),...,F(\theta))) && (3) \end{aligned}$$

where Alt denotes the skew-symmetric part of the form. Notice that, in the case of the finite dimensional matrix groups Gl_n with Lie algebra $\mathfrak{gl}_n$, the set $\mathfrak{Pol}(P)$ is generated by the polynomials $A \in \mathfrak{gl}_n \mapsto \mathrm{tr}(A^k)$, for $k \in 0,...,n$. This leads to classical definition of Chern forms. However, in the case of infinite dimensional structure groups, most situations are still unknown and we do not know how to define a set of generators for $\mathfrak{Pol}(P)$.

Lemma 33. *Let $f \in \mathfrak{Pol}^k(P)$. Then*

$$\begin{aligned} f([a_1,v],a_2,...,a_k) + f(a_1,[a_2,v],...,a_k) + \\ \dots \\ + f(a_1,a_2,...,[a_k,v]) &= 0 \end{aligned}$$

Proof. Let us notice first that f is symmetric. Let $v \in \mathfrak{g}$, and c_t a path in G such that $\{\frac{d}{dt}c_t\}_{t=0} = v$. Let $a_1,...,a_k \in \mathfrak{g}^k$.

$$\begin{aligned} \{\frac{d}{dt}\{f(ad_{c_t^{-1}}a_1,...,ad_{c_t^{-1}}a_k)\}_{t=0} &= f([a_1,v],a_2,...,a_k) + f(a_1,[a_2,v],...,a_k) + \\ &\dots \\ &+ f(a_1,a_2,...,[a_k,v]) \end{aligned}$$

Since f in G-invariant, we get:

$$\begin{aligned} f([a_1,v],a_2,...,a_k) + f(a_1,[a_2,v],...,a_k) + \\ \dots \\ + f(a_1,a_2,..,[a_k,v]) &= 0 \end{aligned}$$

Lemma 34. *Let* $f \in \mathfrak{Pol}^k(P)$ *such that* f*, as a smooth map* $P \times \mathfrak{g}^k \to \mathbb{C}$*, satifies* $d^M f = 0$ *on a system of local trivializations of* P*. Then, the map*

$$Ch^f : \theta \in \mathcal{C}(P) \mapsto Ch^f(\theta) = Ch(\theta, f) \in \Omega^*(P, \mathbb{C})$$

takes values into closed forms on P*. Moreover,*

(i) *it is vanishing on vertical vectors and defines a closed form on* M*.*
(ii) *the cohomology class of this form does not depend on the choice of the chosen connexion* θ *on* P*.*

Proof. The proof runs as in the finite dimensional case, see e.g., [30]. First, it is vanishing on vertical vectors and G-invariant because the curvature of a connexion vanishes on vertical forms and is G-covariant for the coadjoint action. Let us now fix $f \in \mathfrak{Pol}^k(P)$. We compute $df(F(\theta), ..., F(\theta))$. We notice first that it vanishes on vertical vectors trivially. Let us fix $Y_1^h, ..., Y_{2k}^h, X^h$ $2k+1$ horizontal vectors on P at $p \in P$. On a local trivialization of P around p, these vectors read as:

$$\begin{aligned} Y_1^h &= Y_1 - \tilde{\theta}(Y_1) \\ &(...) \\ Y_{2k}^h &= Y_{2k} - \tilde{\theta}(Y_{2k}) \\ X^h &= X - \tilde{\theta}(X) \end{aligned}$$

where $\tilde{\theta}$ stands here for the expression of θ in the local trivilization, and $Y_1, ..., Y_{2k}, X$ $2k+1$ tangent vectors on M at $\pi(p) \in M$. We extend these vector fields on a neighborhood of p

- by the action of G in the vertical directions
- setting the vectors fields constant on $U \times p$, where U is a local chart on M around $\pi(p)$.

Then, we have:

$$f(F(\theta), ..., F(\theta))(Y_1^h, ..., Y_{2k}^h) = f(F(\theta), ..., F(\theta))(Y_1, ..., Y_{2k})$$

since $F(\theta)$ is vanishing on vertical vectors.

Then, on a local trivialization with the notations defined before (the sign Alt is omitted for easier reading), and writing d^M for the differential of forms on any open subset of M,

$$\begin{aligned} d^M f(F(\tilde{\theta}), ..., F(\tilde{\theta})) &= \sum_{i=1}^{k} f(d^M F(\tilde{\theta}), F(\tilde{\theta}), ..., F(\tilde{\theta})) + f(F(\tilde{\theta}), d^M F(\tilde{\theta}), ..., F(\tilde{\theta})) + \\ &\quad ... + f(F(\tilde{\theta}), F(\tilde{\theta}), ..., d^M F(\tilde{\theta})) \end{aligned}$$

and then, using Lemma 33,

$$\begin{aligned} \nabla^\theta f(F(\tilde{\theta}), ..., F(\tilde{\theta})) &= \sum_{i=1}^{k} f(\nabla^\theta F(\tilde{\theta}), F(\tilde{\theta}), ..., F(\tilde{\theta})) + f(F(\tilde{\theta}), \nabla^\theta F(\tilde{\theta}), ..., F(\tilde{\theta})) + \\ &\quad ... + f(F(\tilde{\theta}), F(\tilde{\theta}), ..., \nabla^\theta F(\tilde{\theta})) \end{aligned}$$

Then, by Bianchi identity, we get that:

$$\begin{aligned} d^M Ch(f, \theta) &= \nabla^\theta Ch(f, \theta) \\ &= 0 \end{aligned}$$

This proves (i) Then, following e.g., [30], if θ and θ' are connections, fix $\mu = \theta' - \theta$ and $\theta_t = \theta + t\nu$ for $t \in [0;1]$. We have:

$$\frac{dF(\theta_t)}{dt} = \nabla^{\theta^t}\mu$$

Moreover, μ is G-invariant and vanishes on vertical vectors. Thus,

$$\begin{aligned}\frac{dCh(f,\theta_t)}{dt} &= kf(F(\theta_t),...,F(\theta_t),\nabla^{\theta_t}\mu) \\ &= kd^M(f(F(\theta_t),...,F(\theta_t),\mu))\end{aligned}$$

Integrating in the t-variable, we get:

$$Ch(f,\theta_0) - Ch(f,\theta_1) = -kd^M\int_0^1 f(F(\theta_t),...,F(\theta_t),\mu)dt$$

Even if these computations are local, the two sides are global objects and do not depend on the chosen trivialization, which ends the proof. □

Important Remark. The condition $d^M f = 0$ is a **local** condition, checked in an (adequate) system of trivializations of the principal bundle, because it has to be checked on the vector bundle $Ad(P)^{\times k}$. This is in particular the case when we can find a 0-curvature connection θ on P such that:

$$[\nabla^\theta, f] = 0$$

In that case, since the structure group G is regular, we can find a system of local trivializations of P defined by θ and such that, on any local trivialization, $\nabla^\theta = d^M$ (see e.g., [23,40] for the technical tools that are necessary for this).

This technical remark can appear rather unsatisfactory first because it restricts the ability of application of the previous lemma, secondly because we need have a local (and rather unelegant) condition. This is why we give the following theorem, from Lemma 34.

Theorem 35. *Let $f \in \mathfrak{Pol}(P)$ for which there exists $\theta \in C(P)$ such that $[\nabla^\theta, f] = 0$. We shall note this set of polynomials by $\mathfrak{Pol}_{reg}(P)$. Then, the map:*

$$Ch^f : \theta \in C(P) \mapsto Ch^f(\theta) = Ch(\theta, f) \in \Omega^*(P,\mathbb{C})$$

takes values into closed forms on P. Moreover,

(i) it is vanishing on vertical vectors and defines a closed form on M.
(ii) The cohomology class of this form does not depend on the choice of the chosen connexion θ on P.

Moreover, $\forall(\theta, f) \in C(P) \times \mathfrak{Pol}_{reg}(P), [\nabla^\theta, f] = 0$.

Proof. Let $f \in \mathfrak{Pol}_{reg}(P)$ and let $\theta \in C(P)$ such that $[\nabla^\theta, f] = 0$. Let $\theta' \in \rfloor(P)$ and let $\nu = \theta' - \theta \in \Omega^1(M,\mathfrak{g})$. Let $(\alpha_1, ..., \alpha_k) \in (\Omega^2(M,\mathfrak{g}))^k$.

$$\begin{aligned}[\nabla^{\theta'}, f](\alpha_1,...\alpha_k) &= [\nabla^\theta, f](\alpha_1,...\alpha_k) + f([\alpha_1,\nu],...,\alpha_n) + \\ &\quad ... + f(\alpha_1,...,[\alpha_n,\nu]) \\ &= f([\alpha_1,\nu],...,\alpha_n) + ... + f(\alpha_1,...,[\alpha_n,\nu]) \\ &= 0\end{aligned}$$

Then, $\forall(\theta, f) \in \mathcal{C}(P) \times \mathfrak{Pol}_{reg}(P)$, $[\nabla^\theta, f] = 0$. By the way, $\forall \theta' \in \mathcal{C}(P)$,

$$d^M f(\alpha_1, ..., \alpha_k) = f(\nabla^{\theta'} \alpha_1, ..., \alpha_k) + ... + f(\alpha_1, ..., \nabla^{\theta'} \alpha_k)$$

Applying this to $\alpha_1 = ... = \alpha_k = F(\theta')$, we get:

$$dCh(f, \theta') = f(\nabla^{\theta'} F(\theta'), ..., F(\theta')) + ... + f(F(\theta'), ..., \nabla^{\theta'} F(\theta')) = 0$$

by Bianchi identity. Thus $Ch(f, \theta')$ is closed. Then, mimicking the end of the proof of Lemma 34, we get that the difference $Ch(f, \theta) - Ch(f, \theta')$ is an exact form, which ends the proof.

Proposition 36. *Let $\phi : \mathfrak{g}^k \to \mathbb{C}$ be a $k-$linear, symmetric, Ad$-$invariant form. Let $f : P \times \mathfrak{g}^k \to \mathbb{C}$ be the map induced by ϕ by the formula: $f(x, g) = \phi(g)$. Then $f \in \mathfrak{Pol}_{reg}$.*

Proof. Obsiously, $f \in \mathfrak{Pol}$. Let $\varphi : U \times G \to P$ and $\varphi' : U \times G \to P$ be a local trivialisations of P, where U is an open subset of M. Then there exists a smooth map $g : U \to G$ such that $\varphi'(x, e_G) = \varphi(x, e_G).g(x)$. Then we remark that $\varphi^* f = \varphi'^* f$ is a constant map on horizontal slices since ϕ is Ad-invariant. Moreover, since $\varphi^* f$ in a constant (polynomial-valued) map on $\varphi(x, e_G)$ we get that $[\nabla^\theta, f] = 0$ for the (flat) connection θ such that $T\varphi(x, e_G)$ spans the horizontal bundle over U. □

6.2. Application to $Emb(M, N)$

Mimicking the approach of [6], the cohomology classes of Chern-Weil forms should give rise to homotopy invariants. Applying Theorem 35, we get:

Theorem 37. *The Chern-Weil forms Ch^f is a $H^*(B(M, N))-$valued invariant of the homotopy class of an embedding, $\forall k \in \mathbb{N}^*$.*

When $M = S^1$, $Emb(S^1, N)$ is the space of (parametrized) smooth knots on N, and $B(S^1, N)$ is the space of non parametrized knots. Its connected components are the homotopy classes of the knots, through classical results of differential topology, see e.g., [41]. We now apply the material of the previous section to manifolds of embeddings. For this, we can define invariant polynomials of the type of those obtained in [6] (for mapping spaces) by a field of linear functionnal λ with "good properties" that ensures that:

$$A \mapsto \lambda(A^k) \in \mathfrak{Pol}^k_{reg}$$

This approach is a straightforward generalization of the description of Chern-Weil forms on finite dimensional principal bundles where polynomials are generated by functionnals of the type $A \mapsto \mathrm{tr}(A^k)$ (tr is the classical trace) but as we guess that we can consider other classes of polynomials for spaces of embeddings. In this paper, let us describe how to replace the classical trace of matrices tr by a renormalized trace tr^Q. In the most general case, it is not so easy to define a family of weights $f \in Emb(M, N) \mapsto Q_f$ which satisfy the good properties. Indeed, we have two examples of constructions which match the necessary assumptions for $\mathfrak{Pol}_{reg}$ when $M = S^1$, and the first one is derived from the following example:

Knot Invariant Through Kontsevich and Vishik Trace

The Kontsevich and Vishik trace is a renormalized trace for which $tr^Q([A, B]) = 0$ for each differential operator A, B and does not depend on the weight chosen in the odd class. For example, one can choose $Q = Id + \nabla^* \nabla$, where ∇ is a connection induced on $\mathcal{N}_f$ by the Riemannian metric, as described in [6]. It is an order 2 injective elliptic differential operator (in the odd class), and the coadjoint action of $Aut(\mathcal{N}_f)$ will give rise to another order 2 injective elliptic differential operator [7]. When $Q = Id + \nabla^*\nabla$, this only changes ∇ into another connection on E. Thus, setting:

$$\phi(A, ..., A) = tr^Q(A^k)$$

we have:

$$f \in \mathfrak{Pol}_{reg}$$

Let us now consider a connected component of $B(M, N)$, *i.e.*, a homotopy class of an embedding among the space of embeddings. We apply now the construction to $M = S^1$. The polynomial:

$$\phi : A \mapsto tr^Q(A^k)$$

is $Diff(S^1)-$invariant, and gives rise to an invariant of non oriented knots, *i.e.*, a Chern form on the base manifold:

$$B(S^1, N) = Emb(S^1, N)/Diff(S^1)$$

by theorem 37. This approach can be extended to invariant of embeddings, replacing S^1 by another odd-dimensional manifold.

7. Conclusions

We have given here some groundbreaking properties for a theory of differential invariants on non-linear grassmannians. A work in progress intends to describe such Chern-Weil, or Chern-Simons, or Cheeger-Simons invariant which could lead to non trivial knot invariants.

Conflicts of Interest: The author declares no conflict of interest.

Appendix

Renormalized Traces of PDOs

E is equipped this an Hermitian products $< .,. >$, which induces the following L^2-inner product on sections of E:

$$\forall u, v \in C^\infty(S^1, E), \quad (u, v)_{L^2} = \int_{S^1} < u(x), v(x) > dx$$

where dx is the Riemannian volume. The main references are [42,43], see e.g., [6].

Definition A1. Q is a **weight** of order $s > 0$ on E if and only if Q is a classical, elliptic, admissible pseudo-differential operator acting on smooth sections of E, with an admissible spectrum.

Recall that, under these assumptions, the weight Q has a real discrete spectrum, and that all its eigenspaces are finite dimensional. For such a weight Q of order q, one can define the complex powers of Q [44], see e.g., [4] for a fast overview of technicalities. The powers Q^{-s} of the weight Q are defined for $Re(s) > 0$ using with a contour integral,

$$Q^{-s} = \int_\Gamma \lambda^s (Q - \lambda Id)^{-1} d\lambda$$

where Γ is an "angular" contour around the spectrum of Q. Let A be a log-polyhomogeneous pseudo-differential operator. The map $\zeta(A, Q, s) = s \in \mathbb{C} \mapsto \text{tr}\,(AQ^{-s}) \in \mathbb{C}$, defined for $Re(s)$ large, extends on $\mathbb{C}$ to a meromorphic function with a pole of order $q+1$ at 0 ([45]). When A is classical, $\zeta(A, Q, .)$ has a simple pole at 0 with residue $\frac{1}{q}\text{res}A$, where res is the Wodzicki residue ([46], see also [29]). Notice that the Wodzicki residue extends the Adler trace [47] on formal symbols. Following [45], we define the renormalized trace, see e.g., [4,48] for the renormalized trace of classical operators.

Definition A2. $tr^Q A = lim_{z\to 0}(\mathrm{tr}(AQ^{-z}) - \frac{1}{qz} res A)$.

On the other hand, the operator e^{-tQ} is a smoothing operator for each $t > 0$, which shows that $trAe^{-tQ}$ is well-defined and finite for $t > 0$. From the function $t \mapsto trAe^{-tQ}$, we recover the function $z \mapsto \mathrm{tr}(AQ^{-z})$ by the Mellin transform (see e.g., [42], pp. 115–116), which shows the following lemma:

Lemma A1. *Let A, A' be classical pseudo-differential operators, let Q, Q' be weights.*

$$\forall t > 0, trAe^{-tQ} = trA'e^{-tQ'} \Rightarrow \begin{cases} tr^Q(A) &= tr^{Q'}(A') \\ res(A) &= res(A') \end{cases}$$

If A is trace class, $\mathrm{tr}^Q(A) = \mathrm{tr}(A)$. The functional tr^Q is of course not a trace on $Cl(M, E)$. Notice also that, if A and Q are pseudo-differential operators acting on sections on a real vector bundle E, they also act on $E \otimes \mathbb{C}$. The Wodzicki residue res and the renormalized traces tr^Q have to be understood as functionals defined on pseudo-differential operators acting on $E \otimes \mathbb{C}$. In order to compute $\mathrm{tr}^Q[A, B]$ and to differentiate $\mathrm{tr}^Q A$, in the topology of classical pseudo-differential operators, we need the following ([4], see also [49] for the first point):

Proposition A1.

(i) *Given two (classical) pseudo-differential operators A and B, given a weight Q,*

$$\mathrm{tr}^Q[A, B] = -\frac{1}{q}\mathrm{res}(A[B, \log Q]) \tag{A1}$$

(ii) *Given a differentiable family A_t of pseudo-differential operators, given a differentiable family Q_t of weights of constant order q,*

$$\frac{d}{dt}\left(tr^{Q_t} A_t\right) = tr^{Q_t}\left(\frac{d}{dt}A_t\right) - \frac{1}{q}\mathrm{res}\left(A_t(\frac{d}{dt}\log Q_t)\right) \tag{A2}$$

The following "covariance" property of tr^Q ([4,48]) will be useful to define renormalized traces on bundles of operators,

Proposition A2. *Under the previous notations, if C is a classical elliptic injective operator of order 0, $tr^{C^{-1}QC}(C^{-1}AC)$ is well-defined and equals $\mathrm{tr}^Q A$.*

We moreover have specific properties for weighted traces of a more restricted class of pseudo-differential operators (see [4,50,51]), called odd class pseudo-differential operators following [50,51] :

Definition A3. A classical pseudo-differential operator A is called odd class if and only if:

$$\forall n \in \mathbb{Z}, \forall (x, \xi) \in T^*M, \sigma_n(A)(x, -\xi) = (-1)^n \sigma_n(A)(x, \xi).$$

We note this class Cl_{odd}.

Such a definition is consistent for pseudo-differential operators on smooth sections of vector bundles, and applying the local formula for Wodzicki residue, one can prove [4]:

Proposition A3. *If M is an odd dimensional manifold, A and Q lie in the odd class, then $f(s) = tr(AQ^{-s})$ has no pole at $s = 0$. Moreover, if A and B are odd class pseudo-differential operators, $\mathrm{tr}^Q([A, B]) = 0$ and $\mathrm{tr}^Q A$ does not depend on Q.*

This trace was first defined in the papers [50,51] by Kontesevich and Vishik. We remark that it is in particular a trace on $DO(M, E)$ when M is odd-dimensional.

Let us now describe a class of operators which is, in some sense, complementary to odd class:

Definition A4. A classical pseudo-differential operator A is called even class if and only if:

$$\forall n \in \mathbb{Z}, \forall (x, \xi) \in T^*M, \sigma_n(A)(x, -\xi) = (-1)^{n+1}\sigma_n(A)(x, \xi)$$

We note this class Cl_{even}.

Very easy properties are the following:

Proposition A4.
$Cl_{even} \circ Cl_{odd} = Cl_{odd} \circ Cl_{even} = Cl_{even}$ and
$Cl_{even} \circ Cl_{even} = Cl_{odd} \circ Cl_{odd} = Cl_{odd}$.

Now, following [6], we explore properties of tr^Q on Lie brackets.

Definition A5. Let E be a vector bundle over M, Q a weight and $a \in \mathbb{Z}$. We define :

$$\mathcal{A}_a^Q = \{B \in Cl(M, E); [B, \log Q] \in Cl^a(M, E)\}$$

Theorem A1. [6]

(i) *$\mathcal{A}_a^Q \cap Cl^0(M, E)$ is a subalgebra of $Cl(M, E)$ with unit.*
(ii) *Let $B \in Ell^*(M, E)$, $B^{-1}\mathcal{A}_a^Q B = \mathcal{A}_a^{B^{-1}QB}$, where B^{-1} is the parametrix.*
(iii) *Let $A \in Cl^b(M, E)$, and $B \in \mathcal{A}^Q_{-dimM-b-1}$, then $\mathrm{tr}^Q[A, B] = 0$.*
(iv) *For $a < -\frac{dimM}{2}$, $\mathcal{A}_a^Q \cap Cl^{\frac{-dimM}{2}}(M, E)$ is an algebra on which the renormalized trace is a trace* (i.e., *vanishes on the brackets).*

We now produce non trivial examples of operators that are in $\mathcal{A}_a^Q$ when Q is scalar, and secondly we give a formula for some non vanishing renormalized traces of a bracket.

Lemma A2. *Let Q be a weight on $C_0^\infty(M, V)$ and let B be a classical pseudo-differential operator of order b. If B or Q is scalar, then $[B, \log Q]$ is a classical pseudo-differential operator of order $b - 1$.*

Proposition A5. *Let Q be a scalar weight on $C_0^\infty(M, V)$. Then*

$$Cl^{a+1}(M, V) \subset \mathcal{A}_a^Q$$

Consequently,

(i) *if $ord(A) + ord(B) = -dimM$, $\mathrm{tr}^Q[A, B] = 0$.*
(ii) *when $M = S^1$, if A and B are classical pseudo-differential operators, if A is compact and B is of order 0, $\mathrm{tr}^Q[A, B] = 0$.*

Lemma A3. *Let Q be a scalar weight on $C_0^\infty(M,V)$, and A, B two pseudo-differential operators of orders a and b on $C_0^\infty(M,V)$, such that $a+b=-m+1$ ($m = dim\ M$). Then*

$$\mathrm{tr}^Q[A,B] = -\frac{1}{q}\mathrm{res}\,(A[B,\log Q]) = -\frac{1}{q(2\pi)^n}\int_M\int_{|\xi|=1} tr(\sigma_a(A)\sigma_{b-1}([B,\log Q]))$$

References

1. Pressley, A.; Segal, G. *Loop Groups*; OUP: Oxford, UK, 1988.
2. Magnot, J.P.; Renormalized traces and cocycles on the algebra of S^1-pseudo-differential operators. *Lett. Math. Phys.* **2006**, *75*, 111–127.
3. Magnot, J.P. The Schwinger cocycle on algebras with unbounded operators. *Bull. Sci. Math.* **2008**, *132*, 112–127.
4. Cardona, A.; Ducourtioux, C.; Magnot, J.P.; Paycha, S. Weighted traces on pseudo-differential operators and geometry on loop groups. *Infin. Dimens. Anal. Quant. Probab. Relat. Top.* **2002**, *5*, 503–541.
5. Freed, D. The geometry of loop groups. *J. Differ. Geom.* **1988**, *28*, 223–276.
6. Magnot, J.P. Chern forms on mapping spaces. *Acta Appl. Math.* **2006**, *91*, 67–95.
7. Gilkey, P. *Invariance Theory, the Heat Equation and the Atiyah-Singer Index Theorem.* Publish or Perish Press: Houston, TX, USA, 1985; p. 349.
8. Bokobza-Haggiag, J. Opérateurs pseudo-différentiels sur une variété différentiable. *Ann. Inst. Fourier Grenoble* **1969**, *19*, 125–177.
9. Widom, H. A complete symbolic calculus for pseudo-differential operators. *Bull. Sci. Math.* **1980**, *104*, 19–63.
10. Adams, M.; Ratiu, T.; Schmidt, R. A Lie group structure for pseudo-differential operators. *Math. Ann.* **1986**, *273*, 529–551.
11. Adams, M.; Ratiu, T.; Schmidt, R. A Lie group structure for Fourier integral operators. *Math. Ann.* **1986**, *276*, 19–41.
12. Omori, H. *Infinite Dimensional Lie Groups*; Math. Surveys and Monographs AMS: Providence, RI, USA, 1997; Volume 158.
13. Hörmander, L. Fourier integral operators. I. *Acta Math.* **1971**, *127*, 79–189.
14. Omori, H.; Maeda, Y.; Yoshioka, A. On regular Fréchet Lie groups, I; Some differential geometric expressions of Fourier integral operators on a Riemannian manifold. *Tokyo J. Math.* **1980**, *3*, 353–390.
15. Omori, H; Maeda, Y; Yoshioka, A. On regular Fréchet Lie groups II; Composition rules of Fourier integral operatorson a Riemannian manifold. *Tokyo J. Math.* **1981**, *4*, 221–253.
16. Omori, H; Maeda, Y; Yoshioka, A.; Kobayashi, O. On regular Fréchet Lie groups III; A second cohomology class related to theLie algebra of pseudo-differential oprators of order 1. *Tokyo J. Math.* **1981**, *4*, 255–277.
17. Omori, H; Maeda, Y; Yoshioka, A.; Kobayashi, O. On regular Fréchet Lie groups IV; Definition and fundamental theorems. *Tokyo J. Math.* **1981**, *5*, 365–398.
18. Omori, H; Maeda, Y; Yoshioka, A.; Kobayashi, O. On regular Fréchet Lie groups V; Several basic properties. *Tokyo J. Math.* **1983**, *6*, 39–64.
19. Omori, H.; Maeda, Y.; Yoshioka, A.; Kobayashi, O. Infinite dimensional Lie groups that appear in general relativity. *Tokyo J. Math.* **1983**, *6*, 217–246.
20. Omori, H; Maeda, Y; Yoshioka, A.; Kobayashi, O. On regular Fréchet Lie groups VII; The group generated by pseudo-differential operators of negative order. *Tokyo J. Math.* **1984**, *7*, 315–336.
21. Omori, H; Maeda, Y; Yoshioka, A.; Kobayashi, O. Primordial operators and Fourier integral operators. *Tokyo J. Math.* **1985**, *8*, 1–47.
22. Ratiu, T.; Schmid, R. The differentiable structure of three remarkable diffeomorphism groups. *Math. Z.* **1981**, *177*, 81–100.
23. Kriegl, A.; Michor, P.W. *The Convenient Setting for Global Analysis.* Math. Surveys and Monographs AMS: Providence, RI, USA, 1997; Volume 53.
24. Magnot, J.P. Ambrose-Singer theorem on diffeological bundles and complete integrability of the KP equation. *Int. J. Geom. Methods Mod. Phys.* **2013**, *10*, doi:10.1142/S0219887813500436.

25. Savin, A.Y.; Sternin, B.Y. Uniformization of nonlocal elliptic operators and KK-theory. *Russ. J. Math. Phys.* **2013**, *20*, 345–359.
26. Alekseevsky, D; Michor, P.; Ruppert, W. Extensions of Lie algebras. **2004**, arXiv:math/0005042v3.
27. Abbati, M.C.; Cirelli, R.; Mania, A.; Michor, P. The Lie group of automorphisms of a principal bundle. *J. Geom. Phys.* **1989**, *6*, 215–235.
28. Magnot, J.P. The Kähler form on the loop group and the Radul cocycle on Pseudo-differential Operators. In *GROUP'24: Physical and Mathematical aspects of symmetries*, Proceedings of the 24th International Colloquium on Group Theorical Methods in Physics, Paris, France, 15–20 July 2002; Institut of Physic Conferences Publishing: Bristol, UK, 2003, Volume 173, pp. 671–675.
29. Kassel, C. Le résidu non commutatif (d'après M. Wodzicki) Séminaire Bourbaki, *Astérisque* **1988**, *89*, 177–178.
30. Kobayashi, S.; Nomizu, K. *Fundations of Differential Geometry I, II*; Wiley: Hoboken, NJ, USA, 1963.
31. Schwinger, J. Field theory of commutators. *Phys. Rev. Lett.* **1959**, *3*, 296–297.
32. Cederwall, M.; Ferretti, G.; Nilsson, B; Westerberg, A. Schwinger terms and cohomology of pseudo-differential operators *Commun. Math. Phys.* **1996**, *175*, 203–220.
33. Kravchenko, O.S.; Khesin, B.A. A central extension of the algebra of pseudo-differential symbols *Funct. Anal. Appl.* **1991**, *25*, 152–154.
34. Radul, O.A. Lie albegras of differential operators, their central extensions, and W-algebras *Funct. Anal. Appl.* **1991**, *25*, 25–39.
35. Binz, H.R. Fischer: The manifold of embeddings of a closed manifold. In *Proceedings of Differential Geometric Methods in Theoretical Physics, Clausthal 1978, Lecture Notes in Physics 139*; Springer-Verlag: Berlin, Germany, 1981; pp. 310–329.
36. Gay-Balmaz, F.; Vizman, C. Principal bundles of embeddings and non linear grassmannians. **2014**, arXiv:1402.1512v1.
37. Eells, J. A setting for global analysis. *Bull. Am. Math. Soc.* **1966**, *72*, 751–807.
38. Haller, S.; Vizman, C. Non-linear grassmannian as coadjoint orbits. *Math. Ann.* **2004**, *139*, 771–785.
39. Motilor, M. La Grassmannienne Non-linéaire comme Variété Fréchétique Homogène. *J. Lie Theory* **2008**, *18*, 523–539.
40. Magnot, J.P. Structure groups and holonomy in infinite dimensions. *Bull. Sci. Math.* **2004**, *128*, 513–529.
41. Hirsch, M. *Differential Topology*; Springer: New York, NY, USA, 1976; Volume 33.
42. Scott, S. *Traces and Determinants of Pseudodifferential Operators*; OUP: Oxford, UK, 2010.
43. Paycha, S. *Regularised Integrals, Sums and Traces. an Analytic Point of View*; University Lecture Series; AMS: Providence, RI, USA, 2012; Volume 59.
44. Seeley, R.T. Complex powers of an elliptic operator. *AMS Proc. Symp. Pure Math.* **1968**, *10*, 288–307.
45. Lesch, M. On the non commutative residue for pseudo-differential operators with log-polyhomogeneous symbol. *Ann. Glob. Anal. Geom.* **1998**, *17*, 151–187.
46. Wodzicki, M. Local invariants in spectral asymmetry. *Invent. Math.* **1984**, *75*, 143–178.
47. Adler, M. On a trace functionnal for formal pseudo-differential operators and the symplectic structure of Korteweg-de Vries type equations *Invent. Math.* **1979**, *50*, 219–248.
48. Paycha, S. Renormalized traces as a looking glass into infinite dimensional geometry. *Infin. Dimens. Anal. Quant. Probab. Relat. Top.* **2001**, *4*, 221–266.
49. Melrose, R.; Nistor, V. Homology of pseudo-differential operators I. manifolds without boundary, **1996**, arXiv:funct-an/9606005.
50. Kontsevich, M.; Vishik, S. *Determinants of Elliptic Pseudo-Differential Operators*; Max Plank Institut fur Mathematik: Bonn, Germany, 1994.
51. Kontsevich, M.; Vishik, S. Geometry of determinants of elliptic operators. In *Functional Analysis on the Eve of the 21st Century*; Birkhauser Boston: New Brunswick, NJ, USA, 1993; Volume 1.

2

Entropic Uncertainty Relations for Successive Generalized Measurements

Kyunghyun Baek [1] **and Wonmin Son** [1,2,*]

[1] Department of Physics, Sogang University, Mapo-gu, Shinsu-dong, Seoul 121-742, Korea; baek1013@gmail.com
[2] Department of Physics, University of Oxford, Parks Road, Oxford OX1 3PU, UK
* Correspondence: sonwm@physics.org

Academic Editors: Paul Busch, Takayuki Miyadera and Teiko Heinosaari

Abstract: We derive entropic uncertainty relations for successive generalized measurements by using general descriptions of quantum measurement within two distinctive operational scenarios. In the first scenario, by merging two successive measurements into one we consider successive measurement scheme as a method to perform an overall composite measurement. In the second scenario, on the other hand, we consider it as a method to measure a pair of jointly measurable observables by marginalizing over the distribution obtained in this scheme. In the course of this work, we identify that limits on one's ability to measure with low uncertainty via this scheme come from intrinsic unsharpness of observables obtained in each scenario. In particular, for the Lüders instrument, disturbance caused by the first measurement to the second one gives rise to the unsharpness at least as much as incompatibility of the observables composing successive measurement.

Keywords: entropic uncertainty relations; successive measurements; unsharpness; disturbance

1. Introduction

Ever since Heisenberg proposed uncertainty principle under consideration of γ-ray microscope in [1], the uncertainty principle has become one of the most central concepts in quantum physics. Till now, there have been concatenated debates to find uncertainty relations quantitatively well-formulated to reflect underlying meanings of the uncertainty principle [2]. Among the uncertainty relations, one of the most widely known forms of uncertainty relations may be Robertsons's relation formulated in terms of statistical variances [3]. This relation was discovered by generalizing Kennard's relation [4] for a pair of arbitrary observables, which indicates limitations on one's ability to prepare system being well-localized in position and momentum spaces simultaneously, so-called *preparation relation*. However, underlying meaning of it is not equivalent to Heisenberg's first insight that there should be a trade-off between imprecision of an instrument measuring a particle's position and disturbance of its momentum, which is so-called *error-disturbance relation*. From the Heisenberg's perspective, various forms of error-disturbance relation were derived based on state-dependent and state-independent quantifications of error and disturbance in [5–8], respectively. Here, the point to note in the course of the quantifications is that successive measurement scheme has played major roles in clarifying meaning of error and disturbance, and with increasing experimental ability to control quantum systems these relations were proved [9,10] by applying this scheme. Nevertheless, uncertainty relations for successive measurements have received less attentions than they deserve, as discussed in [11,12].

In the field of quantum information theory, uncertainty relations have been formulated in terms of information-theoretic quantities such as entropy, since they are well-defined as a measure of uncertainty in the sense that they are invariant under relabeling of outcomes and concave functions (refer to [13] for further discussions). This information-theoretic approach has been conducted in both preparation and error-disturbance relations. From the point of preparation relation, entropic uncertainty relations were suggested in [14] and then improved by Massen-Uffink [15]. A generalized version of it is written in the form of [16]

$$H_\rho(A) + H_\rho(B) \geq c \tag{1}$$

where we measure observables A and B described by positive-operator-valued measures (POVM) $\{\hat{A}_i\}$ and $\{\hat{B}_j\}$ on a quantum system $\hat{\rho}$, respectively. In the relation, the Shannon entropy is denoted by $H_\rho(A) = -\sum_{i=1}^{n_A} p_A(i) \log p_A(i)$, where $p_A(i) = \mathrm{Tr}[\hat{A}_i\hat{\rho}]$ is the probability to obtain i-th outcome of a measurement of A. The lower bound representing incompatibility between A and B is given by

$$c = -\log \max_{i,j} \|\sqrt{\hat{A}_i}\sqrt{\hat{B}_j}\|^2 \tag{2}$$

where the operator norm is denoted by $\|\hat{C}\|$ meaning the maximal singular value of $\hat{C}$. Here and in the following, we will take logarithm in base 2 according to the information-theoretic convention. From the point of error-disturbance relation, there have been recent works to formulate the relation in terms of entropy in order to obtain operationally meaningful formulation, based on state-dependent [17] and state-independent quantifications [18]. (see [19,20] for more details.)

Inspired by the Heisenberg's first insight, entropic uncertainty relations for successive projective measurements were considered in an information-theoretic approach [12,21]. Subsequently, this approach was developed based on Rényi's entropies [22] and Tsallis' entropies [23] for a pair of qubit observables. However, the concept of generalized measurements have not been considered in successive measurement scenario. Therefore, the main purpose of the present work is to generalize the entropic uncertainty relations for the case of POVMs. More specifically, we will focus on deriving entropic uncertainty relations for successive generalized measurements with respect to two scenarios. In the first scenario, we consider a statistical distribution of probabilities $p_{AB}(i,j)$ to obtain sequentially measurement outcomes i and j of the first and the second measurements A, B, respectively. In the second scenario, we analyze the marginal distributions $p_A(i)$ and $p_{B'}(j)$ associated with jointly measurable observables A and B', respectively. In particular, it was argued that the second scenario can be considered as a general method to measure any pair of jointly measurable quantum observables [24], and further it has special usefulness due to so-called universality of successive measurement [25]. In this regard, the range of its applications becomes broader (see the references in [25]). Additionally, in both scenarios, the effect of unsharpness of observables on entropic uncertainty relations will be discussed by using the quantification of unsharpness previously defined in [26].

This paper is organized in the following way. In Section 2 we will introduce a quantity defined as a measure of unsharpness and clarify explicit mathematical expressions of measuring process. Subsequently, based on these mathematical descriptions, entropic uncertainty relations in the first scenario of successive measurement scheme will be derived in Section 3 with specific examples. In Section 4, the second scenario will be considered to derive entropic uncertainty relations. Finally, we will highlight important points of the results, in Section 5.

2. Preliminaries

In this section, we introduce the basic concepts necessary to generalize the entropic uncertainty relations for the case of POVMs in successive measurement scenarios.

2.1. Measure of Unsharpness

Here we introduce the measure of unsharpness which was derived based on entropy in [26]. To begin with, let us clarify notations and terminologies as follows. For a finite d-dimensional Hilbert space $\mathcal{H}_d$, we denote the vector space of all linear operators on $\mathcal{H}_d$ by $\mathcal{L}(\mathcal{H}_d)$. Any observable A then is generally described by POVM $\{\hat{A}_i\}$ which is a set of positive operators $\hat{A}_i \in \mathcal{L}(\mathcal{H}_d)$ obeying the completeness relation, $\sum_{i=1}^{n_A} \hat{A}_i = \hat{I}$ with the number of outcomes of the measurement n_A. In a particular case that all POVM elements are given as projections, A is a projection-valued measure (PVM). In this case, the observable A is commonly considered as the description of a measurement with perfect accuracy, and thus is called a *sharp observable*. On the other hand, if A is not a PVM, it is called an *unsharp observable*.

To clarify the distinction between the concepts of sharp and unsharp observables, we consider $\hat{A}_i$ in the form of spectral decomposition

$$\hat{A}_i = \sum_{k=1}^{d} a_i^k |a_i^k\rangle\langle a_i^k| \quad (3)$$

where $0 \leq a_i^k \leq 1$ is an eigenvalue corresponding to an eigenvector $|a_i^k\rangle$. By means of this expression, one can find that the condition for sharp observables is equivalent to the statement that all eigenvalues of $\hat{A}_i$ are given by either 0 or 1, *i.e.*, $\forall a_i^k \in \{0, 1\}$, as discussed in [27]. Otherwise, we can say that it is unsharp. This statement gives us the idea that the unsharpness measure should be defined as a function of a_i^k vanishing only when a_i^k is 0 or 1. Reflecting this idea, the function can be selected in the form of $h(a_i^k) = -a_i^k \log a_i^k$, and by averaging it over all POVM elements the measure of unsharpness is defined as [26]

$$D_\rho(A) = \sum_{i=1}^{n_A} \sum_{k=1}^{d} \langle a_i^k|\hat{\rho}|a_i^k\rangle h(a_i^k) \quad (4)$$

which is so-called *device uncertainty*, where a measurement of A is performed on a quantum system $\hat{\rho}$. As a measure of unsharpness, this quantity possesses essential properties such that $D_\rho(A) = 0$ for all states if and only if A is a PVM, and $D_\rho(A) = H_\rho(A)$ for all states if and only if $\hat{A}_i = \lambda_i \hat{I}$ for all i with $0 \leq \lambda_i \leq 1$ satisfying $\sum_{i=1}^{n_A} \lambda_i = 1$. With additional properties, the validity of this quantity for the unsharpness measure has been verified on the lines of the previous work [28] in which the unsharpness is characterized based on statistical variance. In particular, an important point is that the device uncertainty gives us a nontrivial lower bound of entropy by itself such that

$$H_\rho(A) \geq D_\rho(A) \geq \min_\rho D_\rho(A) \geq -\log \max_i \|\hat{A}_i\| \quad (5)$$

due to the concavity of entropy [29]. Moreover, the minimal device uncertainty can be obtained by diagonalizing $\sum_{i=1}^{n_A} \sum_{k=1}^{d} h(a_i^k)|a_i^k\rangle\langle a_i^k|$ and taking the lowest eigenvalue, which is stronger than $-\log \max_i \|\hat{A}_i\|$ proposed in [16]. In other words, this method makes it available for us to generally find stronger state-independent bounds, and thus will play key roles in deriving entropic uncertainty relations for successive measurements. The detailed properties of the device uncertainty $D_\rho(A)$ has been studied in [26].

2.2. *General Description of Successive Measurements*

In the present work, by a *successive measurement*, we mean a scheme where two measurements are performed one after the other successively. In particular, the second measurement is assumed to be performed immediately on an output state conditionally transformed according to an outcome of the first measurement. Thus, in order to consider successive measurement scheme, we should clarify how input state is transformed to output states conditioned on the measurement outcome. For this purpose, however, the concept of POVM is not enough to fully describe the state transformation. For general description of successive measurements, therefore, we need the concept of an *instrument* [30], which is a mapping $\mathcal{I} : i \to \mathcal{I}_i$ such that each $\mathcal{I}_i$ is a completely positive linear map on $\mathcal{L}(\mathcal{H}_d)$ satisfying $\sum_{i=1}^{n} \mathrm{tr}[\mathcal{I}_i(\hat{\rho})] = 1$ for all states $\hat{\rho}$.

A general description of a quantum measurement is given by a pair of an observable A and an instrument. However, a notable point is that for a given A not all instruments are compatible with

A. For the description to be valid, an instrument should obey the condition that each i-th completely positive linear map $\mathcal{I}_i^A$ satisfies

$$\mathrm{tr}[\mathcal{I}_i^A(\hat{\rho})] = \mathrm{tr}[\hat{A}_i\hat{\rho}] \tag{6}$$

for all states $\hat{\rho}$, and in this case we say that an instrument $\mathcal{I}^A$ is A-compatible. Accordingly, A-compatible instrument illustrates that a measurement outcome i is obtained with the probability $p_i^A = \mathrm{tr}[\hat{A}_i\hat{\rho}]$ for an input state $\hat{\rho}$, and a normalized output state $\mathcal{I}_i^A(\hat{\rho})/p_i^A$ is generated as depicted in Figure 1a. Among A-compatible instruments, the most common instruments occurring in applications may be the *Lüders instrument*, defined by

$$\mathcal{I}_i^{A_L}(\hat{\rho}) = \sqrt{\hat{A}_i}\hat{\rho}\sqrt{\hat{A}_i} \tag{7}$$

The Lüders instrument is the generalized version of projective measurements for general POVM, in the sense that for sharp observables it illustrates the same with state transformation of projective measurement.

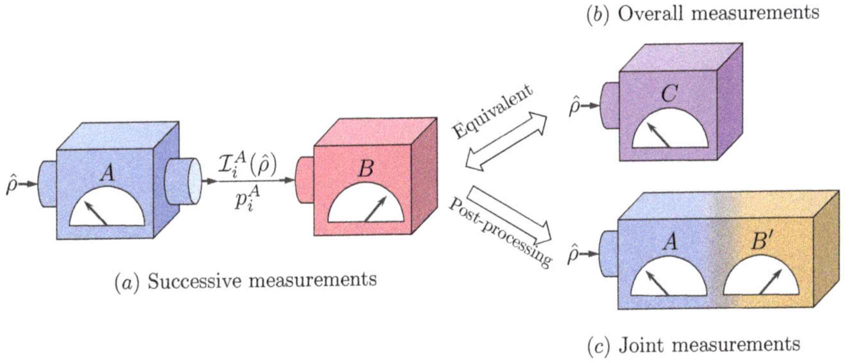

Figure 1. Relations among measurement schemes. (**a**) Successive measurement of observables A and B, where the first measurement A gives rise to output state $\mathcal{I}_i^A(\rho)/p_A(i)$ conditioned on its outcome i; (**b**) Overall measurement of C obtained by performing the successive measurement of A and B; (**c**) Joint measurements obtained in successive measurement scheme by considering the marginal distributions.

Additionally, the concept of measuring process deserves consideration in order to describe successive measurement in a more specific way, of which a description is known to be consistent with the description of instrument [31]. A *measuring process* is defined to be a quadruple $(\mathcal{K}, |\xi\rangle, \hat{U}, F)$ consisting of a Hilbert space $\mathcal{K}$ associated with a probe system, a state vector $|\xi\rangle$ on $\mathcal{K}$, a unitary operator $\hat{U}$ on $\mathcal{H}\otimes\mathcal{K}$, and an observable F on $\mathcal{K}$. The quadruple $(\mathcal{K}, |\xi\rangle, \hat{U}, F)$ is compatible with an observable A if it satisfies

$$\mathrm{tr}[\hat{A}_i\hat{\rho}] = \mathrm{tr}[\hat{U}(\hat{\rho}\otimes|\xi\rangle\langle\xi|)\hat{U}^\dagger(\hat{I}\otimes\hat{F}_i)] \tag{8}$$

for all i and states. It means that the measuring process for A gives rise to the same with a probability distribution obtained by performing a measurement of A directly on the system. In this case, an instrument $\mathcal{I}^A$ is related to the measuring process $(\mathcal{K}, |\xi\rangle, \hat{U}, F)$ in the following manner

$$\mathcal{I}_i^A(\hat{\rho}) = \mathrm{tr}_{\mathcal{K}}[\hat{U}(\hat{\rho}\otimes|\xi\rangle\langle\xi|)\hat{U}^\dagger(\hat{I}\otimes\hat{F}_i)] \tag{9}$$

for all i and states. In this way, each measuring process defines a unique instrument and conversely for every instrument there exists a measuring process explicitly describing the same state transformation [31]. We refer to [32] for more details.

Based on the above mathematical descriptions, there are two scenarios to investigate statistical properties of a distribution obtained in the successive measurement scheme. Now, let us consider the first scenario in which we sequentially perform two measurements A and B as depicted in Figure 1a, where the numbers of measurement outcomes are n_A and n_B, respectively. Then, this scenario can be seen as a method to obtain the *overall observable* C described by POVM $\{\hat{C}_{ij}\}$ obeying

$$\mathrm{tr}[\hat{C}_{ij}\hat{\rho}] = \mathrm{tr}[\mathcal{I}_i^A(\hat{\rho})\hat{B}_j] = p_{AB}(i,j) \tag{10}$$

for all i, j and all states $\hat{\rho}$. In the Heisenberg picture, equivalently, it can be rewritten as

$$\hat{C}_{ij} = \mathcal{I}_i^{A*}(\hat{B}_j) \tag{11}$$

where $\mathcal{I}_i^{A*}$ denotes the adjoint map of $\mathcal{I}_i^A$. As illustrated in Figure 1b, namely, the successive measurements A, B are merged into C having $n_A n_B$ outcomes. Consequently, a task to analyze statistical properties in the successive measurements can be accomplished by exploring the overall observable C without loss of generality. In this approach, we will investigate the entropic uncertainty relation in Section 3.

On the other hand, in the second scenario we take into account the marginal distributions obtained as $\sum_{j=1}^{n_B} p_{AB}(i,j) = p_A(i)$ and $\sum_{i=1}^{n_A} p_{AB}(i,j) = p_{B'}(j)$. Namely, the successive measurement scheme is considered as a strategy to perform a joint measurement of A and B' as depicted in Figure 1c, where the observables A and B' are described by

$$\hat{A}_i = \sum_{j=1}^{n_B} \hat{C}_{ij} \quad \text{and} \quad \hat{B}'_j = \sum_{i=1}^{n_A} \hat{C}_{ij} \tag{12}$$

for all i, j, respectively. It is worth noting that performing the second measurement B is effectively equivalent to perform the measurement B' on the initial system, since the first measurement disturbs the second one. However, it has been shown that, despite of this disturbance, any jointly measurable pair of quantum observables can be measured by means of successive measurement scheme [24]. Therefore, considering the second scenario is a general way to explicitly explore the concept of jointly measurable observables. Entropic uncertainty relations in this scenario will be taken into account in Section 4.

3. Generalized Version of Entropic Uncertainty Relation for Successive Measurements

In the previous section, the mathematical methods have been presented, which are necessary to describe successive measurement in general. Based on the methods, we derive entropic uncertainty relations within the first scenario. As mentioned in Section 2.2, we can consider performing successive measurement of A and B as a method to implement the corresponding overall measurement of C. This fact implies that uncertainty existing in this scenario can be equivalently characterized as entropy of C,

$$H_\rho(A,B) = H_\rho(C) \tag{13}$$

since $p_{AB}(i,j) = p_C(i,j)$ for all i, j. In other words, our goal to analyze uncertainty existing in the first scenario can be achieved under consideration of the overall observable C. From this point of view, one can identify that the reason why we cannot avoid uncertainty in this scheme originates from intrinsic unsharpness of the overall observable C. By quantitatively formulating this fact, as described in Equation (5), we obtain entropic form of uncertainty relation lower bounded by device uncertainty characterizing unsharpness of C such that

$$H_\rho(A,B) \geq D_\rho(C) \geq \min_\rho D_\rho(C) \equiv \mathcal{D}_1 \tag{14}$$

where state-independent bound $\mathcal{D}_1$ is obtained by minimizing device uncertainty of C over all states. An important point here is that sequentially measuring incompatible observables may give rise to unavoidable unsharpness as the second measurement is disturbed by performing the first one. We can clearly observe the phenomena in the following cases.

3.1. Projective Measurement Model

In order to examine how much unsharpness emerges due to the incompatibility in the first scenario, let us consider successive projective measurements of observables A and B described by orthonormal bases $\{|a_i\rangle\}$ and $\{|b_j\rangle\}$ in $\mathcal{H}_d$, respectively. Then, according to the state transformation $\mathcal{I}_i^A(\rho) = \langle a_i|\hat{\rho}|a_i\rangle |a_i\rangle\langle a_i|$, this successive measurement can be considered as the overall measurement of C, which is described by

$$\hat{C}_{ij} = |\langle a_i|b_j\rangle|^2 |a_i\rangle\langle a_i| \tag{15}$$

for all i, j. In this case, by calculating the minimal value of device uncertainty defined in Equation (5) we obtain

$$H_\rho(A, B) \geq \min_i \left(-\sum_{j=1}^{d} |\langle a_i|b_j\rangle|^2 \log |\langle a_i|b_j\rangle|^2 \right) \tag{16}$$

where the lower bound was proposed in [12]. Here, it is notable that this bound is stronger than $-\log \max_{i,j} |\langle a_i|b_j\rangle|^2$, which is widely known as a measure of incompatibility [15]. Thus, successively performing projective measurements of incompatible sharp observables induces unavoidable unsharpness which gives limits on one's ability to measure with arbitrarily low uncertainty.

3.2. Lüders Instrument

As a generalized version of projective measurement model, let us assume that we implement the Lüders instrument for an unsharp observable A at first and later a measurement of B in the first scenario. In this case, each map $\mathcal{I}_i^{A_L}$ is fully determined by POVM element $\hat{A}_i$ as defined in Equation (7), so that by applying adjoint map of $\mathcal{I}_i^{A_L}$ to each POVM element of B such as Equation (11), we obtain the explicit form of the overall observable C described by

$$\hat{C}_{ij} = \sqrt{\hat{A}_i} B_j \sqrt{\hat{A}_i} \tag{17}$$

for all i, j. Then, it is straightforward to formulate relations among the concepts of uncertainty, unsharpness and incompatibility by directly using the relations in Equation (5)

$$H_\rho(A, B) \geq \mathcal{D}_1 \geq -\log \max_{i,j} \|\sqrt{\hat{A}_i} B_j \sqrt{\hat{A}_i}\| = c \tag{18}$$

Here, the last inequality in Equation (18) means that the minimal value of device uncertainty gives rise to a stronger bound than the incompatibility c defined in Equation (2). Therefore, as observed in the case of projective measurement model, one can identify that measuring incompatible observables by means of the Lüders instrument imposes the unavoidable unsharpness. In the following examples, we will analyze the relationships in Equation (18).

3.3. Examples in Spin $\frac{1}{2}$ System

In order to clarify the relationship between $\mathcal{D}_1$ and c and verify the validity of $\mathcal{D}_1$ for lower bound of uncertainty relation (18), let us consider an example of successively measuring two spin observables Z at first and $X(\theta)$ later in $\mathcal{H}_2$ described by

$$\hat{Z}_{\pm} = \frac{\hat{I} \pm s\hat{\sigma}_z}{2} \quad \text{and} \quad \hat{X}_{\pm}(\theta) = \frac{\hat{I} \pm t(\sin\theta\hat{\sigma}_x + \cos\theta\hat{\sigma}_z)}{2} \tag{19}$$

respectively, where $\hat{\sigma}_x$ and $\hat{\sigma}_z$ are the Pauli spin matrices and unsharp parameters are denoted by $0 \leq s, t \leq 1$. Then the incompatibility of the observables is determined by θ which is angle between directions of measurement components. Additionally, we assume that a Z-compatible instrument is induced by a measuring process $(\mathcal{H}_2, |\phi\rangle, \hat{U}_{CNOT}, \sigma_z)$, where an initial state of the probe system is $|\phi\rangle = \sqrt{(1+s)/2}|0\rangle + \sqrt{(1-s)/2}|1\rangle$ and a unitary operator $\hat{U}_{CNOT}$ gives rise to a CNOT gate controlled by eigenstates of $\hat{\sigma}_z$, $|0\rangle$ and $|1\rangle$. Namely, this measuring process leads to the Lüders instrument for Z. In this case, the successive measurement scheme is equivalent to perform the overall measurement of S defined in terms of four POVM elements

$$\hat{S}_{\pm\pm} = \frac{1}{4}\Big((1 + st\cos\theta)\hat{I} \pm \sqrt{1-s^2}t\sin(\theta)\ \hat{\sigma}_x \pm (s + t\cos\theta)\hat{\sigma}_z\Big) \tag{20}$$

$$\hat{S}_{\pm\mp} = \frac{1}{4}\Big((1 - st\cos\theta)\hat{I} \mp \sqrt{1-s^2}t\sin(\theta)\ \hat{\sigma}_x \pm (s - t\cos\theta)\hat{\sigma}_z\Big) \tag{21}$$

Calculating the device uncertainty for the overall observable S, we obtain state-independent lower bounds

$$\mathcal{D}_1 = \sum_{\substack{\mu=\pm1 \\ \nu=\pm1}} h\left(\frac{1}{4}\left(1 + \mu st\cos\theta + \nu\sqrt{s^2 + t^2 + \mu 2st\cos\theta + s^2t^2(\cos^2\theta - 1)}\right)\right) \tag{22}$$

We plot the lower bounds $\mathcal{D}_1$ and $c = -\log\left(\frac{1}{4}(1 + st|\cos\theta| + \sqrt{s^2 + t^2 + 2st|\cos\theta| + s^2t^2(\cos^2\theta - 1)})\right)$ presented in Equation (14) versus the angle θ. As a result, we can check that $\mathcal{D}_1$ gives rise to strictly stronger bound than c except when the spin components are mutually parallel or perpendicular in Figure 2a. With increasing unsharpness of the observables, it becomes evident that $\mathcal{D}_1$ is well-formulated to reflect the effect of unsharpness as observed in Figure 2b,c. In the examples, we analytically confirm the validity of $\mathcal{D}_1$ for a lower bound by comparing with c, and, as a result, observe that measuring incompatible observables gives rise to unavoidable unsharpness originating from the incompatibility on the assumption that we implement the Lüders instrument in the first scenario.

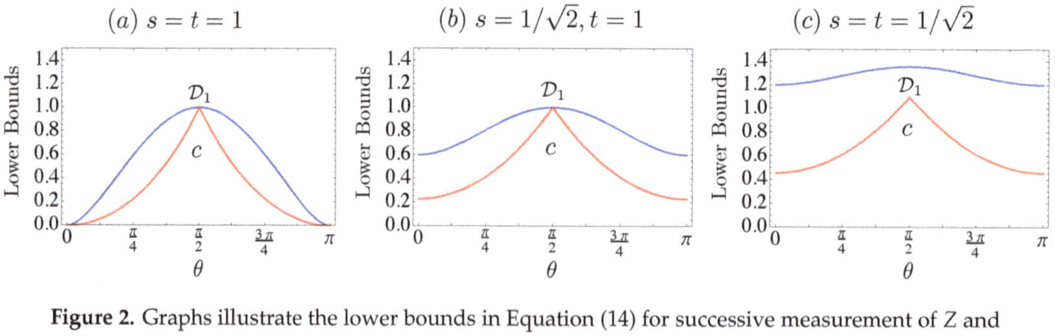

Figure 2. Graphs illustrate the lower bounds in Equation (14) for successive measurement of Z and $X(\theta)$ with respect to angle θ. Varying unsharpness of the observables, we present three cases. (**a**) At $s = t = 1$, both observables are sharp; (**b**) At $s = 1/\sqrt{2}, t = 1$, only the first one is unsharp; (**c**) At $s = t = 1/\sqrt{2}$, both are unsharp.

4. Entropic Uncertainty Relations for a Jointly Measurable Pair of Observables Obtained via Successive Measurement Scheme

In this section, we consider entropic uncertainty relations for a pair of jointly measurable observables obtained via successive measurement scheme within the second scenario. As a first step, let us clarify the concept of a joint measurement. Given observables A and B, they are *jointly measurable* if and only if there exists a *joint observable* M composed of $n_A n_B$-elements of POVM satisfying [33]

$$\sum_{j=1}^{n_B} \hat{M}_{ij} = \hat{A}_i \quad \text{and} \quad \sum_{i=1}^{n_A} \hat{M}_{ij} = \hat{B}_j \tag{23}$$

Specifically, in the second scenario, the overall observable C can be seen as a joint observable of A and B' by the definitions (12). Moreover, as mentioned in Section 2.2, this scheme generally provides a method to implement joint measurements of any jointly measurable observables [24]. Hence, entropic uncertainty relations obtained within this scenario is applicable to any pair of them.

Both observables A and B' obtained via the second scenario may have their own unsharpness, so that an amount of uncertainties about A and B' may not vanish due to the unsharpness of them. As formulating this fact, we obtain entropic uncertainty relations within the second scenario in the form of

$$H_\rho(A) + H_\rho(B') \geq D_\rho(A) + D_\rho(B') \geq \min_\rho \left[D_\rho(A) + D_\rho(B')\right] \equiv \mathcal{D}_2 \tag{24}$$

where the sum of device uncertainties is written as

$$D_\rho(A) + D_\rho(B') = \text{tr}\left[\hat{\rho}\left(\sum_{i=1}^{n_A}\sum_{k=1}^{d} h(a_i^k)|a_i^k\rangle\langle a_i^k| + \sum_{j=1}^{n_B}\sum_{l=1}^{d} h(b'^l_j)|b'^l_j\rangle\langle b'^l_j|\right)\right] \tag{25}$$

and thus minimizing it over all states can be accomplished by diagonalizing $\left(\sum_{i=1}^{n_A}\sum_{k=1}^{d} h(a_i^k)|a_i^k\rangle\langle a_i^k| + \sum_{j=1}^{n_B}\sum_{l=1}^{d} h(b'^l_j)|b'^l_j\rangle\langle b'^l_j|\right)$ and taking the lowest eigenvalue. An important point, here, is that the second measurement B may be perturbed to be B' because of disturbance caused by the first measurement A, while A is preserved. This fact implies that even when both observables A and B applied in this scheme are sharp, it is possible for the perturbed one B' to become unsharp. In particular, this behavior is apparently observed when applying a pair of incompatible observables to this scenario, since measuring one of a pair of incompatible observables disturbs the other, according to the Heisenberg's insight. From the point of view that the incompatibility imposes unavoidable unsharpness on the observables A and B', we will discuss more details in the following specific measurement models.

4.1. Projective Measurement Model

In the same way in Section 3.1, let us consider projective measurements of A and B described in $\mathcal{H}_d$ by orthonormal bases $\{|a_i\rangle\}$ and $\{|b_j\rangle\}$, respectively. Then, in the second scenario, the first one A remains itself, while the second one B is disturbed to be B' described by

$$\hat{B}'_j = \sum_{i=1}^{d} |\langle a_i|b_j\rangle|^2 |a_i\rangle\langle a_i| \tag{26}$$

for all j. According to the Lüders theorem [34], B is not disturbed if and only if all elements of A and B commute each other. In this case, thus, even though we perform two sharp observables sequentially, the second one involves the unavoidable unsharpness originating from the incompatibility between them. This behavior can be quantitatively formulated in the form of

$$H_\rho(A) + H_\rho(B') \geq D_\rho(B') \geq \min_\rho D_\rho(B') = \min_i \left(-\sum_{j=1}^{d} |\langle a_i|b_j\rangle|^2 \log |\langle a_i|b_j\rangle|^2\right) \tag{27}$$

where its lower bound was proposed in [12]. A generalized version of it for POVMs will be taken into account in the following.

4.2. Lüders Instruments

As a next step, in the same manner as Section 3.2, let us assume we sequentially implement the Lüders instrument of an observable A at first and later another one of B, in the second scenario. Then, according to Equation (7), it is equivalent to implement joint measurement of a pair of observables A and B' described by

$$\hat{A}_i \quad \text{and} \quad \hat{B}'_j = \sum_{i=1}^{n_A} \sqrt{\hat{A}_i}\hat{B}_j\sqrt{\hat{A}_i} \tag{28}$$

for all i, j, respectively. In this case, likewise as discussed above, performing the Lüders instrument of A gives rise to disturbance to B, in a way for B to become B'. Then, by applying the relations in Equation (5) directly, we obtain

$$H_\rho(A) + H_\rho(B') \geq D_\rho(A) + D_\rho(B') \geq \min_\rho \left[D_\rho(A) + D_\rho(B')\right] \geq -\log \max_j \| \sum_{i=1}^{n_A} \sqrt{\hat{A}_i}\hat{B}_j\sqrt{\hat{A}_i} \| \tag{29}$$

where the last inequality follows from applying the third inequality in Equation (5) solely to the observable B' and its bound is a new form of incompatibility larger than $-\log \max_{i,j} \| \sqrt{\hat{A}_i}\sqrt{\hat{B}_j} \|^2$, which was conjectured in [35] and proved later in [36]. A distinct point from projective measurement model is that there is a possibility for A to be unsharp, and loosing sharpness of A may decrease disturbance to B caused by A. Namely, a trade-off between the unsharpness of A and B can be observed. In the following examples, this phenomenon will be examined.

4.3. Examples in Spin $\frac{1}{2}$ System

As an example in the second scenario, we assume to implement the Lüders instrument of Z induced by the measuring process $(\mathcal{H}_2, |\phi\rangle, \hat{U}_{CNOT}, \sigma_z)$ and a measurement of X successively in $\mathcal{H}_2$ described by

$$\hat{Z}_\pm = \frac{\hat{I} \pm s\hat{\sigma}_z}{2} \quad \text{and} \quad \hat{X}_\pm = \frac{\hat{I} \pm \hat{\sigma}_x}{2} \tag{30}$$

respectively. Here we restrict ourselves for X to be sharp, in order to observe more clearly unsharpness appearing due to the first measurement. In this case, we can consider it as a method to measure a pair of jointly measurable observables Z and X', where the perturbed observable X' is given as [37]

$$\hat{X}'_\pm = \frac{\hat{I} \pm t\hat{\sigma}_x}{2} \tag{31}$$

with the unsharp parameter $t = \sqrt{1 - s^2}$. Consequently, this scheme provides an optimized method to jointly measure incompatible observables σ_z and σ_x, in the sense that the unsharp parameters s and t saturate the inequality $s^2 + t^2 \leq 1$, which is a necessary and sufficient condition for Z and X' to be jointly measurable [38]. However, even in the optimized method, we can not avoid unsharpness, and there is the trade-off between the unsharpness of Z and X' such that the more sharpness of Z, the more unsharpness of X'. This behavior can be examined quantitatively by considering entropic uncertainty relations (4) given in the form of

$$H_\rho(Z) + H_\rho(X') \geq D(Z) + D(X') = H_{bin}\left(\frac{1+s}{2}\right) + H_{bin}\left(\frac{1+\sqrt{1-s^2}}{2}\right) \tag{32}$$

where binary entropy is denoted by $H_{bin}(q) = -h(q) - h(1-q)$. The trade-off between the unsharpness of them is illustrated in Figure 3, and the total unsharpness characterized by $D(Z) + D(X')$ is maximized when unsharpness equally distributed $s = t = 1/\sqrt{2}$, while minimized at extreme points $s = 1$ or $s = 0$.

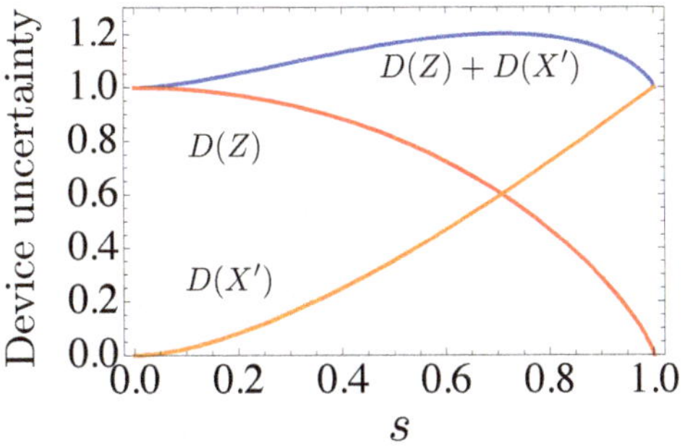

Figure 3. For a pair of jointly measurable observables Z and X' obtained via successive measurement scheme, we plot device uncertainties of them and their summation denotes by $D(Z)$, $D(X')$ and $D(Z) + D(X')$, respectively, versus the unsharp parameter s. We can observe trade-off relation between $D(Z)$ and $D(X')$ characterizing unsharpness of Z and X'.

5. Conclusions

The main purpose of present work is to suggest entropic uncertainty relations for successive generalized measurement within two distinctive scenarios. Before deriving the relations, in Section 2, we have introduced device uncertainty as a measure of unsharpness and its mathematical properties in order to investigate the effect of unsharpness on entropic uncertainty relations in successive measurement scheme. Subsequently, we have explicitly explained general description of successive measurement scheme with respect to two scenarios. In the first scenario, we have considered this scheme as a method to implement an overall measurement, and, as a result, observed that unsharpness of the overall observable gives limits on one's ability to measure it with arbitrarily low uncertainty as formulated in Equation (14). Assuming to perform the Lüders instrument as the first measurement in this scenario, it is clearly shown that this unsharpness comes from disturbance caused by the first measurement to the second one. The amount of unsharpness appears at least as much as the incompatibility of a pair of observables composing successive measurement, as formulated in Equation (18). In the second scenario, on the other hand, this scheme has been considered as a method to measure a pair of jointly measurable observables. Consequently, we have figured out that total unsharpness in both observables is a major factor that gives rise to unavoidable uncertainty as observed in Equation (24). Also, under the assumption for the first measurement to be described by the Lüders instrument, it becomes clear that the first measurement leads to unsharpness of the second one by disturbing it at least as much as incompatibility as shown in Equation (29). It is notable that this form of uncertainty relations is applicable to any pair of jointly observables, since Heinosaari *et al.* have proved that we can obtain any pair of jointly measurable observables via successive measurement scheme in [24].

Acknowledgments: This work was done with support of ICT R&D program of MSIP/IITP (No. 2014-044-014-002), the R&D Convergence Program of NST (National Research Council of Science and Technology) of Korea (Grant No. CAP-15-08-KRISS) and National Research Foundation (NRF) grant (No. NRF- 2013R1A1A2010537).

Author Contributions: K. Baek and W. Son contribute to the manuscript equally.

Conflicts of Interest: The authors declare no conflict of interest.

Abbreviations

The following abbreviations are used in this manuscript:

POVM: Positive-Operator-Valued Measure
PVM: Projection-Valued Measure

References

1. Heisenberg, W. Über den anschulichen Inhalt der quantentheoretischen Kinematik und Mechanik. *Z. Phys.* **1927**, *43*, 172–198.
2. Busch, P.; Heinosaari, T.; Lahti, P. Heisenberg's uncertainty principle. *Phys. Rep.* **2007**, *452*, 155–176.
3. Robertson, H.P. The uncertainty principle. *Phys. Rev.* **1929**, *34*, 163.
4. Kennard, E.H. Zur Quantenmechanik einfacher Bewegungstypen. *Z. Phys.* **1927**, *44*, 326–352.
5. Ozawa, M. Universally valid reformulation of the Heisenberg uncertainty principle on noise and disturbance in measurement. *Phys. Rev. A* **2003**, *67*, 042105.
6. Branciard, C. Error-tradeoff and error-disturbance relations for incompatible quantum measurements. *Proc. Nat. Acad. Sci. USA* **2013**, *110*, 6742–6747.
7. Busch, P.; Lahti, P.; Werner, R.F. Proof of Heisenberg's error-disturbance relation. *Phy. Rev. Lett.* **2013**, *111*, 160405.
8. Busch, P.; Lahti, P.; Werner, R.F. Heisenberg uncertainty for qubit measurements. *Phy. Rev. A* **2014**, *89*, 012129.
9. Erhart, J.; Spona, S.; Sulyok, G.; Badurek, G.; Ozawa, M.; Yuji, H. Experimental demonstration of a universally valid error-disturbance uncertainty relation in spin measurements. *Nat. Phys.* **2012**, *8*, 185–189.
10. Rozema, L.A.; Darabi, A.; Mahler, D.H.; Hayat, A.; Soudagar, Y.; Steinberg, A.M. Violation of Heisenberg's measurement-disturbance relationship by weak measurements. *Phys. Rev. Lett.* **2012**, *109*, 100404.
11. Distler, J.; Paban, S. Uncertainties in successive measurements. *Phy. Rev. A* **2013**, *87*, 062112.
12. Srinivas, M.D. Optimal entropic uncertainty relation for successive measurements in quantum information theory. *Paramana-J. Phys.* **2002**, *60*, 1137–1152.
13. Uffink, J.B.M. Measures of Uncertainty and the Uncertainty Principle. Ph.D. thesis, University of Utrecht, Utrecht, The Netherlands, 1990.
14. Deutsch, D. Uncertainty in Quantum Measurements. *Phy. Rev. Lett.* **1983**, *50*, 631.
15. Maasen, H.; Uffink, J.B.M. Generalized entropic uncertainty relations. *Phy. Rev. Lett.* **1988**, *60*, 1103–1106.
16. Krishna, M.; Parthasarathy, K.R. An entropic uncertainty principle for quantum measurements. *Indian J. Stat. Ser. A* **2002**, *64*, 842.
17. Coles, P.J.; Furrer, F. State-dependent approach to entropic measurement-disturbance relations. *Phys. Lett. A* **2015**, *379*, 105–112.
18. Buscemi, F.; Hall, M.J.W.; Ozawa, M.; Wilde, M.M. Noise and disturbance in quantum measurements: An information-theoretic approach. *Phys. Rev. Lett.* **2014**, *112*, 050401.
19. Wehner, S.; Winter, A. Entropic uncertainty relations—A survey. *New J. Phys.* **2010**, *12*, 025009.
20. Coles, P.J.; Berta, M.; Tomamichel, M.; Wehner, S. Entropic uncertainty relations and their applications. 2015. Available online: http://arxiv.org/abs/1511.04857 (accessed on 2 June 2016).
21. Baek, K.; Farrow, T.; Son, W. Optimized entropic uncertainty for successive projective measurements. *Phys. Rev. A* **2014**, *89*, 032108.
22. Zhang, L.; Zhang, Y.; Yu, C. Rényi entropy uncertainty relation for successive projective measurements. *Quantum Inf. Process.* **2015**, *14*, 2239–2253.
23. Rastegin, A.E. Uncertainty and certainty relations for successive projective measurements of a qubit in terms of Tsallis' tntropies. *Commun. Theor. Phys.* **2015**, *63*, 687.
24. Heinosaari, T.; Wolf, M. Non-disturbing quantum measurements. *J. Math. Phys.* **2010**, *51*, 092201.
25. Heinosaari, T.; Miyadera, T. Universality of sequential quantum measurements. *Phys. Rev. A* **2015**, *91*, 022110.
26. Baek, K.; Son, W. Unsharpness of generalized measurement and its effects in entropic uncertainty relations. *Sci. Rep.*, under review, 2016.
27. Busch, P. On the sharpness and bias of quantum effects. *Found. Phys.* **2009**, *39*, 712.
28. Massar, S. Uncertainty relations for positive-operator-valued measures. *Phys. Rev. A* **2007**, *76*, 042114.

29. Cover, T.M.; Thomas, J.A. *Elements of Information Theory*; Wiley: New York, NY, USA, 1991.
30. Davies, E.B. *Quantum Theory of Open Systems*; Academic Press: London, UK, 1976.
31. Ozawa, M. Conditional expectation and repeated measurements of continuous quantum observables. *J. Math. Phys.* **1984**, *25*, 79–87.
32. Heinosaari, T.; Ziman, M. *The Mathematical Language of Quantum Theory from Uncertainty to Entanglement*; Cambridge University Press: Cambridge, UK, 2012.
33. Lahti, P.; Pulmannová, S. Coexistent observables and effects in quantum mechanics. *Rep. Math. Phys.* **1997**, *39*, 339.
34. Lüders, G. Über die Zustandsänderung durch den Messprozess. *Ann. Phys.* **1951**, *8*, 322–328.
35. Tomamichel, M. A Framework for Non-Asymptotic Quantum Information Theory. Ph.D. Thesis, ETH Zurich, Zurich, Switzerland, 2012.
36. Coles, P.J.; Piani, M. Improved entropic uncertainty relations and information exclusion relations. *Phys. Rev. A* **2014**, *89*, 022112.
37. Carmeli, C.; Heinosaari, T.; Toigo, A. Informationally complete joint measurements on finite quantum systems. *Phys. Rev. A* **2012**, *85*, 012109.
38. Busch, P. Unsharp reality and joint measurements for spin observables. *Phys. Rev. D* **1986**, *33*, 2253.

3

Conformal Maps, Biharmonic Maps, and the Warped Product

Seddik Ouakkas * and Djelloul Djebbouri

Laboratory of Geometry, Analysis, Control and Applications, University de Saida, BP138, En-Nasr, 20000 Saida, Algeria; ddjebbouri20@gmail.com

* Correspondence: seddik.ouakkas@gmail.com

Academic Editor: Sadayoshi Kojima

Abstract: In this paper we study some properties of conformal maps between equidimensional manifolds, we construct new example of non-harmonic biharmonic maps and we characterize the biharmonicity of some maps on the warped product manifolds.

Keywords: biharmonic map; conformal map; warped product

Mathematics Subject Classifications (2000): 31B30, 58E20, 58E30

---

1. Introduction.

Let $\phi : (M^m, g) \to (N^n, h)$ be a smooth map between Riemannian manifolds. Then ϕ is said to be harmonic if it is a critical point of the energy functional :

$$E(\phi) = \frac{1}{2}\int_K |d\phi|^2 dv_g \tag{1}$$

for any compact subset $K \subset M$. Equivalently, ϕ is harmonic if it satisfies the associated Euler-Lagrange equations :

$$\tau(\phi) = Tr_g \nabla d\phi = 0, \tag{2}$$

and $\tau(\phi)$ is called the tension field of ϕ. One can refer to [1–4] for background on harmonic maps. In the context of harmonic maps, the stress-energy tensor was studied in details by Baird and Eells in [5]. The stress-energy tensor for a map $\phi : (M^m, g) \longrightarrow (N^n, h)$ defined by

$$S(\phi) = e(\phi)g - \phi^* h$$

and the relation between $S(\phi)$ and $\tau(\phi)$ is given by

$$divS(\phi) = -h(\tau(\phi), d\phi).$$

The map ϕ is said to be biharmonic if it is a critical point of the bi-energy functional :

$$E_2(\phi) = \frac{1}{2}\int_M |\tau(\phi)|^2 dv_g \tag{3}$$

Equivalently, ϕ is biharmonic if it satisfies the associated Euler-Lagrange equations :

$$\tau_2(\phi) = -Tr_g \left(\nabla^\phi\right)^2 \tau(\phi) - Tr_g R^N(\tau(\phi), d\phi)d\phi = 0, \tag{4}$$

where ∇^{ϕ} is the connection in the pull-back bundle $\phi^{-1}(TN)$ and, if $(e_i)_{1\leq i\leq m}$ is a local orthonormal frame field on M, then

$$Tr_g\left(\nabla^{\phi}\right)^2\tau(\phi)=\left(\nabla^{\phi}_{e_i}\nabla^{\phi}_{e_i}-\nabla^{\phi}_{\nabla_{e_i}e_i}\right)\tau(\phi),$$

where we sum over repeated indices. We will call the operator $\tau_2(\phi)$, the bi-tension field of the map ϕ. In analogy with harmonic maps, Jiang In [6] has constructed for a map ϕ the stress bi-energy tensor defined by

$$S_2(\phi)=\left(\frac{-1}{2}\left|\tau(\phi)\right|^2+divh\left(\tau(\phi),d\phi\right)\right)g-2symh\left(\nabla\tau(\phi),d\phi\right),$$

where

$$symh\left(\nabla\tau(\phi),d\phi\right)(X,Y)=\frac{1}{2}\left\{h\left(\nabla_X\tau(\phi),d\phi\left(Y\right)\right)+h\left(\nabla_Y\tau(\phi),d\phi\left(X\right)\right)\right\},$$

for any $X,Y\in\Gamma\left(TM\right)$. The stress bi-energy tensor was also studied in [7] and those results could be useful when we study conformal maps. The stress bi-energy tensor of ϕ satisfies the following relationship

$$divS_2(\phi)=h\left(\tau_2(\phi),d\phi\right).$$

Clearly any harmonic map is biharmonic, therefore it is interesting to construct non-harmonic biharmonic maps. In [8] the authors found new examples of biharmonic maps by conformally deforming the domain metric of harmonic ones. While in [9] the author analyzed the behavior of the biharmonic equation under the conformal change the domain metric, she obtained metrics $\widetilde{g}=e^{2\gamma}$ such that the idendity map $Id:(M,g)\longrightarrow(M,\widetilde{g})$ is biharmonic non-harmonic. Moreover, in [10] the author gave some extensions of the result in [9] together with some further constructions of biharmonic maps. The author in [11] deform conformally the codomain metric in order to render a semi-conformal harmonic map biharmonic. In [12] the authors studied the case where $\phi:(M^n,g)\longrightarrow(N^n,h)$ is a conformal mapping between equidimensional manifolds where they show that a conformal mapping ϕ is biharmonic if and only if the gradient of its dilation satisfies a second order elliptic partial differential equation. We can refer the reader to [13], for a survey of biharmonic maps. In the first section of this paper, we present some properties for a conformal mapping $\phi:(M^n,g)\longrightarrow(N^n,h)$, we prove that the stress bi-energy tensor depend only on the dilation (Theorem 1) and we calculate the bitension field of ϕ (Theorem 2). In the last section we study the biharmonicity of some maps on the warped product (Theorem 4 and 5), with this setting we obtain new examples of biharmonic non-harmonic maps.

2. Some properties for conformal maps.

We study conformal maps between equidimensional manifolds of the same dimension $n\geq 3$. Note that by a result in [12], any such map can have no critical points and so is a local conformal diffeomorphism. Recall that a mapping $\phi:(M^n,g)\to(N^n,h)$ is called conformal if there exist a C^∞ function $\lambda:M\to\mathbb{R}_+^*$ such that for any $X,Y\in\Gamma(TM)$:

$$h(d\phi(X),d\phi(Y))=\lambda^2g(X,Y).$$

The function λ is called the dilation for the map ϕ. The tension field and the stress energy tensor for a conformal map are given by (see [14]):

Proposition 1. *Let $\phi:(M^n,g)\to(N^n,h)$ be a conformal map of dilation λ, we have*

$$(i)\ \ divS(\phi)=(n-2)\lambda^2d\ln\lambda, \tag{5}$$

$$(ii)\ \ divh(\tau(\phi),d\phi)=(2-n)\left(2\lambda^2\left|grad\ln\lambda\right|^2+\lambda^2\Delta\ln\lambda\right). \tag{6}$$

$$(iii)\ \ \tau(\phi)=(2-n)d\phi(grad\ln\lambda). \tag{7}$$

$$(iv)\ \ \left|\tau(\phi)\right|^2=(2-n)^2\lambda^2\left|grad\ln\lambda\right|^2. \tag{8}$$

Note that the conformal map $\phi : (M^n, g) \to (N^n, h)$ of dilation λ is harmonic if and only if $n = 2$ or the dilation λ is constant.

In the first, wa calculate the stress bi-energy tensor for a conformal map ϕ when we prove that $S_2(\phi)$ depend only the dilation.

Theorem 1. *Let $\phi : (M^n, g) \to (N^n, h)$ be a conformal map with dilation λ, then we have*

$$S_2(\phi) = (2-n)\,\lambda^2 \left\{ \left(\frac{n-2}{2} \left| grad \ln \lambda \right|^2 + \Delta \ln \lambda \right) g - 2\nabla d \ln \lambda \right\}, \tag{9}$$

and the trace of $S_2(\phi)$ is given by

$$TrS_2(\phi) = -\,(2-n)^2\,\lambda^2 \left\{ \frac{n}{2} \left| grad \ln \lambda \right|^2 + \Delta \ln \lambda \right\}. \tag{10}$$

To prove Theorem 1, we need the following Lemma :

Lemma 1. *Let $\phi : (M^n, g) \to (N^n, h)$ be a conformal map with dilation λ, then for any function $f \in C^\infty(M)$ and for any $X, Y \in \Gamma(TM)$, we have*

$$\begin{aligned} h\left(\nabla_X d\phi\left(gradf\right), d\phi\left(Y\right)\right) &= \lambda^2 \left(X\left(\ln\lambda\right) Y\left(f\right) - Y\left(\ln\lambda\right) X\left(f\right)\right) \\ &\quad + \lambda^2 \nabla df\left(X, Y\right) + \lambda^2 d\ln\lambda\left(gradf\right) g\left(X, Y\right). \end{aligned} \tag{11}$$

Proof of Lemma 1. Let $f \in C^\infty(M)$, for any $X, Y \in \Gamma(TM)$, we have

$$\begin{aligned} h\left(\nabla_X d\phi\left(gradf\right), d\phi\left(Y\right)\right) &= X\left(\lambda^2 g\left(gradf, Y\right)\right) - h\left(d\phi\left(gradf\right), \nabla_X d\phi\left(Y\right)\right) \\ &= X\left(\lambda^2\right) g\left(gradf, Y\right) + \lambda^2 g\left(\nabla_X gradf, Y\right) + \lambda^2 g\left(gradf, \nabla_X Y\right) \\ &\quad - h\left(d\phi\left(gradf\right), \nabla d\phi\left(X, Y\right)\right) - h\left(d\phi\left(gradf\right), d\phi\left(\nabla_X Y\right)\right) \\ &= X\left(\lambda^2\right) g\left(gradf, Y\right) + \lambda^2 g\left(\nabla_X gradf, Y\right) + \lambda^2 g\left(gradf, \nabla_X Y\right) \\ &\quad - h\left(d\phi\left(gradf\right), \nabla d\phi\left(X, Y\right)\right) - \lambda^2 g\left(gradf, \nabla_X Y\right). \end{aligned}$$

Note that

$$g\left(\nabla_X gradf, Y\right) = \nabla df\left(X, Y\right),$$

then we obtain

$$h\left(\nabla_X d\phi\left(gradf\right), d\phi\left(Y\right)\right) = 2\lambda^2 X\left(\ln\lambda\right) Y\left(f\right) + \lambda^2 \nabla df\left(X, Y\right) - h\left(d\phi\left(gradf\right), \nabla d\phi\left(X, Y\right)\right).$$

By similary, we have

$$h\left(\nabla_Y d\phi\left(gradf\right), d\phi\left(X\right)\right) = 2\lambda^2 Y\left(\ln\lambda\right) X\left(f\right) + \lambda^2 \nabla df\left(X, Y\right) - h\left(d\phi\left(gradf\right), \nabla d\phi\left(X, Y\right)\right).$$

Then, we deduce that

$$\begin{aligned} h\left(\nabla_X d\phi\left(gradf\right), d\phi\left(Y\right)\right) &= h\left(d\phi\left(X\right), \nabla_Y d\phi\left(gradf\right)\right) \\ &\quad + 2\lambda^2 \left(X\left(\ln\lambda\right) Y\left(f\right) - Y\left(\ln\lambda\right) X\left(f\right)\right). \end{aligned} \tag{12}$$

For the term $h\left(d\phi\left(X\right),\nabla_{Y}d\phi\left(gradf\right)\right)$, we have

$$\begin{aligned} h\left(\nabla_{Y}d\phi\left(gradf\right),d\phi\left(X\right)\right) &= h\left(\nabla d\phi\left(gradf,Y\right),d\phi\left(X\right)\right)+\lambda^{2}g\left(\nabla_{Y}gradf,X\right) \\ &= h\left(\nabla_{gradf}d\phi\left(Y\right),d\phi\left(X\right)\right)-\lambda^{2}g\left(\nabla_{gradf}Y,X\right) \\ &+\lambda^{2}g\left(\nabla_{Y}gradf,X\right) \\ &= gradf\left(\lambda^{2}g\left(X,Y\right)\right)-h\left(\nabla_{gradf}d\phi\left(X\right),d\phi\left(Y\right)\right) \\ &-\lambda^{2}g\left(\nabla_{gradf}Y,X\right)+\lambda^{2}g\left(\nabla_{Y}gradf,X\right) \\ &= 2\lambda^{2}d\ln\lambda\left(gradf\right)g\left(X,Y\right)-h\left(\nabla d\phi\left(X,gradf\right),d\phi\left(Y\right)\right) \\ &+\lambda^{2}g\left(\nabla_{Y}gradf,X\right). \end{aligned}$$

We deduce that

$$\begin{aligned} h\left(\nabla_{Y}d\phi\left(gradf\right),d\phi\left(X\right)\right) &= -h\left(\nabla_{X}d\phi\left(gradf\right),d\phi\left(Y\right)\right)+2\lambda^{2}\nabla df\left(X,Y\right) \\ &+2\lambda^{2}d\ln\lambda\left(gradf\right)g\left(X,Y\right). \end{aligned} \tag{13}$$

Finally, if we replace (13) in (12), we obtain

$$\begin{aligned} h\left(\nabla_{X}d\phi\left(gradf\right),d\phi\left(Y\right)\right) &= \lambda^{2}\left(X\left(\ln\lambda\right)Y\left(f\right)-Y\left(\ln\lambda\right)X\left(f\right)\right) \\ &+\lambda^{2}\nabla df\left(X,Y\right)+\lambda^{2}d\ln\lambda\left(gradf\right)g\left(X,Y\right). \end{aligned}$$

This completes the proof of Lemma 1.

Remark 1. *Let $\phi:(M^{n},g)\to(N^{n},h)$ be a conformal map with dilation λ, then if we consider $f=\ln\lambda$, the equation (11) gives*

$$h\left(\nabla_{X}d\phi\left(grad\ln\lambda\right),d\phi\left(Y\right)\right)=\lambda^{2}\left(\nabla d\ln\lambda\left(X,Y\right)+\left|grad\ln\lambda\right|^{2}g\left(X,Y\right)\right).$$

Proof of Theorem 1. By definition, the stress bi-energy tensor is given by :

$$S_{2}(\phi)=\left(-\frac{1}{2}\left|\tau(\phi)\right|^{2}+divh\left(\tau(\phi),d\phi\right)\right)g-2symh\left(\nabla\tau(\phi),d\phi\right). \tag{14}$$

Using the equations (2) et (4) for the Proposition 1, we have

$$-\frac{1}{2}\left|\tau(\phi)\right|^{2}+divh\left(\tau(\phi),d\phi\right)=(2-n)\lambda^{2}\left(\frac{n+2}{2}\left|grad\ln\lambda\right|^{2}+\Delta\ln\lambda\right). \tag{15}$$

Calculate now $symh\left(\nabla\tau(\phi),d\phi\right)$, we have by definition for any $X,Y\in\Gamma\left(TM\right)$

$$\begin{aligned} symh\left(\nabla\tau(\phi),d\phi\right)(X,Y) &= \frac{1}{2}\left(h\left(\nabla_{X}\tau\left(\phi\right),d\phi\left(Y\right)\right)+h\left(\nabla_{Y}\tau\left(\phi\right),d\phi\left(X\right)\right)\right) \\ &= \frac{2-n}{2}\left(h\left(\nabla_{X}d\phi\left(grad\ln\lambda\right),d\phi\left(Y\right)\right)+h\left(\nabla_{Y}\left(grad\ln\lambda\right),d\phi\left(X\right)\right)\right). \end{aligned}$$

By Lemma 1, we have

$$h\left(\nabla_{X}d\phi\left(grad\ln\lambda\right),d\phi\left(Y\right)\right)=\lambda^{2}\left(\nabla d\ln\lambda\left(X,Y\right)+\left|grad\ln\lambda\right|^{2}g\left(X,Y\right)\right)$$

and

$$h\left(\nabla_{Y}d\phi\left(grad\ln\lambda\right),d\phi\left(X\right)\right)=\lambda^{2}\left(\nabla d\ln\lambda\left(X,Y\right)+\left|grad\ln\lambda\right|^{2}g\left(X,Y\right)\right),$$

then

$$symh\left(\nabla\tau(\phi),d\phi\right)(X,Y)=(2-n)\,\lambda^2\left(\nabla d\ln\lambda\,(X,Y)+|grad\ln\lambda|^2 g\,(X,Y)\right). \tag{16}$$

If we substitute (15) and (16) in (14), we conclude that

$$S_2(\phi)=(2-n)\,\lambda^2\left\{\left(\frac{n-2}{2}|grad\ln\lambda|^2+\Delta\ln\lambda\right)g-2\nabla d\ln\lambda\right\}$$

Calculate now the trace of stress bi-energy tensor. Let $(e_i)_{1\leq i\leq n}$ be an orthonormal frame on M, we have

$$\begin{aligned}Tr_g S_2(\phi)&=S_2(\phi)(e_i,e_i)\\&=(2-n)\,\lambda^2\left(\frac{n-2}{2}|grad\ln\lambda|^2+\Delta\ln\lambda\right)g\,(e_i,e_i)\\&-2\,(2-n)\,\lambda^2\nabla d\ln\lambda\,(e_i,e_i)\\&=(2-n)\,n\lambda^2\left(\frac{n-2}{2}|grad\ln\lambda|^2+\Delta\ln\lambda\right)\\&-2\,(2-n)\,\lambda^2\,(\Delta\ln\lambda)\\&=(2-n)\,\lambda^2\left\{\frac{n\,(n-2)}{2}|grad\ln\lambda|^2+(n-2)\,\Delta\ln\lambda\right\}.\end{aligned}$$

Then

$$TrS_2(\phi)=-\,(2-n)^2\,\lambda^2\left\{\frac{n}{2}|grad\ln\lambda|^2+\Delta\ln\lambda\right\}.$$

By calculating the Laplacian of the function $\lambda^{\frac{n}{2}}$ and by using

$$\Delta\lambda^{\frac{n}{2}}=\frac{n}{2}\lambda^{\frac{n}{2}}\left(\frac{n}{2}|grad\ln\lambda|^2+\Delta\ln\lambda\right),$$

we obtain immediately the following corollary

Corollary 1. *Let $\phi:(M^n,g)\to(N^n,h)$, $(n\neq 2)$ to be a conformal map of dilation λ, then the trace of $S_2(\phi)$ is zero if and only if the function $\lambda^{\frac{n}{2}}$ is harmonic.*

The bi-tension field of the conformal map is given by

Theorem 2. *Let $\phi:(M^n,g)\to(N^n,h)$, $(n\geq 3)$ to be a conformal map of dilation λ, then bi-tension field of ϕ is given by :*

$$\tau_2(\phi)=(n-2)\,d\phi\,(H)$$

where

$$\begin{aligned}H=grad\Delta\ln\lambda&-\frac{(n-6)}{2}grad\left(|grad\ln\lambda|^2\right)+2Ricci^M\,(grad\ln\lambda)\\&-\left(2\,(\Delta\ln\lambda)+(n-2)\,|grad\ln\lambda|^2\right)grad\ln\lambda.\end{aligned} \tag{17}$$

Remark 2. *A. Balmus in [9] studied the case where $\phi=Id_M$, she obtained the biharmonicity of the identity map from (M,g) onto $(M,\lambda^2 g)$, this case was also studied in [15].*

To prove the Theorem 2, we need two Lemmas. In the first Lemma, we give a simple formula of the term $Tr_g\left(\nabla^\phi\right)^2 d\phi\,(grad\gamma)$ for a conformal map $\phi:(M^n,g)\to(N^n,h)$ $(n\geq 3)$ of dilation λ and for any function $\gamma\in C^\infty(M)$.

Lemma 2. *Let $\phi:(M^n,g)\to(N^n,h)$ $(n\geq 3)$ to be a conformal map of dilation λ, then for any function $\gamma\in C^\infty(M)$, we have*

$$\begin{aligned}Tr_g\left(\nabla^\phi\right)^2 d\phi\,(grad\gamma)&=d\phi\,(grad\Delta\gamma)+4d\phi\left(\nabla_{grad\ln\lambda}grad\gamma\right)+d\phi\left(Ricci^M\,(grad\gamma)\right)\\&+(\Delta\ln\lambda)\,d\phi\,(grad\gamma)-2\,(\Delta\gamma)\,d\phi\,(grad\ln\lambda)\\&-(n-2)\,d\ln\lambda\,(grad\gamma)\,d\phi\,(grad\ln\lambda).\end{aligned} \tag{18}$$

Proof of Lemma 2. Let $\gamma \in C^{\infty}(M)$, by definition, we have

$$Tr_g\left(\nabla^{\phi}\right)^2 d\phi\left(grad\gamma\right) = \nabla^{\phi}_{e_i}\nabla^{\phi}_{e_i}d\phi\left(grad\gamma\right) - \nabla^{\phi}_{\nabla_{e_i}e_i}d\phi\left(grad\gamma\right). \tag{19}$$

(Here henceforth we sum over repeated indices.) Let us start with the calculation of the term $\nabla^{\phi}_{e_i}\nabla^{\phi}_{e_i}d\phi\left(grad\gamma\right)$, we have

$$\nabla^{\phi}_{e_i}d\phi\left(grad\gamma\right) = \nabla d\phi\left(e_i, grad\gamma\right) + d\phi\left(\nabla_{e_i}grad\gamma\right).$$

It is known that (see [16])

$$\nabla d\phi\left(e_i, grad\gamma\right) = e_i\left(\ln\lambda\right)d\phi\left(grad\gamma\right) + d\ln\lambda\left(grad\gamma\right)d\phi\left(e_i\right) - e_i\left(\gamma\right)d\phi\left(grad\ln\lambda\right),$$

then

$$\begin{aligned}\nabla^{\phi}_{e_i}d\phi\left(grad\gamma\right) &= e_i\left(\ln\lambda\right)d\phi\left(grad\gamma\right) + d\ln\lambda\left(grad\gamma\right)d\phi\left(e_i\right)\\ &\quad - e_i\left(\gamma\right)d\phi\left(grad\ln\lambda\right) + d\phi\left(\nabla_{e_i}grad\gamma\right).\end{aligned} \tag{20}$$

It follows that

$$\begin{aligned}\nabla^{\phi}_{e_i}\nabla^{\phi}_{e_i}d\phi\left(grad\gamma\right) &= \nabla^{\phi}_{e_i}\left\{e_i\left(\ln\lambda\right)d\phi\left(grad\gamma\right)\right\} + \nabla^{\phi}_{e_i}\left\{d\ln\lambda\left(grad\gamma\right)d\phi\left(e_i\right)\right\}\\ &\quad - \nabla^{\phi}_{e_i}\left\{e_i\left(\gamma\right)d\phi\left(grad\ln\lambda\right)\right\} + \nabla^{\phi}_{e_i}d\phi\left(\nabla_{e_i}grad\gamma\right).\end{aligned} \tag{21}$$

We will study term by term the right-hand of this expression. For the first term $\nabla^{\phi}_{e_i}\left\{e_i\left(\ln\lambda\right)d\phi\left(grad\gamma\right)\right\}$, we have

$$\nabla^{\phi}_{e_i}\left\{e_i\left(\ln\lambda\right)d\phi\left(grad\gamma\right)\right\} = e_i\left(\ln\lambda\right)\nabla^{\phi}_{e_i}d\phi\left(grad\gamma\right) + e_i\left(e_i\left(\ln\lambda\right)\right)d\phi\left(grad\gamma\right).$$

By using the equation (20), we deduce that

$$\begin{aligned}\nabla^{\phi}_{e_i}\left\{e_i\left(\ln\lambda\right)d\phi\left(grad\gamma\right)\right\} &= e_i\left(\ln\lambda\right)e_i\left(\ln\lambda\right)d\phi\left(grad\gamma\right) + e_i\left(\ln\lambda\right)d\ln\lambda\left(grad\gamma\right)d\phi\left(e_i\right)\\ &\quad - e_i\left(\ln\lambda\right)e_i\left(\gamma\right)d\phi\left(grad\ln\lambda\right) + e_i\left(\ln\lambda\right)d\phi\left(\nabla_{e_i}grad\gamma\right)\\ &\quad + e_i\left(e_i\left(\ln\lambda\right)\right)d\phi\left(grad\gamma\right),\end{aligned}$$

then, we obtain

$$\begin{aligned}\nabla^{\phi}_{e_i}\left\{e_i\left(\ln\lambda\right)d\phi\left(grad\gamma\right)\right\} &= \left|grad\ln\lambda\right|^2 d\phi\left(grad\gamma\right) + d\phi\left(\nabla_{grad\ln\lambda}grad\gamma\right)\\ &\quad + e_i\left(e_i\left(\ln\lambda\right)\right)d\phi\left(grad\gamma\right).\end{aligned} \tag{22}$$

For the second term $\nabla^{\phi}_{e_i}\left\{d\ln\lambda\left(grad\gamma\right)d\phi\left(e_i\right)\right\}$, a similar calculation gives

$$\begin{aligned}\nabla^{\phi}_{e_i}\left\{d\ln\lambda\left(grad\gamma\right)d\phi\left(e_i\right)\right\} &= d\ln\lambda\left(grad\gamma\right)\nabla^{\phi}_{e_i}d\phi\left(e_i\right) + e_i\left\{g\left(grad\ln\lambda, grad\gamma\right)\right\}d\phi\left(e_i\right)\\ &= d\ln\lambda\left(grad\gamma\right)\nabla^{\phi}_{e_i}d\phi\left(e_i\right) + g\left(\nabla_{e_i}grad\ln\lambda, grad\gamma\right)d\phi\left(e_i\right)\\ &\quad + g\left(grad\ln\lambda, \nabla_{e_i}grad\gamma\right)d\phi\left(e_i\right)\\ &= d\ln\lambda\left(grad\gamma\right)\nabla^{\phi}_{e_i}d\phi\left(e_i\right) + g\left(\nabla_{grad\gamma}grad\ln\lambda, e_i\right)d\phi\left(e_i\right)\\ &\quad + g\left(\nabla_{grad\ln\lambda}grad\gamma, e_i\right)d\phi\left(e_i\right),\end{aligned}$$

it follows that

$$\nabla^{\phi}_{e_i}\left\{d\ln\lambda\left(grad\gamma\right)d\phi\left(e_i\right)\right\} = d\ln\lambda\left(grad\gamma\right)\nabla^{\phi}_{e_i}d\phi\left(e_i\right) + d\phi\left(\nabla_{grad\gamma}grad\ln\lambda\right) + d\phi\left(\nabla_{grad\ln\lambda}grad\gamma\right). \tag{23}$$

For the third term $\nabla^{\phi}_{e_i}\left\{e_i\left(\gamma\right)d\phi\left(grad\ln\lambda\right)\right\}$, by using the same calculation method and the equation (20), we have

$$\begin{aligned}\nabla^{\phi}_{e_i}\left\{e_i\left(\gamma\right)d\phi\left(grad\ln\lambda\right)\right\} &= e_i\left(\gamma\right)\nabla^{\phi}_{e_i}d\phi\left(grad\ln\lambda\right) + e_i\left(e_i\left(\gamma\right)\right)d\phi\left(grad\ln\lambda\right)\\ &= e_i\left(\gamma\right)e_i\left(\ln\lambda\right)d\phi\left(grad\ln\lambda\right) + e_i\left(\gamma\right)d\ln\lambda\left(grad\ln\lambda\right)d\phi\left(e_i\right)\\ &- e_i\left(\gamma\right)e_i\left(\ln\lambda\right)d\phi\left(grad\ln\lambda\right) + e_i\left(\gamma\right)d\phi\left(\nabla_{e_i}grad\ln\lambda\right)\\ &+ e_i\left(e_i\left(\gamma\right)\right)d\phi\left(grad\ln\lambda\right),\end{aligned}$$

which gives us

$$\nabla^{\phi}_{e_i}\left\{e_i\left(\gamma\right)d\phi\left(grad\ln\lambda\right)\right\} = \left|grad\ln\lambda\right|^2 d\phi\left(grad\gamma\right) + d\phi\left(\nabla_{grad\gamma}grad\ln\lambda\right) + e_i\left(e_i\left(\gamma\right)\right)d\phi\left(grad\ln\lambda\right). \tag{24}$$

Now let us look at the last term $\nabla^{\phi}_{e_i}d\phi\left(\nabla_{e_i}grad\gamma\right)$, a simple calculation gives

$$\begin{aligned}\nabla^{\phi}_{e_i}d\phi\left(\nabla_{e_i}grad\gamma\right) &= e_i\left(\ln\lambda\right)d\phi\left(\nabla_{e_i}grad\gamma\right) + d\ln\lambda\left(\nabla_{e_i}grad\gamma\right)d\phi\left(e_i\right)\\ &- g\left(e_i,\nabla_{e_i}grad\gamma\right)d\phi\left(grad\ln\lambda\right) + d\phi\left(\nabla_{e_i}\nabla_{e_i}grad\gamma\right)\\ &= 2d\phi\left(\nabla_{grad\ln\lambda}grad\gamma\right) - \left(\Delta\gamma\right)d\phi\left(grad\ln\lambda\right)\\ &+ d\phi\left(\nabla_{e_i}\nabla_{e_i}grad\gamma\right),\end{aligned}$$

then

$$\nabla^{\phi}_{e_i}d\phi\left(\nabla_{e_i}grad\gamma\right) = d\phi\left(\nabla_{e_i}\nabla_{e_i}grad\gamma\right) + 2d\phi\left(\nabla_{grad\ln\lambda}grad\gamma\right) - \left(\Delta\gamma\right)d\phi\left(grad\ln\lambda\right). \tag{25}$$

If we replace (22), (23), (24) and (25) in (21), we obtain

$$\begin{aligned}\nabla^{\phi}_{e_i}\nabla^{\phi}_{e_i}d\phi\left(grad\gamma\right) &= 4d\phi\left(\nabla_{grad\ln\lambda}grad\gamma\right) + e_i\left(e_i\left(\ln\lambda\right)\right)d\phi\left(grad\gamma\right)\\ &+ d\ln\lambda\left(grad\gamma\right)\nabla^{\phi}_{e_i}d\phi\left(e_i\right) - e_i\left(e_i\left(\gamma\right)\right)d\phi\left(grad\ln\lambda\right)\\ &+ d\phi\left(\nabla_{e_i}\nabla_{e_i}grad\gamma\right) - \left(\Delta\gamma\right)d\phi\left(grad\ln\lambda\right).\end{aligned} \tag{26}$$

To complete the proof, it remains to investigate the term $\nabla^{\phi}_{\nabla_{e_i}e_i}d\phi\left(grad\gamma\right)$, we have

$$\nabla^{\phi}_{\nabla_{e_i}e_i}d\phi\left(grad\gamma\right) = \nabla d\phi\left(\nabla_{e_i}e_i, grad\gamma\right) + d\phi\left(\nabla_{\nabla_{e_i}e_i}grad\gamma\right),$$

Therefore, by using the equation (20), we obtain

$$\nabla^{\phi}_{\nabla_{e_i}e_i}d\phi\left(grad\gamma\right) = \nabla_{e_i}e_i\left(\ln\lambda\right)d\phi\left(grad\gamma\right) + d\ln\lambda\left(grad\gamma\right)d\phi\left(\nabla_{e_i}e_i\right) - \nabla_{e_i}e_i\left(\gamma\right)d\phi\left(grad\ln\lambda\right) + d\phi\left(\nabla_{\nabla_{e_i}e_i}grad\gamma\right). \tag{27}$$

By substituting (26) and (27) in (19), we deduce

$$\begin{aligned} Tr_g\left(\nabla^{\phi}\right)^2 d\phi\left(grad\gamma\right) &= \nabla^{\phi}_{e_i}\nabla^{\phi}_{e_i}d\phi\left(grad\gamma\right) - \nabla^{\phi}_{\nabla_{e_i}e_i}d\phi\left(grad\gamma\right) \\ &= d\phi\left(Tr_g\nabla^2 grad\gamma\right) + 4d\phi\left(\nabla_{grad\ln\lambda}grad\gamma\right) \\ &+ \left(\Delta\ln\lambda\right)d\phi\left(grad\gamma\right) + d\ln\lambda\left(grad\gamma\right)\tau\left(\phi\right) \\ &- 2\left(\Delta\gamma\right)d\phi\left(grad\ln\lambda\right). \end{aligned}$$

Finally, using the fact that (see [11])

$$Tr_g\nabla^2 grad\gamma = grad\Delta\gamma + Ricci^M\left(grad\gamma\right)$$

and

$$\tau\left(\phi\right) = \left(2-n\right)d\phi\left(grad\ln\lambda\right),$$

we conclude that

$$\begin{aligned} Tr_g\left(\nabla^{\phi}\right)^2 d\phi\left(grad\gamma\right) &= d\phi\left(grad\Delta\gamma\right) + 4d\phi\left(\nabla_{grad\ln\lambda}grad\gamma\right) + d\phi\left(Ricci^M\left(grad\gamma\right)\right) \\ &+ \left(\Delta\ln\lambda\right)d\phi\left(grad\gamma\right) - 2\left(\Delta\gamma\right)d\phi\left(grad\ln\lambda\right) \\ &- \left(n-2\right)d\ln\lambda\left(grad\gamma\right)d\phi\left(grad\ln\lambda\right). \end{aligned}$$

This completes the proof of Lemma 2. Now, in the second Lemma, we will calculate $Tr_g R^N\left(d\phi\left(grad\gamma\right), d\phi\right)d\phi$ for a conformal maps $\phi : (M^n, g) \to (N^n, h)$ $(n \geq 3)$ of dilation λ and for any function $\gamma \in C^{\infty}(M)$

Lemma 3. *Let $\phi : (M^n, g) \to (N^n, h)$ $(n \geq 3)$ to be a conformal map of dilation λ, then for any function $\gamma \in C^{\infty}(M)$, we have*

$$\begin{aligned} Tr_g R^N\left(d\phi\left(grad\gamma\right), d\phi\right)d\phi &= d\phi\left(Ricci^M\left(grad\gamma\right)\right) - \left(n-2\right)d\phi\left(\nabla_{grad\gamma}grad\ln\lambda\right) \\ &- \left(\Delta\ln\lambda + \left(n-2\right)\left|grad\ln\lambda\right|^2\right)d\phi\left(grad\gamma\right) \\ &+ \left(n-2\right)d\ln\lambda\left(grad\gamma\right)d\phi\left(grad\ln\lambda\right) \end{aligned} \tag{28}$$

Proof of Lemma 3. Let $\gamma \in C^{\infty}(M)$, by definition we have

$$Tr_g R^N\left(d\phi\left(grad\gamma\right), d\phi\right)d\phi = R^N\left(d\phi\left(grad\gamma\right), d\phi\left(e_i\right)\right)d\phi\left(e_i\right) \tag{29}$$

but we know that (see [16])

$$\begin{aligned} Ric^N\left(d\phi\left(X\right), d\phi\left(Y\right)\right) &= Ric^M\left(X, Y\right) + \left(n-2\right)X\left(\ln\lambda\right)Y\left(\ln\lambda\right) \\ &- \left(n-2\right)\left|grad\ln\lambda\right|^2 g\left(X, Y\right) \\ &- \left(n-2\right)\nabla d\ln\lambda\left(X, Y\right) - \left(\Delta\ln\lambda\right)g\left(X, Y\right). \end{aligned}$$

Then

$$\begin{aligned} Ric^N\left(d\phi\left(grad\gamma\right), d\phi\left(e_i\right)\right) &= Ric^M\left(grad\gamma, e_i\right) + \left(n-2\right)grad\gamma\left(\ln\lambda\right)e_i\left(\ln\lambda\right) \\ &- \left(n-2\right)\left|grad\ln\lambda\right|^2 g\left(grad\gamma, e_i\right) \\ &- \left(n-2\right)\nabla d\ln\lambda\left(grad\gamma, e_i\right) - \left(\Delta\ln\lambda\right)g\left(grad\gamma, e_i\right) \end{aligned}$$

it follows that

$$\begin{aligned} Ric^{N}\left(d\phi\left(grad\gamma\right),d\phi\left(e_{i}\right)\right) &= Ric^{M}\left(grad\gamma,e_{i}\right)+\left(m-2\right)d\ln\lambda\left(grad\gamma\right)e_{i}\left(\ln\lambda\right) \\ &- \left(n-2\right)\left|grad\ln\lambda\right|^{2}e_{i}\left(\gamma\right)-\left(n-2\right)\nabla d\ln\lambda\left(grad\gamma,e_{i}\right) \\ &- \left(\Delta\ln\lambda\right)e_{i}\left(\gamma\right). \end{aligned} \tag{30}$$

If we replace (30) in (29), we deduce that

$$\begin{aligned} Tr_{g}R^{N}\left(d\phi\left(grad\gamma\right),d\phi\right)d\phi &= R^{N}\left(d\phi\left(grad\gamma\right),d\phi\left(e_{i}\right)\right)d\phi\left(e_{i}\right) \\ &= d\phi\left(Ricci^{M}\left(grad\gamma\right)\right)+\left(n-2\right)d\ln\lambda\left(grad\gamma\right)d\phi\left(grad\ln\lambda\right) \\ &- \left(n-2\right)\left|grad\ln\lambda\right|^{2}d\phi\left(grad\gamma\right)-\left(n-2\right)\nabla d\ln\lambda\left(grad\gamma,e_{i}\right)d\phi\left(e_{i}\right) \\ &- \left(\Delta\ln\lambda\right)d\phi\left(grad\gamma\right). \end{aligned}$$

To complete the proof, we will simplify the term $\nabla d\ln\lambda\left(grad\gamma,e_{i}\right)d\phi\left(e_{i}\right)$, we obtain

$$\begin{aligned} \nabla d\ln\lambda\left(grad\gamma,e_{i}\right)d\phi\left(e_{i}\right) &= \left\{e_{i}\left(g\left(grad\ln\lambda,grad\gamma\right)\right)-d\ln\lambda\left(\nabla_{e_{i}}grad\gamma\right)\right\}d\phi\left(e_{i}\right) \\ &= g\left(\nabla_{e_{i}}grad\ln\lambda,grad\gamma\right)d\phi\left(e_{i}\right) \\ &= g\left(\nabla_{grad\gamma}grad\ln\lambda,e_{i}\right)d\phi\left(e_{i}\right) \\ &= d\phi\left(\nabla_{grad\gamma}grad\ln\lambda\right), \end{aligned}$$

which finally gives

$$\begin{aligned} Tr_{g}R^{N}\left(d\phi\left(grad\gamma\right),d\phi\right)d\phi &= d\phi\left(Ricci^{M}\left(grad\gamma\right)\right)-\left(n-2\right)d\phi\left(\nabla_{grad\gamma}grad\ln\lambda\right) \\ &- \left(\Delta\ln\lambda+\left(n-2\right)\left|grad\ln\lambda\right|^{2}\right)d\phi\left(grad\gamma\right) \\ &+ \left(n-2\right)d\ln\lambda\left(grad\gamma\right)d\phi\left(grad\ln\lambda\right). \end{aligned}$$

This completes the proof of Lemma 3. We are now able to prove Theorem 2.
Proof of Theorem 2. By definition, the bi-tension field is given by

$$\tau_{2}\left(\phi\right)=-Tr_{g}\left(\nabla^{\phi}\right)^{2}\tau\left(\phi\right)-Tr_{g}R^{N}\left(\tau\left(\phi\right),d\phi\right)d\phi.$$

The tension field of the conformal map ϕ is given by

$$\tau\left(\phi\right)=\left(2-n\right)d\phi\left(grad\ln\lambda\right),$$

it follows that

$$\tau_{2}\left(\phi\right)=\left(n-2\right)\left(Tr_{g}\left(\nabla^{\phi}\right)^{2}d\phi\left(grad\ln\lambda\right)+Tr_{g}R^{N}\left(d\phi\left(grad\ln\lambda\right),d\phi\right)d\phi\right). \tag{31}$$

By Lemma 2, we have

$$\begin{aligned} Tr_{g}\left(\nabla^{\phi}\right)^{2}d\phi\left(grad\ln\lambda\right) &= d\phi\left(grad\Delta\ln\lambda\right)+2d\phi\left(grad\left(\left|grad\ln\lambda\right|^{2}\right)\right) \\ &- \left(\Delta\ln\lambda\right)d\phi\left(grad\ln\lambda\right)+d\phi\left(Ricci^{M}\left(grad\ln\lambda\right)\right) \\ &- \left(n-2\right)\left|grad\ln\lambda\right|^{2}d\phi\left(grad\ln\lambda\right). \end{aligned} \tag{32}$$

By using lemma 3 and the fact that $\nabla_{grad \ln \lambda} grad \ln \lambda = \frac{1}{2} grad \left(|grad \ln \lambda|^2 \right)$

$$Tr_g R^N \left(d\phi \left(grad \ln \lambda \right), d\phi \right) d\phi = d\phi \left(Ricci^M \left(grad \ln \lambda \right) \right) - \left(\Delta \ln \lambda \right) d\phi \left(grad \ln \lambda \right) - \frac{(n-2)}{2} d\phi \left(grad \left(|grad \ln \lambda|^2 \right) \right). \tag{33}$$

If we replace (32) and (33) in (31), we deduce that

$$\begin{aligned} \tau_2 (\phi) &= (n-2)\, d\phi \left(grad \Delta \ln \lambda \right) - \frac{(n-2)(n-6)}{2} d\phi \left(grad \left(|grad \ln \lambda|^2 \right) \right) \\ &- (n-2) \left(2 \left(\Delta \ln \lambda \right) + (n-2) |grad \ln \lambda|^2 \right) d\phi \left(grad \ln \lambda \right) \\ &+ 2 (n-2)\, d\phi \left(Ricci^M \left(grad \ln \lambda \right) \right). \end{aligned}$$

Then the bi-tension field of ϕ is given by :

$$\tau_2(\phi) = (n-2)\, d\phi\, (H)$$

where

$$\begin{aligned} H &= grad \Delta \ln \lambda - \frac{(n-6)}{2} grad \left(|grad \ln \lambda|^2 \right) + 2 Ricci^M \left(grad \ln \lambda \right) \\ &- \left(2 \left(\Delta \ln \lambda \right) + (n-2) |grad \ln \lambda|^2 \right) grad \ln \lambda. \end{aligned}$$

The proof of Theorem 2 is complete. By application of Theorem 2, we get the following result (see [15]).

Theorem 3. *([12]) Let $\phi : (M^n, g) \to (N^n, h)$ $(n \geq 3)$ to be a conformal map of dilation λ, then ϕ is biharmonic if and only if the dilation λ satisfies*

$$\begin{aligned} & grad \left(\Delta \ln \lambda \right) - \left(2 \left(\Delta \ln \lambda \right) + (n-2) |grad \ln \lambda|^2 \right) grad \ln \lambda \\ & + \frac{6-n}{2} grad \left(|grad \ln \lambda|^2 \right) + 2 Ricci^M (grad \ln \lambda) = 0. \end{aligned}$$

In particular, we prove that the biharmonicity of the conformal map $\phi : (\mathbb{R}^n, g) \to (N^n, h)$ $(n \geq 3)$ where the dilation λ is radial $(\ln \lambda = \alpha(r), r = |x|$ and $\alpha \in C^\infty(\mathbb{R}, \mathbb{R}))$ is equivalent to an ordinary differential equation of the second order. More precisely, we have

Corollary 2. *Let $\phi : (\mathbb{R}^n, g) \to (N^n, h)$ $(n \geq 3)$ to be a conformal map of dilation λ when we suppose that $\ln \lambda$ is radial $(\ln \lambda = \alpha(r), r = |x|$ and $\alpha \in C^\infty(\mathbb{R}, \mathbb{R}))$. Then ϕ is biharmonic if and only if $\beta = \alpha'$ satisfies the following ordinary differential equation :*

$$\beta'' - (n-4)\beta\beta' + \frac{n-1}{r}\beta' - \frac{n-1}{r^2}\beta - \frac{2(n-1)}{r}\beta^2 - (n-2)\beta^3 = 0. \tag{34}$$

Proof of Corollary 2 Let $\phi : (\mathbb{R}^n, g) \to (N^n, h)$ $(n \geq 3)$ to be a conformal map of dilation λ such that $\ln \lambda = \alpha(r)$. By Theorem 3, ϕ is biharmonic if and only if the dilation λ satisfies

$$\begin{aligned} & grad \left(\Delta \ln \lambda \right) - \left(2 \left(\Delta \ln \lambda \right) + (n-2) |grad \ln \lambda|^2 \right) grad \ln \lambda \\ & + \frac{6-n}{2} grad \left(|grad \ln \lambda|^2 \right) + 2 Ricci^M (grad \ln \lambda) = 0. \end{aligned}$$

A direct calculation gives

$$grad \ln \lambda = \alpha' \frac{\partial}{\partial r},$$

$$|grad \ln \lambda|^2 = (\alpha')^2,$$

$$grad\left(|grad \ln \lambda|^2\right) = 2\alpha'\alpha''\frac{\partial}{\partial r},$$

$$\Delta \ln \lambda = \alpha'' + \frac{n-1}{r}\alpha'$$

and

$$grad\left(\Delta \ln \lambda\right) = \left(\alpha''' + \frac{n-1}{r}\alpha'' - \frac{n-1}{r^2}\alpha'\right)\frac{\partial}{\partial r}.$$

Therefore ϕ is biharmonic if and only if the function α satisfies the following differential equation

$$\alpha''' - (n-4)\,\alpha'\alpha'' + \frac{n-1}{r}\alpha'' - \frac{n-1}{r^2}\alpha' - \frac{2(n-1)}{r}(\alpha')^2 - (n-2)(\alpha')^3 = 0.$$

If we denote $\beta = \alpha'$, the biharmonicity of ϕ is equivalent to the differential equation

$$\beta'' - (n-4)\,\beta\beta' + \frac{n-1}{r}\beta' - \frac{n-1}{r^2}\beta - \frac{2(n-1)}{r}\beta^2 - (n-2)\,\beta^3 = 0.$$

As a consequence of the Corollary 2, We will present some remarks which we give a particular solutions of the equation (34) that allows us to construct a biharmonic non-harmonic maps.

Remark 3. . *Looking for particular solutions of type $\beta = \frac{a}{r}$ ($a \in \mathbb{R}^*$). By (34), we deduce that $\phi : (\mathbb{R}^n, g) \to (N^n, h)$ ($n \geq 3$) is biharmonic if and only if a is a solution of the algebraic equation*

$$(n-2)\,a^2 + (n+2)\,a + 2n - 2 = 0.$$

This equation has real solutions if and only if $n \in \{3, 4\}$.

1. *If $n = 3$, we find $a = \frac{-5+\sqrt{17}}{2}$ or $a = \frac{-5-\sqrt{17}}{2}$, so $\lambda = Cr^{-\left(\frac{5-\sqrt{17}}{2}\right)}$ or $\lambda = Cr^{-\left(\frac{5+\sqrt{17}}{2}\right)}$ ($C \in \mathbb{R}_+^*$). It follows that any conformal map $\phi : (\mathbb{R}^3, g) \to (N^3, h)$ of dilation $\lambda = Cr^{-\left(\frac{5-\sqrt{17}}{2}\right)}$ or $\lambda = Cr^{-\left(\frac{5+\sqrt{17}}{2}\right)}$ is biharmonic non-harmonic.*
2. *If $n = 4$, we find $a = -1$ or $a = -2$, so $\lambda = \frac{C}{r^2}$ or $\lambda = \frac{C}{r}$ ($C \in \mathbb{R}_+^*$). Then, in this case any conformal map $\phi : (\mathbb{R}^4, g) \to (N^4, h)$ of dilation $\lambda = \frac{C}{r^2}$ or $\lambda = \frac{C}{r}$ is biharmonic non-harmonic. For example, the inversion $\phi : (\mathbb{R}^n \backslash \{0\}, g_{\mathbb{R}^n}) \longrightarrow (\mathbb{R}^n \backslash \{0\}, g_{\mathbb{R}^n})$ definded by $\phi(x) = \frac{x}{|x|^2}$ is a conformal map with dilation $\lambda = \frac{1}{r^2}$. By (34), the inversion is biharmonic non-harmonic if and only if $n = 4$.*

Remark 4. . *Looking for particular solutions of type $\beta = \frac{ar}{1+r^2}$ ($a \in \mathbb{R}^*$). By (34), $\phi : (\mathbb{R}^n, g) \to (N^n, h)$ ($n \geq 3$) is biharmonic if and only we have*

$$(n-2)\,a^2 + (n+2)\,a + 2n - 2 = 0$$

and

$$3(n-2)\,a + 2n + 4 = 0.$$

These two equations gives $a = -2$ and $n = 4$, it follows that the dilation is equal to $\lambda = \frac{C}{r^2+1}$ ($C \in \mathbb{R}_+^*$). Then, all conformal maps $\phi : (\mathbb{R}^4, g) \to (N^4, h)$ of dilation $\lambda = \frac{C}{r^2+1}$ are biharmonic non-harmonic. For example, the inverse of the stereographic projection of the sphere $\phi : \mathbb{R}^n \longrightarrow S^n$ definded by $\phi(x) = \frac{1}{|x|^2+1}\left(|x|^2 - 1, 2x\right)$ is a conformal map with dilation $\lambda = \frac{2}{r^2+1}$. By (34), the inverse of the stereographic projection is biharmonic non-harmonic if and only if $n = 4$.

The last part of this paper is devoted to the study of biharmonic maps between warped product manifolds, these maps were also studied in [17]. We will give some results of the biharmonicity in other particular cases.

3. Biharmonic maps and the warped product

Let (M^m, g) and (N^n, h) two Riemannian manifolds and let $f \in C^\infty(M)$ be a positive function. The warped product $M \times_f N$ is the product manifolds $M \times N$ endowed with the Riemannian metric G_f defined, for $X, Y \in \Gamma(T(M \times N))$, by

$$G_f(X,Y) = g(d\pi(X), d\pi(Y)) + (f \circ \pi)^2 h(d\eta(X), d\eta(Y)),$$

where $\pi : M \times N \longrightarrow M$ and $\eta : M \times N \longrightarrow N$ are respectively the first and the second projection. The function f is called the warping function of the warped product. Let $X, Y \in \Gamma(T(M \times N))$, $X = (X_1, X_2)$, $Y = (Y_1, Y_2)$. Denote by ∇ the Levi-Civita connection on the Riemannian product $M \times N$. The Levi-Civita connection $\widetilde{\nabla}$ of the warped product $M \times_f N$ is given by

$$\widetilde{\nabla}_X Y = \nabla_X Y + X_1(\ln f)(0, Y_2) + Y_1(\ln f)(0, X_2) - f^2 h(X_2, Y_2)(grad \ln f, 0). \tag{35}$$

In the first, we consider a smooth map $\phi : (M^m, g) \longrightarrow (P^p, k)$ and we defined the map $\widetilde{\phi} : \left(M^m \times_f N^n, G_f\right) \longrightarrow (P^p, k)$ by $\widetilde{\phi}(x,y) = \phi(x)$. We will study the biharmonicity of $\widetilde{\phi}$. By calculating the tension field of $\widetilde{\phi}$, we obtain the following result:

Proposition 2. *Let $\phi : (M^m, g) \longrightarrow (P^p, k)$ be a smooth map. The tension field of the map $\widetilde{\phi} : \left(M^m \times_f N^n, G_f\right) \longrightarrow (P^p, k)$ defined by $\widetilde{\phi}(x,y) = \phi(x)$ is given by*

$$\tau(\widetilde{\phi}) = \tau(\phi) + n d\phi(grad \ln f) \tag{36}$$

Proof of Proposition 2. Let us choose $\{e_i\}_{1 \le i \le m}$ to be an orthonormal frame on M and $\{f_j\}_{1 \le j \le n}$ to be an orthonormal frame on N. An orthonormal frame on $M \times_f N$ is given by $\left\{(e_i, 0), \frac{1}{f}(0, f_j)\right\}$. Note that in this case we have $d\widetilde{\phi}(X,Y) = (d\phi(X), 0)$ for any $X \in \Gamma(TM)$ and $Y \in \Gamma(TN)$. By definition to the tension field, we have

$$\begin{aligned} \tau(\widetilde{\phi}) &= Tr_{G_f} \nabla d\widetilde{\phi} \\ &= \nabla^{\widetilde{\phi}}_{(e_i,0)} d\widetilde{\phi}(e_i, 0) + \frac{1}{f^2} \nabla^{\widetilde{\phi}}_{(0,f_j)} d\widetilde{\phi}(0, f_j) \\ &\quad - d\widetilde{\phi}\left(\widetilde{\nabla}_{(e_i,0)}(e_i, 0)\right) - \frac{1}{f^2} d\widetilde{\phi}\left(\widetilde{\nabla}_{(0,f_j)}(0, f_j)\right). \end{aligned}$$

A simple calculation gives

$$\nabla^{\widetilde{\phi}}_{(e_i,0)} d\widetilde{\phi}(e_i, 0) = \nabla^{\phi}_{e_i} d\phi(e_i)$$

and

$$\nabla^{\widetilde{\phi}}_{(0,f_j)} d\widetilde{\phi}(0, f_j) = 0,$$

By using the equation (35), we deduce that

$$\widetilde{\nabla}_{(e_i,0)}(e_i, 0) = (\nabla_{e_i} e_i, 0)$$

and

$$\widetilde{\nabla}_{(0,f_j)}(0, f_j) = \left(0, \nabla_{f_j} f_j\right) - n f^2 (grad \ln f, 0).$$

It follows that

$$\tau(\widetilde{\phi}) = \nabla^{\phi}_{e_i} d\phi(e_i) - d\phi\left(\nabla^{M}_{e_i} e_i\right) + nd\phi(grad \ln f),$$

then, we obtain

$$\tau(\widetilde{\phi}) = \tau(\phi) + nd\phi(grad \ln f).$$

Remark 5. *If $\phi : (M^m, g) \longrightarrow (P^m, k)$ $(m \geq 3)$ is a conformal map with dilation λ, the tension field of $\widetilde{\phi}$ is given by*

$$\tau(\widetilde{\phi}) = (2-m)\, d\phi(grad \ln \lambda) + nd\phi(grad \ln f) = d\phi\left(grad \ln\left(\lambda^{2-m} f^n\right)\right).$$

Then $\widetilde{\phi}$ is harmonic if and only if the function $\lambda^{2-m} f^n$ is constant.

We will now calculate the bitension field of the map $\widetilde{\phi} : \left(M^m \times_f N^n, G_f\right) \longrightarrow (P^p, k)$.

Theorem 4. *Let $\phi : (M^m, g) \longrightarrow (P^p, k)$ be a smooth map. The bitension field of the map $\widetilde{\phi} : \left(M^m \times_f N^n, G_f\right) \longrightarrow (P^p, k)$ defined by $\widetilde{\phi}(x, y) = \phi(x)$ is given by*

$$\begin{aligned} \tau_2(\widetilde{\phi}) = \tau_2(\phi) - n&\left(Tr_g \nabla^2 d\phi(grad \ln f) + Tr_g R^P(d\phi(grad \ln f), d\phi)\, d\phi\right) \\ &- n\nabla_{grad \ln f}\tau(\phi) - n^2 \nabla_{grad \ln f} d\phi(grad \ln f). \end{aligned} \tag{37}$$

Proof of Theorem 4. By definition of the bi-tension field, we have

$$\tau_2(\widetilde{\phi}) = -Tr_{G_f}\left(\nabla^{\widetilde{\phi}}\right)^2 \tau(\widetilde{\phi}) - Tr_{G_f} R^P\left(\tau(\widetilde{\phi}), d\widetilde{\phi}\right) d\widetilde{\phi} \tag{38}$$

For the first term $Tr_{G_f}\left(\nabla^{\widetilde{\phi}}\right)^2 \tau(\widetilde{\phi})$, we have

$$\begin{aligned} Tr_{G_f}\left(\nabla^{\widetilde{\phi}}\right)^2 \tau(\widetilde{\phi}) = &\nabla^{\widetilde{\phi}}_{(e_i,0)} \nabla^{\widetilde{\phi}}_{(e_i,0)} \tau(\widetilde{\phi}) + \frac{1}{f^2} \nabla^{\widetilde{\phi}}_{(0,f_j)} \nabla^{\widetilde{\phi}}_{(0,f_j)} \tau(\widetilde{\phi}) \\ &- \nabla^{\widetilde{\phi}}_{\widetilde{\nabla}_{(e_i,0)}(e_i,0)} \tau(\widetilde{\phi}) - \frac{1}{f^2} \nabla^{\widetilde{\phi}}_{\widetilde{\nabla}_{(0,f_j)}(0,f_j)} \tau(\widetilde{\phi}). \end{aligned}$$

We will study term by term the right-hand of this expression. A simple calculation gives

$$\begin{aligned} \nabla^{\widetilde{\phi}}_{(e_i,0)} \nabla^{\widetilde{\phi}}_{(e_i,0)} \tau(\widetilde{\phi}) &= \nabla^{\widetilde{\phi}}_{(e_i,0)} \nabla^{\widetilde{\phi}}_{(e_i,0)} \tau(\phi) + n\nabla^{\widetilde{\phi}}_{(e_i,0)} \nabla^{\widetilde{\phi}}_{(e_i,0)} d\phi(grad \ln f) \\ &= \nabla^{\phi}_{e_i} \nabla^{\phi}_{e_i} \tau(\phi) + n\nabla^{\phi}_{e_i} \nabla^{\phi}_{e_i} d\phi(grad \ln f) \end{aligned}$$

and

$$\nabla^{\widetilde{\phi}}_{(0,f_j)} \nabla^{\widetilde{\phi}}_{(0,f_j)} \tau(\widetilde{\phi}) = 0.$$

By using the equation (35), we obtain

$$\nabla^{\widetilde{\phi}}_{\widetilde{\nabla}_{(e_i,0)}(e_i,0)} \tau(\widetilde{\phi}) = \nabla^{\phi}_{\nabla^{M}_{e_i} e_i} \tau(\phi) + n\nabla^{\phi}_{\nabla^{M}_{e_i} e_i} d\phi(grad \ln f),$$

and

$$\nabla^{\widetilde{\phi}}_{\widetilde{\nabla}_{(0,f_j)}(0,f_j)} \tau(\widetilde{\phi}) = -nf^2 \nabla^{\phi}_{grad \ln f} \tau(\phi) - n^2 f^2 \nabla^{\phi}_{grad \ln f} d\phi(grad \ln f).$$

Then, we deduce that

$$\begin{aligned} Tr_{G_f}\left(\nabla^{\widetilde{\phi}}\right)^2 \tau(\widetilde{\phi}) = &Tr_g\left(\nabla^{\phi}\right)^2 \tau(\phi) + nTr_g\left(\nabla^{\phi}\right)^2 d\phi(grad \ln f) \\ &+ n\nabla^{\phi}_{grad \ln f} \tau(\phi) + n^2 \nabla^{\phi}_{grad \ln f} d\phi(grad \ln f). \end{aligned} \tag{39}$$

To complete the proof, we will simplify the term $Tr_{G_f} R^P \left(\tau\left(\widetilde{\phi}\right), d\widetilde{\phi}\right) d\widetilde{\phi}$, we have

$$\begin{aligned} Tr_{G_f} R^P \left(\tau\left(\widetilde{\phi}\right), d\widetilde{\phi}\right) d\widetilde{\phi} &= R^P \left(\tau\left(\widetilde{\phi}\right), d\widetilde{\phi}\left(e_i, 0\right)\right) d\widetilde{\phi}\left(e_i, 0\right) \\ &+ \frac{1}{f^2} R^P \left(\tau\left(\widetilde{\phi}\right), d\widetilde{\phi}\left(0, f_j\right)\right) d\widetilde{\phi}\left(0, f_j\right) \\ &= R^P \left(\tau\left(\widetilde{\phi}\right), d\widetilde{\phi}\left(e_i, 0\right)\right) d\widetilde{\phi}\left(e_i, 0\right) \\ &= R^P \left(\tau\left(\phi\right), d\phi\left(e_i\right)\right) d\phi\left(e_i\right) \\ &+ n R^P \left(d\phi\left(grad \ln f\right), d\phi\left(e_i\right)\right) d\phi\left(e_i\right). \end{aligned}$$

It follows that

$$Tr_{G_f} R^P \left(\tau\left(\widetilde{\phi}\right), d\widetilde{\phi}\right) d\widetilde{\phi} = Tr_g R^P \left(\tau\left(\phi\right), d\phi\right) d\phi + n Tr_g R^P \left(d\phi\left(grad \ln f\right), d\phi\right) d\phi. \tag{40}$$

If we replace (39) and (40) in (38), we obtain

$$\begin{aligned} \tau_2\left(\widetilde{\phi}\right) &= \tau_2\left(\phi\right) - n\left(Tr_g \nabla^2 d\phi\left(grad \ln f\right) + Tr_g R^P \left(d\phi\left(grad \ln f\right), d\phi\right) d\phi\right) \\ &- n \nabla_{grad \ln f} \tau\left(\phi\right) - n^2 \nabla_{grad \ln f} d\phi\left(grad \ln f\right). \end{aligned}$$

The proof of Theorem 4 is complete.

Remark 6. *Theorem 4 is a particular result of generalized warped product manifolds (see [18]).*

As a consequence, if ϕ is harmonic, we have

Corollary 3. *Let $\phi : (M^m, g) \longrightarrow (P^p, k)$ a harmonic map. The map $\widetilde{\phi} : \left(M^m \times_f N^n, G_f\right) \longrightarrow (P^p, k)$ defined by $\widetilde{\phi}(x, y) = \phi(x)$ is biharmonic if and only if*

$$Tr_g \nabla^2 d\phi\left(grad \ln f\right) + Tr_g R^P \left(d\phi\left(grad \ln f\right), d\phi\right) d\phi + n \nabla_{grad \ln f} d\phi\left(grad \ln f\right) = 0.$$

In particular if $\phi = Id_M$, the first projection $P_1 : \left(M^m \times_f N^n, G_f\right) \longrightarrow (M^m, g)$ defined by $P_1(x, y) = x$ is biharmonic if and only if (see [17])

$$grad \Delta \ln f + \frac{n}{2} grad\left(\left|grad \ln f\right|^2\right) + 2 Ricci^M\left(grad \ln f\right) = 0.$$

In the following we shall present an example of biharmonic non-harmonic maps.

Example 1. *Let $\widetilde{\varphi} : \mathbb{R}^m \setminus \{0\} \times_f N^n \longrightarrow \mathbb{R}^m \setminus \{0\}$ defined by $\widetilde{\varphi}(x, y) = \frac{x}{|x|^2}$ when we suppose that $\ln f$ is radial ($\ln f = \alpha(r)$). Then by Theorem 4, we deduce that the map $\widetilde{\varphi} : \mathbb{R}^m \setminus \{0\} \times_f N^n \longrightarrow \mathbb{R}^m \setminus \{0\}$ is biharmonic if and only if the function α satisfies the following differential equation*

$$n\alpha''' + \frac{n(m-5)}{r}\alpha'' - \frac{3n(3m-7)}{r^2}\alpha' + n^2\alpha'\alpha'' - \frac{2n^2}{r}\left(\alpha'\right)^2 - \frac{8(m-2)(m-4)}{r^3} = 0.$$

Let $\beta = \alpha'$, this equation becomes

$$n\beta'' + \frac{n(m-5)}{r}\beta' - \frac{3n(3m-7)}{r^2}\beta + n^2\beta\beta' - \frac{2n^2}{r}\beta^2 - \frac{8(m-2)(m-4)}{r^3} = 0.$$

Looking for particular solutions of type $\beta = \frac{a}{r}$ ($a \in \mathbb{R}^*$), then $\widetilde{\varphi} : \mathbb{R}^m \setminus \{0\} \times_f N^n \longrightarrow \mathbb{R}^m \setminus \{0\}$ is biharmonic if and only if

$$3n^2a^2 + 2n(5m-14)a + 8(m-2)(m-4) = 0.$$

This equation has two solutions $a = \frac{4-2m}{n}$ and $a = \frac{4(4-m)}{3n}$.

1. *For* $a = \frac{4-2m}{n}$, *we obtain* $f(r) = Cr^{\frac{4-2m}{n}}$ *and in this case* $\widetilde{\varphi} : \mathbb{R}^m \setminus \{0\} \times_f N^n \longrightarrow \mathbb{R}^m \setminus \{0\}$ *is harmonic so biharmonic.*
2. *For* $a = \frac{4(4-m)}{3n}$, *we obtain* $f(r) = Cr^{\frac{4(4-m)}{3n}}$ *and in this case* $\widetilde{\varphi} : \mathbb{R}^m \setminus \{0\} \times_f N^n \longrightarrow \mathbb{R}^m \setminus \{0\}$ *is biharmonic non-harmonic.*

Now, we consider a smooth map $\psi : (N^n, g) \longrightarrow (P^p, k)$ and we define the map $\widetilde{\psi} : \left(M^m \times_f N^n, G_f\right) \longrightarrow (P^p, k)$ by $\widetilde{\psi}(x, y) = \psi(y)$. We will study the biharmonicity of $\widetilde{\psi}$, we obtain the following result :

Theorem 5. *Let* $\psi : (N^n, h) \to (P^p, k)$ *be a smooth map, we define* $\widetilde{\psi} : \left(M^m \times_{f2} N^n, G_{f2}\right) \to (P^p, k)$ *by* $\widetilde{\psi}(x, y) = \psi(y)$. *The tension field and the bi-tension field of* $\widetilde{\psi}$ *are given by*

$$\tau(\widetilde{\psi}) = \frac{1}{f^2 \circ \pi} \tau(\psi) \tag{41}$$

and

$$\tau_2(\widetilde{\psi}) = \frac{1}{f^4 \circ \pi} \tau_2(\psi) - \frac{2}{f^2 \circ \pi} \left(\left(\Delta \ln f + (n-2) |grad \ln f|^2\right) \circ \pi\right) \tau(\psi). \tag{42}$$

Proof of Theorem 5. In the first, we calculate the tension field of of $\widetilde{\psi}$. By definition to the tension field, we have

$$\begin{aligned}
\tau(\widetilde{\psi}) &= Tr_{G_f} \nabla d\widetilde{\psi} \\
&= \nabla^{\widetilde{\psi}}_{(e_i,0)} d\widetilde{\psi}(e_i, 0) + \frac{1}{f^2 \circ \pi} \nabla^{\widetilde{\psi}}_{(0,f_j)} d\widetilde{\psi}(0, f_j) \\
&\quad - d\widetilde{\psi}\left(\widetilde{\nabla}_{(e_i,0)}(e_i, 0)\right) - \frac{1}{f^2 \circ \pi} d\widetilde{\psi}\left(\widetilde{\nabla}_{(0,f_j)}(0, f_j)\right).
\end{aligned}$$

By using the equation (35), we obtain

$$\tau(\widetilde{\psi}) = \frac{1}{f^2 \circ \pi} \nabla^{\psi}_{f_j} d\psi(f_j) - \frac{1}{f^2 \circ \pi} d\psi\left(\nabla_{f_j} f_j\right) = \frac{1}{f^2 \circ \pi} \tau(\psi),$$

then

$$\tau(\widetilde{\psi}) = \frac{1}{f^2 \circ \pi} \tau(\psi).$$

By this expression, we deduce that $\widetilde{\psi}$ is harmonic if and only if ψ is harmonic. Now, we will calculate the bi-tension field of $\widetilde{\psi}$. By definition, we have

$$\tau_2(\widetilde{\psi}) = -Tr_{G_f}\left(\nabla^{\widetilde{\psi}}\right)^2 \tau(\widetilde{\psi}) - Tr_{G_f} R^P\left(\tau(\widetilde{\psi}), d\widetilde{\psi}\right) d\widetilde{\psi}. \tag{43}$$

For the first term $Tr_{G_f}\left(\nabla^{\widetilde{\psi}}\right)^2 \tau(\widetilde{\psi})$, we have

$$\begin{aligned}
Tr_{G_f}\left(\nabla^{\widetilde{\psi}}\right)^2 \tau(\widetilde{\psi}) &= \nabla^{\widetilde{\psi}}_{(e_i,0)} \nabla^{\widetilde{\psi}}_{(e_i,0)} \tau(\widetilde{\psi}) + \frac{1}{f^2 \circ \pi} \nabla^{\widetilde{\psi}}_{(0,f_j)} \nabla^{\widetilde{\psi}}_{(0,f_j)} \tau(\widetilde{\psi}) \\
&\quad - \nabla^{\widetilde{\psi}}_{\widetilde{\nabla}_{(e_i,0)}(e_i,0)} \tau(\widetilde{\psi}) - \frac{1}{f^2 \circ \pi} \nabla^{\widetilde{\psi}}_{\widetilde{\nabla}_{(0,f_j)}(0,f_j)} \tau(\widetilde{\psi}).
\end{aligned}$$

A long calculation gives

$$\nabla^{\widetilde{\psi}}_{(e_i,0)} \nabla^{\widetilde{\psi}}_{(e_i,0)} \tau(\widetilde{\psi}) = \frac{2}{f^2 \circ \pi} \left(\left(2 |grad \ln f|^2 - e_i(e_i(\ln f))\right) \circ \pi\right) \tau(\psi)$$

and

$$\frac{1}{f^2 \circ \pi} \nabla^{\widetilde{\psi}}_{(0,f_j)} \nabla^{\widetilde{\psi}}_{(0,f_j)} \tau(\widetilde{\psi}) = \frac{1}{f^4 \circ \pi} \nabla^{\psi}_{f_j} \nabla^{\psi}_{f_j} \tau(\psi).$$

Finally, by (35), we obtain

$$\nabla^{\widetilde{\psi}}_{\widetilde{\nabla}_{(e_i,0)}(e_i,0)}\tau(\widetilde{\psi}) = \frac{2}{f^2\circ\pi}\left(\nabla_{e_i}e_i\left((\ln f)\right)\circ\pi\right)\tau(\psi)$$

and

$$\frac{1}{f^2\circ\pi}\nabla^{\widetilde{\psi}}_{\widetilde{\nabla}_{(0,f_j)}(0,f_j)}\tau(\widetilde{\psi}) = \frac{1}{f^4\circ\pi}\nabla^{\psi}_{\nabla_{f_j}f_j}\tau(\psi) + \frac{2n}{f^2\circ\pi}\left(\left(|grad\ln f|^2\right)\circ\pi\right)\tau(\psi).$$

Which gives us

$$Tr_{G_f}\left(\nabla^{\widetilde{\psi}}\right)^2\tau(\widetilde{\psi}) = \frac{1}{f^4\circ\pi}Tr_h\nabla^2\tau(\psi) - \frac{2}{f^2\circ\pi}\left(\left(\Delta\ln f + (n-2)\,|grad\ln f|^2\right)\circ\pi\right)\tau(\psi) \quad (44)$$

Finally for the first term $Tr_{G_f}R^P(\tau(\widetilde{\psi}), d\widetilde{\psi})\,d\widetilde{\psi}$, it is easy to verify that

$$Tr_{G_f}R^P(\tau(\widetilde{\psi}), d\widetilde{\psi})\,d\widetilde{\psi} = \frac{1}{f^4\circ\pi}Tr_hR^P(\tau(\psi), d\psi)\,d\psi. \quad (45)$$

If we substitute (44) and (45) in (43), we obtain

$$\tau_2(\widetilde{\psi}) = \frac{1}{f^4\circ\pi}\tau_2(\psi) - \frac{2}{f^2\circ\pi}\left(\left(\Delta\ln f + (n-2)\,|grad\ln f|^2\right)\circ\pi\right)\tau(\psi).$$

This completes the proof of Theorem 5. An immediate consequence of Theorem 5 is given by the following corollary :

Corollary 4. *Let $\psi : (N^n, h) \longrightarrow (P^p, k)$ a biharmonic non-harmonic map. The map $\widetilde{\phi}$: $\left(M^m \times_f N^n, G_{f^2}\right) \longrightarrow (P^p, k)$ defined by $\widetilde{\psi}(x,y) = \psi(y)$ is biharmonic if and only if the function f^{n-2} is harmonic.*

Acknowledgments: The authors would like to thank the referee for some useful comments and their helpful suggestions that have improved the quality of this paper.

Author Contributions: The authors provide equal contributions to this paper.

Conflicts of Interest: The authors declare no conflict of interest.

References

1. Eells, J.; Lemaire, L. A report on harmonic maps. *Bull. Lond. Math. Soc.* **1978**, *16*, 1–68.
2. Eells, J.; Lemaire, L. Another report on harmonic maps. *Bull. Lond. Math. Soc.* **1988**, *20*, 385–524.
3. Eells, J.; Lemaire, L. *Selected topics in harmonic maps*; CBMS Regional Conference Series in Mathematics; American Mathematical Society: Providence, RI, USA, 1981; Vol. 150.
4. Eells, J.; Ratto, A. *Harmonic Maps and Minimal Immersions with Symmetries: Methods of Ordinary Differential Equations Applied to Elliptic Variational Problems*; Princeton University Press: Princeton, NJ, USA, 1993; Vol. 130.
5. Baird, P.; Eells, J. *A Conservation Law for Harmonic Maps*; Lecture Notes in Math. 894; Springer : Berlin, Germany, 1981; pp. 1–25.
6. Jiang, G.Y. 2-harmonic maps and their first and second variational formulas. *Chin. Ann. Math. Ser.* **1986**, *A7*, 389–402.
7. Loubeau, E.; Montaldo, S.; Oniciuc, C. The stress-energy tensor for biharmonic maps. *Math. Z.* **2008**, *259*, 503–524.
8. Baird, P.; Kamissoko, D. On constructing biharmonic maps and metrics. *Ann. Glob. Anal. Geom.* **2003**, *23*, 65–75.
9. Balmus, A. Biharmonic properties and conformal changes. *Analele Stiintifice ale Univ. Al.I. Cuza Iasi Mat.* **2004**, *50*, 367–372.

10. Ou, Y.-L. p-harmonic morphisms, biharmonic morphisms, and non-harmonic biharmonic maps. *J. Geom. Phys.* **2006**, *56*, 358–374.
11. Ouakkas, S. Biharmonic maps, conformal deformations and the Hopf maps. *Diff. Geom. Appl.* **2008**, *26*, 495–502.
12. Baird, P.; Fardoun, A.; Ouakkas, S. Conformal and semi-conformal biharmonic maps. *Ann. Glob. Anal. Geom.* **2008**, *34*, 403–414.
13. Montaldo, S.; Oniciuc, C. A short survey of biharmonic maps between Riemannian manifolds. *Rev. Union Mat. Argent.* **2006**, *47*, 1–22.
14. Baird, P. *Harmonic Maps with Symmetry, Harmonic Morphisms and Deformation of Metrics*; Research Notes in Mathematics; CRC Press: London, UK,1983; pp. 27–39.
15. Baird, P.; Loubeau, E.; Oniciuc, C. Harmonic and biharmonic maps from surfaces. In *Harmonic maps and differential geometry*; Amer. Math. Soc.: Providence, RI, USA, 2011; pp. 234–241.
16. Baird, P.; Wood, J.C. *Harmonic Morphisms between Riemannain Manifolds*; London mathematical Society Monographs (N.S.); Oxford University Press: Oxford, UK, 2003.
17. Balmus, A.; Montaldo, S.; Oniciuc, C. Biharmonic maps between warped product manifolds. *J. Geom. Phys.* **2007**, *57*, 449–466.
18. Djaa, N.E.H.; Boulal, A.; Zagane, A. Generalized warped product manifolds and biharmonic maps. *Acta Math. Univ. Comen.* **2012**, *81*, 283–298.

4

Existence of Semi Linear Impulsive Neutral Evolution Inclusions with Infinite Delay in Frechet Spaces

Dimplekumar N. Chalishajar [1,*], Kulandhivel Karthikeyan [2,†] and Annamalai Anguraj [3,†]

1 Department of Applied Mathematics, Virginia Military Institute (VMI), 431 Mallory Hall, Lexington, VA 24450, USA
2 Department of Mathematics, KSR College of Technology, Tiruchengode 637215, India; karthi_phd2010@yahoo.co.in
3 Department of Mathematics, PSG College of Arts and Science, Coimbatore 641014 , India; angurajpsg@yahoo.com
* Correspondence: dipu17370@gmail.com
† These authors contributed equally to this work.

Academic Editor: Hari M. Srivastava

Abstract: In this paper, sufficient conditions are given to investigate the existence of mild solutions on a semi-infinite interval for first order semi linear impulsive neutral functional differential evolution inclusions with infinite delay using a recently developed nonlinear alternative for contractive multivalued maps in Frechet spaces due to Frigon combined with semigroup theory. The existence result has been proved without assumption of compactness of the semigroup. We introduced a new phase space for impulsive system with infinite delay and claim that the phase space considered by different authors are not correct.

Keywords: impulsive differential inclusions; fixed point; Frechet spaces; nonlinear alternative due to Frigon

Mathematics Subject Classification (2000): 34A37, 34G20, 47H20

---

1. Introduction

In recent years, impulsive differential and partial differential equations have become more important in some mathematical models of real phenomena, especially in control, biological and medical domains. In these models, the investigated simulating processes and phenomena usually are subject to short-term perturbations whose duration is negligible in comparison with the duration of the process. Consequently, it is natural to assume that these perturbations act instantaneously in the form of impulses. The theory of impulsive differential equations has seen considerable development, see the monographs of Bainov and Semeonov [1], Lakshimikantham [2] and Perestyuk [3]. Recently, several works reported existence results for mild solutions for impulsive neutral functional differential equations or inclusions, such as ([4–11]) and references therein. However, the results obtained there are only in connection with finite delay. Since many systems arising from realistic models heavily depend on histories (*i.e.*, there is an effect of infinite delay on state equations), there is a real need to discuss partial functional differential systems with infinite delay, where numerous properties of their solutions are studied and detailed bibliographies are given. The literature related to first and second order nonlinear non autonomous neutral impulsive systems with or without state dependent delay is not vast. To the best of our knowledge, this is almost an untreated article in a literature and is one of the main motivations of this paper.

When the delay is infinite, the notion of the phase space plays an important role in the study of both qualitative and quantitative theory. A usual choice is a seminormed space satisfying suitable axioms, introduced by Hale and Kato in [12]; see also Corduneanu and Lakshmikantham [13]; Graef [14] and Baghli and Benchohra [15,16]. Unfortunately, we can not find broad literature about the system involving infinite delay with impulse effects. Henderson and Ouahab [17] discussed existence results for nondensely defined semilinear functional differential inclusions in Frechet spaces. Hernández *et al.* [18] studied existence of solutuions for impulsive partial neutral functional differential equations for first and second order systems with infinite delay. Recently, Arthi and Balachandran [19] proved controllability of the second order impulsive functional differential equations with state dependent delay using fixed point approach and cosine operator theory. It has been observed that the existence or the controllability results proved by different authors are through an axiomatic defination of the phase space given by Hale and Kato [12]. However, as remarked by Hino, Murakami, and Naito [20], it has come to our attention that these axioms for the phase space are not correct for the impulsive system with infinite delay [21,22]. This motivated us to generate a new phase space for the existence of a nonautonomous impulsive neutral inclusion with infinite delay. This is another motivation of this paper. To the best of our knowledge, the result proved in this paper is new and are not available in the literature.

On the other hand, researchers have been proving the controllability results using compactness assumption of semigroups and the family of cosine operators. However, as remarked by Triggiani [23], if X is an infinite dimensional Banach space, then the linear control system is never exactly controllable on the given interval if either B is compact or associated semigroup is compact. According to Triggiani [23], this is a typical case for most control systems governed by parabolic partial differential equations and hence the concept of exact controllability is very limited for many parabolic partial differential equations. Nowadays, researchers are engaged to overcome this problem, refer to ([19,21,22]). Very recently, Chalishajar and Acharya [22] studied the controllability of second order neutral functional differential inclusion, with infinite delay and impulse effect on unbounded domain, without compactness of the family of cosine operators. Ntouyas and O'Regan [24] gave some remarks on controllability of evolution equations in Banach paces and proved a result without compactness assumption.

In the last few years, researchers have diverted to fractional impulsive equations due to their extensive applications in various fields. Fečkan *et al.* [25] have discussed the existence of PC-mild solutions for Cauchy problems and nonlocal problems for impulsive fractional evolution equations involving Caputo fractional derivative by utilizing the theory of operators semigroup, probability density functions via impulsive conditions, a new concept on a PC-mild solution is introduced in their paper. We refer to the readers the book of Zhou [26]. Recently, Fu *et al.* [27] studied the existence of PC-mild solutions for Cauchy and nonlocal problems of impulsive fractional evolution equations for which the impulses are not instantaneous, by using the theory of operator semigroups, probability density functions, and some suitable fixed point theorems.

The rest of this paper is organized as follows: In Section 2, we introduce the system, recall some basic definitions, and preliminary facts that will be used throughout this paper. The existence theorems for semi linear impulsive neutral evolution inclusions with infinite delay, and their proofs are arranged in Section 3. Finally, in Section 4, an example is presented to illustrate the applications of the obtained results.

2. Preliminaries

In this paper, we shall consider the existence of mild solutions for first order impulsive partial neutral functional evolution differential inclusions with infinite delay in a Banach space E

$$\frac{d}{dt}\Big[y(t) - g(t,y_t)\Big] \in A(t)y(t) + F(t,y_t) \quad (1)$$
$$t \in J = [0,+\infty), \quad t \neq t_k, \quad k = 1,2,\dots$$
$$\Delta y|_{t=t_k} = I_k(y(t_k^-)), \quad k = 1,2\dots \quad (2)$$
$$y_0 = \phi \in \mathcal{B}_h \quad (3)$$

where $F : J \times \mathcal{B}_h \to \mathcal{P}(E)$ is a multivalued map with nonempty compact values, $\mathcal{P}(E)$ is the family of all subsets of E, $g : J \times \mathcal{B}_h \to E$ and $I_k : E \to E, k = 1,2,\dots$ are given functions, $\phi \in \mathcal{B}_h$ are given functions and $\{A(t)\}_{0\le t<+\infty}$ is a family of linear closed (not necessarily bounded) operators from E into E that generate an evolution system of operators $\{U(t,s)\}_{(t,s)\in J\times J}$ for $0 \le s \le t < +\infty$. $\Delta y|_{t=t_k} = y(t_k^+) - y(t_k^-)$, $y(t_k^+)$ and $y(t_k^-)$ represent right and left limits of $y(t)$ at $t = t_k$ respectively. For any continuous function y and any $t \ge 0$, we denote by y_t the element of $\mathcal{B}_h$ defined by $y_t(\theta) = y(t+\theta)$ for $\theta \in (-\infty,0]$. We assume that the histories y_t belongs to some abstract phase space $\mathcal{B}_h$ to be specified below.

We present the abstract phase space $\mathcal{B}_h$. Assume that $h : (-\infty,0) \to (0,\infty)$ be a continuous function with $l = \int_{-\infty}^{0} h(s)ds < +\infty$. Define,

$$\mathcal{B}_h := \{\phi : [-\infty,0] \to X \text{ such that, for any } r > 0, \phi(\theta) \text{ is bounded and measurable function on } [-r,0] \text{ and } \int_{-\infty}^{0} h(s) \sup_{s\le\theta\le0}|\phi(\theta)|ds < +\infty\}.$$

Here, $\mathcal{B}_h$ endowed with the norm

$$\|\phi\|_{\mathcal{B}_h} = \int_{-\infty}^{0} h(s) \sup_{s\le\theta\le0} |\phi(\theta)|ds, \ \forall\phi \in \mathcal{B}_h$$

Then, it is easy to show that $(\mathcal{B}_h, \|.\|_{\mathcal{B}_h})$ is a Banach space.

Lemma 1. *Suppose $y \in \mathcal{B}_h$; then, for each $t \in J, y_t \in \mathcal{B}_h$. Moreover,*

$$l|y(t)| \le \|y_t\|_{\mathcal{B}_h} \le l \sup_{s\le\theta\le0} (|y(s)| + \|y_0\|_{\mathcal{B}_h})$$

where $l := \int_{-\infty}^{0} h(s)ds < +\infty$.

Proof. For any $t \in [0,a]$, it is easy to see that, y_t is bounded and measurable on $[-a,0]$ for $a > 0$, and

$$\begin{aligned}
\|y_t\|_{\mathcal{B}_h} &= \int_{-\infty}^{0} h(s) \sup_{\theta\in[s,0]} |y_t(\theta)|ds \\
&= \int_{-\infty}^{-t} h(s) \sup_{\theta\in[s,0]} |y(t+\theta)|ds + \int_{-t}^{0} h(s) \sup_{\theta\in[s,0]} |y(t+\theta)|ds \\
&= \int_{-\infty}^{-t} h(s) \sup_{\theta_1\in[t+s,t]} |y(\theta_1)|ds + \int_{-t}^{0} h(s) \sup_{\theta_1\in[t+s,t]} |y(\theta_1)|ds \\
&\le \int_{-\infty}^{-t} h(s)\left[\sup_{\theta_1\in[t+s,0]} |y(\theta_1)| + \sup_{\theta_1\in[0,t]} |y(\theta_1)|\right]ds + \int_{-t}^{0} h(s) \sup_{\theta_1\in[0,t]} |y(\theta_1)|ds \\
&= \int_{-\infty}^{-t} h(s) \sup_{\theta_1\in[t+s,0]} |y(\theta_1)|ds + \int_{-\infty}^{0} h(s)ds. \sup_{s\in[0,t]} |y(s)| \\
&\le \int_{-\infty}^{-t} h(s) \sup_{\theta_1\in[s,0]} |y(\theta_1)|ds + l. \sup_{s\in[0,t]} |y(s)| \\
&\le \int_{-\infty}^{0} h(s) \sup_{\theta_1\in[s,0]} |y(\theta_1)|ds + l \sup_{s\in[0,t]} |y(s)| \\
&= \int_{-\infty}^{0} h(s) \sup_{\theta_1\in[s,0]} |y_0(\theta_1)|ds + l \sup_{s\in[0,t]} |y(s)| \\
&= l \sup_{s\in[0,t]} |y(s)| + \|y_0\|_{\mathcal{B}_h}
\end{aligned}$$

Since $\phi \in \mathcal{B}_h$, then $y_t \in \mathcal{B}_h$. Moreover,

$$\|y_t\|_{\mathcal{B}_h} = \int_{-\infty}^{0} h(s) \sup_{\theta\in[s,0]} |y_t(\theta)|ds \ge |y_t(\theta)| \int_{-\infty}^{0} h(s)ds = l|y(t)|$$

The proof is complete. □

Next, we introduce definitions, notation and preliminary facts from multi-valued analysis, which are useful for the development of this paper (see [28]).

Let $C([0,b],E)$ denote the Banach space of all continuous functions from $[0,b]$ into E with the norm

$$\|y\|_\infty = \sup\{\|y(t)\| : 0 \le t \le b\}$$

and let $L^1([0,\infty),E)$ be the Banach space of measurable functions $y : [0,\infty) \to E$, that are Lebesgue integrable with the norm

$$\|y\|_{L^1} = \int_0^\infty \|y(t)\|dt \quad \text{for all} \quad y \in L^1([0,\infty),E)$$

Let X be a Frechet space with a family of semi-norms $\{\|\cdot\|_n\}_{n\in\mathbb{N}}$. Let $Y \subset X$, we say that Y is bounded if for every $n \in \mathbb{N}$, there exists $\bar{M}_n > 0$ such that

$$\|y\|_n \le \bar{M}_n \quad \text{for all} \quad y \in Y$$

To X we associate a sequence of Banach spaces $\{(X^n, \|\cdot\|_n)\}$ as follows: for every $n \in \mathbb{N}$, we consider the equivalence relation $\sim_n$ defined by :$x \sim_n y$ if and only if $\|x-y\|_n = 0$ for all $x,y \in X$. We denote $X^n = (X|_{\sim_n}, \|\cdot\|_n)$ the quotient space, the completion of X^n with respect to $\|\cdot\|_n$. To every $Y \subset X$, we associate a sequence the $\{Y^n\}$ of subsets $Y^n \subset X^n$ as follows: for every $x \in X$, we denote $[x]_n$ the equivalence class of x of subset X^n and we define $Y^n = \{[x]_n : x \in Y\}$. We denote

$\bar{Y}_n$, $\text{int}(Y^n)$ and $\partial_n Y^n$, respectively, the closure, the interior, and the boundary of Y^n with respect to $\|\cdot\|$ in X^n. We assume that the family of semi-norms $\{\|\cdot\|_n\}$ verifies:

$$\|x\|_1 \le \|x\|_2 \le \|x\|_3 \le \dots \quad \text{for every} \quad x \in X$$

Let (X, d) be a metric space. We use the following notations:
$\mathcal{P}_{cl}(X) := \{Y \in \mathcal{P}(X) : Y\text{closed}\}, \mathcal{P}_b(X) := \{Y \in \mathcal{P}(X) : Y\text{bounded}\}$
$\mathcal{P}_{cv}(X) := \{Y \in \mathcal{P}(X) : Y\text{convex}\}, \mathcal{P}_{cp}(X) := \{Y \in \mathcal{P}(X) : Y\text{compact}\}$.
Consider $H_d : \mathcal{P}(X) \times \mathcal{P}(X) \to R^+ \cup \{\infty\}$, given by

$$H_d(\mathcal{A}, \mathcal{B}) := \max\{\sup_{a \in \mathcal{A}} d(a, \mathcal{B}), \sup_{b \in \mathcal{B}} d(\mathcal{A}, b)\}$$

where $d(\mathcal{A}, b) := \inf_{a \in \mathcal{A}} d(a, b), d(a, \mathcal{B}) := \inf_{b \in \mathcal{B}} d(a, b)$. Then, $(\mathcal{P}_{b,cl}(X), H_d)$ is a metric space and $(\mathcal{P}_{cl}(X), H_d)$ is a generalized (complete) metric space (see [29]).

Definition 1. *We say that a family $\{A(t)\}_{t \ge 0}$ generates a unique linear evolution system $\{U(t,s)\}_{(t,s) \in \Delta}$ for $\Delta_1 = \{(t,s) \in J \times J : 0 \le s \le t < +\infty\}$ satisfying the following properties:*

(1) $U(t,t) = I$ where I is the identity operator in E,
(2) $U(t,s)U(s,\tau) = U(t,\tau)$ for $0 \le \tau \le s \le t < +\infty$,
(3) $U(t,s) \in B(E)$ the space of bounded linear operators on E, where for every $(t,s) \in \Delta_1$ and for each $y \in E$, the mapping $(t,s) \to U(t,s)y$ is continuous.

More details on evolution systems and their properties could be found in the books of Ahmed [30], Engel and Nagel [31], and Pazy [32].

Definition 2. *A multivalued map $G : J \to \mathcal{P}_{cl}(X)$ is said to be measurable if for each $x \in E$, the function $Y : J \to X$ defined by*

$$Y(t) = d(x, G(t)) = \inf\{|x - z| : z \in G(t)\}$$

is measurable where d is the metric induced by the normed Banach space X.

Definition 3. *A function $F : J \times \mathcal{B}_h \to \mathcal{P}(X)$ is said to be an L^1_{loc}-Caratheodory multivalued map if it satisfies:*
(i) $x \to F(t,y)$ is continuous(with respect to the metric H_d) for almost all $t \in J$;
(ii) $t \to F(t,y)$ is measurable for each $y \in \mathcal{B}_h$;
(iii) for every positive constant k there exists $h_k \in L^1_{loc}(J; R^+)$ such that

$$\|F(t,y)\| \le h_k(t) \quad \textit{for all} \quad \|y\|_{\mathcal{B}_h} \le k \quad \textit{and for almost all} \quad t \in J$$

A multivalued map $G : X \to \mathcal{P}(X)$ has convex(closed) values if $G(x)$ is convex (closed) for all $x \in X$. We say that G is bounded on bounded sets if $G(B)$ is bounded in X for each bounded set B of X, *i.e.*,

$$\sup_{x \in B}\{\sup\{\|y\| : y \in G(x)\}\} < \infty$$

Finally, we say that G has fixed point if there exists $x \in X$ such that $x \in G(x)$.
For each $y \in B_*$, let the set $S_{F,y}$ known as the set of selectors from F defined by

$$S_{F,y} = \{v \in L^1(J; E) : v(t) \in F(t, y_t), \quad \text{a.e.} t \in J\}$$

For more details on multivalued maps, we refer to the books of Aubin and Cellina [33] and Deimling [34], Gorniewicz [35], Hu and Papageorgiou [36], and Tolstonogov [37].

Definition 4. *A multivalued map $F : X \to \mathcal{P}(X)$ is called an admissible contraction with constant $\{k_n\}_{n\in\mathbb{N}}$ if for each $n \in \mathbb{N}$ there exists $k_n \in (0,1)$ such that*
(i) $H_d(F(x), F(y)) \leq k_n\|x-y\|_n$ for all $x, y \in X$.
(ii) For every $x \in X$ and every $\epsilon \in (0,\infty)^n$, there exists $y \in F(x)$ such that

$$\|x-y\|_n \leq \|x - F(x)\|_n + \epsilon_n \quad \text{for every} \quad n \in \mathbb{N}$$

The following nonlinear alternative will be used to prove our main results.
Theorem 2.1 (Nonlinear Alternative of Frigon, [38,39]). Let X be a Frechet space and U an open neighborhood of the origin in X and let $N : \bar{U} \to \mathcal{P}(X)$ be an admissible multivalued contraction. Assume that N is bounded. Then, one of the following statements holds:
(C1) N has a fixed point;
(C2) There exists $\lambda \in [0,1)$ and $x \in \partial U$ such that $x \in \lambda N(x)$.

3. Existence Results

We consider the space

$$\begin{aligned} PC = \{y : (-\infty,\infty) \to E | y(t_k^-) \quad \text{and} \quad y(t_k^-) \quad \text{exist with} \quad y(t_k) = y(t_k^-), \\ y(t) = \phi(t) \quad \text{for} \quad t \in (-\infty,\infty), \quad y_k \in C(J_k, E), k = 1,2,3,\dots\} \end{aligned}$$

where y_k is the restriction of y to $J_k = (t_k, t_k+1], k = 1,2,3,\dots$
Now, we set

$$\begin{aligned} B_* = \{y : (-\infty,\infty) \to E : y \in PC \cap \mathcal{B}_h\} \\ B_k = \{y \in B_* : \sup_{t\in J_k^*} |y(t)| < \infty\}, \quad \text{where} \quad J_k^* = (-\infty, t_k] \end{aligned}$$

Let $\|\cdot\|_k$ be the semi-norm in B_k defined by

$$\|y\|_k = \|y_0\|_{\mathcal{B}_h} + \sup\{|y(s)| : 0 \leq s \leq t_k\}, \quad y \in B_k$$

To prove our existence results for the impulsive neutral functional differential evolution problem with infinite delay $(1)-(3)$. Firstly, we define the mild solution.

Definition 5. *We say that the function $y(\cdot) : (-\infty,+\infty) \to E$ is a mild solution of the evolution system $(1)-(3)$ if $y(t) = \phi(t)$ for all $t \in (-\infty, 0]$, $\Delta y|_{t=t_k} = I_k(y(t_k^-)), k = 1,2,\dots$ and the restriction of $y(\cdot)$ to the interval J is continuous and there exists $f(\cdot) \in L^1(J;E) : f(t) \in F(t,y_t)$ a.e in J such that y satisfies the following integral equation:*

$$\begin{aligned} y(t) &= U(t,0)\Big[\phi(0) - g(0,\phi)\Big] + g(t,y_t) + \int_0^t U(t,s)A(s)g(s,y_s)ds \\ &+ \int_0^t U(t,s)f(s)ds + \sum_{0<t_k<t} U(t,t_k)I_k(y(t_k^-)), \quad \text{for each} \quad t \in [0,+\infty) \end{aligned} \tag{4}$$

We need to introduce the following hypotheses, which are assumed hereafter:
(H1) There exists a constant $M_1 \geq 1$ such that

$$\|U(t,s)\|_{B(E)} \leq M_1 \quad \text{for every} \quad (t,s) \in \Delta_1$$

(H2) The multifunction $F : J \times \mathcal{B}_h \to \mathcal{P}(E)$ is L^1_{loc}-Caratheodory with compact and convex values for each $u \in \mathcal{B}_h$ and there exist a function $p \in L^1_{loc}(J, R^+)$ and a continuous nondecreasing function $\psi : J \to (0, \infty)$ such that

$$\|F(t,u)\|_{\mathcal{P}(E)} \leq p(t)\psi(\|u\|_{\mathcal{B}_h}) \quad \text{for a.e} \quad t \in J \quad \text{and each} \quad u \in \mathcal{B}_h$$

(H3) For all $r > 0$, there exists $l_r \in L^1_{loc}(J; R^+)$ such that

$$H_d(F(t,u) - F(t,v)) \leq l_r(t)\|u - v\|_{\mathcal{B}_h}$$

for each $t \in J$ and for all $u, v \in \mathcal{B}_h$ with $\|u\|_{\mathcal{B}_h} \leq r$ and $\|v\|_{\mathcal{B}_h} \leq r$ and

$$d(0, F(t,0)) \leq l_r(t) \quad \text{a.e} \quad t \in J$$

(H4) There exists a constant $M_0 > 0$ such that

$$\|A^{-1}(t)\|_{B(E)} \leq M_0 \quad \text{for all} \quad t \in J$$

(H5) There exists a constant $d_k > 0, \quad k = 1, 2, \dots$ such that

$$\|I_k(x) - I_k(\bar{x})\| \leq d_k\|x - \bar{x}\| \quad \text{for each} \quad k = 1, 2, \dots \quad \text{for all} \quad x, \bar{x} \in E$$

(H6) There exists a constant $0 < L < \frac{1}{M_0 K_n}$ such that

$$\|A(t)g(t,\phi)\| \leq L(\|\phi\|_{\mathcal{B}_h} + 1) \qquad \text{for all} \qquad t \in J, \quad \phi \in \mathcal{B}_h$$

(H7) There exists a positive constant $c_k, \quad k = 1, 2, \dots$ such that

$$\|I_k(z)\| \leq c_k \quad \text{for all} \quad z \in E, \ \sum_{k=1}^{\infty} c_k < \infty$$

(H8) There exists a constant $L_* > 0$ such that

$$\|A(s)g(s,\phi) - A(\bar{s})g(\bar{s},\bar{\phi})\| \leq L_*(|s - \bar{s}| + \|\phi - \bar{\phi}\|_{\mathcal{B}_h}) \quad \text{for all} \quad s, \bar{s} \in J \quad \text{and} \quad \phi, \bar{\phi} \in \mathcal{B}_h.$$

For every $n \in \mathbb{N}$, let us take here $\bar{l_n}(t) = M_1 K_n \Big[L_* + l_n(t)\Big]$ for the family of semi-norm $\{\| \cdot \|_n\}_{n \in \mathbb{N}}$. In what follows, we fix $\tau > 1$ and assume

$$\Big[M_0 L_* K_n + \frac{1}{\tau} + M_1 \sum_{k=1}^{m} d_k\Big] < 1$$

Theorem 3.1 Suppose that hypotheses (H1)-(H8) are satisfied. Moreover

$$\int_{\delta_n}^{+\infty} \frac{ds}{s + \psi(s)} > \frac{M_1 K_n}{1 - M_0 L K_n} \int_0^n \max(L, p(s))ds \quad \text{for each} \quad n \in \mathbb{N} \tag{5}$$

with

$$\begin{aligned} \delta_n &= (K_n M_1 H + M_n)\|\phi\|_{\mathcal{B}_h} + \frac{K_n}{1 - M_0 L K_n}\Big[(M_1 + 1)M_0 L + M_1 L n \\ &\quad + M_0 L[M_1(K_n H + 1) + M_n]\|\phi\|_{\mathcal{B}_h} + M_1 \sum_{k=1}^{m} c_k\Big] \end{aligned}$$

Then, the impulsive neutral evolution problem $(1)-(3)$ has a mild solution.

Proof. We transform the Problem $(1)-(3)$ into a fixed point problem. Consider an operator $N: B_* \to B_*$ defined by

$$N(y) = h \in B_* : h(t) = \begin{cases} \phi(t) \quad \text{if} \quad t \leq 0 \\ U(t,0)[\phi(0) - g(0,\phi)] + g(t,y_t) + \int_0^t U(t,s)A(s)g(s,y_s)ds \\ + \int_0^t U(t,s)f(s)ds + \sum_{0<t_k<t} U(t,t_k)I_k(y(t_k^-)), t \in J \end{cases}$$

where $f \in S_{F,y} = \{v \in L^1(J,E) : v(t) \in F(t,y_t) \quad \text{for a.e} \quad t \in J\}$. Clearly, the fixed points of the operator N are mild solutions of the Problem $(1)-(3)$. We remark also that, for each $y \in B_*$, the set $S_{F,y}$ is nonempty since F has a measurable selection by (H2) (see [40], Theorem III.6).

For $\phi \in \mathcal{B}_h$, we will define the function $x(\cdot) : (-\infty, +\infty) \to E$ by

$$x(t) = \begin{cases} \phi(t), & \text{if} \quad t \in (-\infty, 0] \\ U(t,0)\phi(0), & \text{if} \quad t \in J \end{cases}$$

Then, $x_0 = \phi$. For each function, $z \in C([0,\infty), E)$. We can decompose y into $y(t) = z(t) + x(t)$. Let $B_*^k = \{z \in B_k : z_0 = 0\}$. Then, for any $z \in B_*^k$, we have

$$\|z\|_k = \|z_0\|_{\mathcal{B}_h} + \sup\{\|z(s)\| : 0 \leq s \leq t_k\| = \sup\{\|z(s)\| : 0 \leq s \leq t_k\}$$

Thus $(B_*^k, \|\cdot\|_k)$is a Banach space, if we set $C_1 = \{z \in B_*; z_0 = 0\}$ with the Bielecki-norm on B_*^k defined by

$$\|z\|_{B_*^k} = \max\{\|z(t)\|e^{-\tau L_n^*(t)} : t \in [0,t_k]\}$$

where $L_n^*(t) = \int_0^t \bar{l}_n(s)ds, \bar{l}_n(t) = M_1 K_n l_n(t)$ and $\tau > 0$ is a constant. Then C_1 is a Frechet space with family of seminorms $\|\cdot\|_{B_*^k}$. It is obvious that y satisfies (4) if and only if z satisfies $z_0 = 0$ and

$$\begin{aligned} z(t) &= g(t, z_t + x_t) - U(t,0)g(0,\phi) + \int_0^t U(t,s)A(s)g(s, z_s + x_s)ds + \int_0^t U(t,s)f(s)ds \\ &+ \sum_{0<t_k<t} U(t,t_k)I_k(z(t_k^+) + x(t_k^-)) \end{aligned}$$

where $f(t) \in F(t, z_t + x_t)$ a.e $t \in J$.

Let us define a multivalued operator $\mathcal{F} : C_1 \to C_1$ by

$$\begin{aligned} \mathcal{F}(z) &= h \in B_*^k : h(t) = g(t, z_t + x_t) - U(t,0)g(0,\phi) + \int_0^t U(t,s)A(s)g(s, z_s + x_s)ds \\ &+ \int_0^t U(t,s)f(s)ds + \sum_{0<t_k<t} U(t,t_k)I_k(z(t_k^-), x(t_k^-)), \quad t \in J \end{aligned}$$

where $f \in S_{F,z} = \{v \in L^1(J,E) : v(t) \in F(t, z_t + x_t) \quad \text{for a.e } t \in J\}$. Obviously, the operator inclusion N has a fixed point is equivalent to the operator inclusion $\mathcal{F}$ has one, so it turns to prove that $\mathcal{F}$ has a fixed point.

Let $z \in B_*^k$ be a possible fixed point of the operator $\mathcal{F}$. Given $n \in \mathbb{N}$, then z should be solution of the inclusion $z \in \lambda\mathcal{F}(z)$ for some $\lambda \in (0,1)$ and there exists $f \in S_{F,z} \Leftrightarrow f(t) \in F(t, z_t + x_t)$ such that, for each $t \in [0,n]$, we have

$$\begin{aligned}
\|z(t)\| &= \|A^{-1}(t)\|_{B(E)}\|A(t)g(t,z_t+x_t)\| + \|U(t,0)\|_{B(E)}\|A^{-1}(0)\|_{B(E)}\|A(0)g(0,\phi)\| \\
&\quad + \int_0^t \|U(t,s)\|_{B(E)}\|A(s)g(s,z_s+x_s)\|ds + \int_0^t \|U(t,s)\|_{B(E)}\|f(s)\|ds \\
&\quad + \sum_{0<t_k<t} \|U(t,t_k)\|_{B(E)}\|I_k(z(t_k^-)+x(t_k^-))\| \\
&\leq M_0L(\|z_t+x_t\|_{\mathcal{B}_h}+1) + M_1M_0L(\|\phi\|_{\mathcal{B}_h}+1) + M_1\int_0^t L(\|z_s+x_s\|_{\mathcal{B}_h}+1)ds \\
&\quad +M_1\int_0^t p(s)\psi(\|z_s+x_s\|_{\mathcal{B}_h})ds + M_1\sum_{k=1}^m c_k \\
&\leq M_0L\|z_t+x_t\|_{\mathcal{B}_h} + M_0L(1+M_1) + M_1Ln + M_1M_0L\|\phi\|_{\mathcal{B}_h} + M_1\int_0^t L\|z_s+x_s\|_{\mathcal{B}_h}ds \\
&\quad +M_1\int_0^t p(s)\psi(\|z_s+x_s\|_{\mathcal{B}_h})ds + M_1\sum_{k=1}^m c_k
\end{aligned}$$

Assumption $(A1)$ gives

$$\begin{aligned}
\|z_s+x_s\|_{\mathcal{B}_h} &\leq \|z_s\|_{\mathcal{B}_h} + \|x_s\|_{\mathcal{B}_h} \\
&\leq K(s)\|z(s)\| + M(s)\|z_0\|_{\mathcal{B}_h} + K(s)\|x(s)\| + M(s)\|x_0\|_{\mathcal{B}_h} \\
&\leq K_n\|z(s)\| + K_n\|U(s,0)\|_{B(E)}\|\phi(0)\| + M_n\|\phi\|_{\mathcal{B}_h} \\
&\leq K_n\|z(s)\| + K_nM_1\|\phi(0)\| + M_n\|\phi\|_{\mathcal{B}_h} \\
&\leq K_n\|z(s)\| + K_nM_1H\|\phi\|_{\mathcal{B}_h} + M_n\|\phi\|_{\mathcal{B}_h} \\
&\leq K_n\|z(s)\| + (K_nM_1H + M_n)\|\phi\|_{\mathcal{B}_h}
\end{aligned}$$

Set

$$C_n = (K_nM_1H + M_n)\|\phi\|_{\mathcal{B}_h}$$

then we have

$$\|x_s+z_s\|_{\mathcal{B}_h} \leq K_n\|z(s)\| + C_n \tag{6}$$

Using the inequality Equation (6) and the nondecreasing character of ψ, we obtain,

$$\begin{aligned}
\|z(t)\| &\leq M_0L(K_n\|z(t)\| + C_n) + M_0L(1+M_1) + M_1Ln + M_1M_0L\|\phi\|_{\mathcal{B}_h} \\
&\quad +M_1\int_0^t L(K_n\|z(s)\| + C_n)ds + M_1\int_0^t p(s)\psi(K_n\|z(s)\| + C_n)ds \\
&\quad +M_1\sum_{k=1}^m c_k \\
&\leq M_0LK_n\|z(t)\| + M_0L(1+M_1) + M_1Ln + M_0LC_n + M_1M_0L\|\phi\|_{\mathcal{B}_h} \\
&\quad +M_1\Big[\int_0^t L(K_n\|z(s)\| + C_n)ds + \int_0^t p(s)\psi(K_n\|z(s)\| + C_n)ds\Big] \\
&\quad +M_1\sum_{k=1}^m c_k
\end{aligned}$$

Then

$$(1-M_0LK_n)\|z(t)\| \leq (M_1+1)M_0L+MLn+M_0LC_n+M_1M_0L\|\phi\|_{\mathcal{B}_h}+M_1\sum_{k=1}^{m}c_k$$
$$+M_1\Big[\int_0^t L(K_n\|z(s)\|+C_n)ds+\int_0^t p(s)\psi(K_n\|z(s)\|+C_n)ds\Big]$$

Set

$$\delta_n = C_n + \frac{K_n}{1-M_0LK_n}\Big[(M_1+1)M_0L+MLn+M_0LC_n+M_1M_0L\|\phi\|_{\mathcal{B}_h}+M_1\sum_{k=1}^{m}c_k\Big]$$

Thus,

$$K_n\|z(t)\|+C_n \leq \delta_n + \frac{M_1K_n}{1-M_0LK_n}\Big[\int_0^t L(K_n\|z(s)\|+C_n)ds+\int_0^t p(s)\psi(K_n\|z(s)\|+C_n)ds\Big]$$

We consider the function μ defined by

$$\mu(t)=\sup\{K_n\|z(s)\|+C_n : 0\leq s\leq t\}, 0\leq t<+\infty$$

Let $t^*\in[0,t]$ be such that $\mu(t)=K_n\|z(t^*)\|+C_n$. By the previous inequality, we have

$$\mu(t)\leq \delta_n+\frac{M_1K_n}{1-M_0LK_n}\Big[\int_0^t L\mu(s)ds+\int_0^t p(s)\psi(\mu(s))ds\Big] \quad \text{for} \quad t\in[0,n]$$

Let us take the right-hand side of the above inequality as $v(t)$. Then, we have $\mu(t)\leq v(t)$ for all $t\in[0,n]$. From the definition of v, we have $v(0)=\delta_n$ and

$$v'(t)=\frac{M_1K_n}{1-M_0LK_n}[L\mu(t)+p(t)\psi(\mu(t))] \quad \text{a.e} \quad t\in[0,n]$$

Using the nondecreasing character of ψ, we get

$$v'(t)=\frac{M_1K_n}{1-M_0LK_n}[Lv(t)+p(t)\psi(v(t))] \quad \text{a.e} \quad t\in[0,n]$$

This implies that for each $t\in[0,n]$and using the condition (5), we get

$$\begin{aligned}\int_{\delta_n}^{v(t)}\frac{ds}{s+\psi(s)} &\leq \frac{M_1K_n}{1-M_0LK_n}\int_0^t \max(L,p(s))ds\\ &\leq \frac{M_1K_n}{1-M_0LK_n}\int_0^n \max(L,p(s))ds\\ &< \int_{\delta_n}^{+\infty}\frac{ds}{s+\psi(s)}\end{aligned}$$

Thus, for every $t\in[0,n]$, there exists a constant Λ_n such that $v(t)\leq\Lambda_n$ and hence $\mu(t)\leq\Lambda_n$. Since $\|z\|_{B_*^k}\leq\mu(t)$, we have $\|z\|_{B_*^k}\leq\Lambda_n$.

Set $\Omega=\{z\in B_*^k : \sup\{\|z(t)\| : 0\leq t\leq n\}<\Lambda_n+1 \quad \text{for all} \quad n\in\mathbb{N}\}$.

Clearly, Ω is an open subset of B_*^k. We shall show that $\mathcal{F} : \bar{\Omega} \to \mathcal{P}(B_*^k)$ is a contraction and an admissible operator. First, we prove that $\mathcal{F}$ is a contraction. Let $z, \bar{z} \in B_*^k$ and $h \in \mathcal{F}(\bar{z})$. Then, there exists $f(t) \in F(t, z_t + x_t)$ such that for each $t \in [0, n]$,

$$\begin{aligned} h(t) &= g(t, z_t + x_t) - U(t,0)g(0,\phi) + \int_0^t U(t,s)A(s)g(s, z_s + x_s)ds \\ &\quad + \int_0^t U(t,s)f(s)ds + \sum_{0<t_k<t} U(t,t_k)I_k(z(t_k^-) + x(t_k^-)) \end{aligned}$$

From (H3) it follows that,

$$H_d(F(t, z_t + x_t), F(t, \bar{z}_t + x_t)) \le l_n(t)\|z_t - \bar{z}_t\|_{\mathcal{B}_h}$$

Hence, there is $\rho \in F(t, \bar{z}_t + x_t)$ such that

$$\|f(t) - \rho\| \le l_n(t)\|z_t - \bar{z}_t\|_{\mathcal{B}_h} \quad t \in [0, n]$$

Consider $\Omega_* : [0, n] \to \mathcal{P}(E)$, given by

$$\Omega_* = \{\rho \in E : \|f(t) - \rho\| \le l_n(t)\|z_t - \bar{z}_t\|\}$$

Since the multivalued operator $\mathcal{V}(t) = \Omega_*(t) \bigcap F(t, \bar{z}_t + x_t)$ is measurable (in [40], see proposition III, 4), there exists a function $\bar{f}(t)$, which is a measurable selection for $\mathcal{V}$. Thus, $\bar{f}(t) \in F(t, \bar{z}_t + x_t)$ and using (A1), we obtain for each $t \in [0, n]$

$$\begin{aligned} \|f(t) - \bar{f}(t)\| &\le l_n(t)\|z_t - \bar{z}_t\|_{\mathcal{B}_h} \\ &\le l_n(t)\Big[K(t)\|z(t) - \bar{z}(t)\| + M(t)\|z_0 - \bar{z}_0\|_{\mathcal{B}_h}\Big] \\ &\le l_n(t)K_n\|z(t) - \bar{z}(t)\| \end{aligned} \tag{7}$$

Let us define, for each $t \in [0, n]$

$$\begin{aligned} \bar{h}(t) &= g(t, \bar{z}_t + x_t) - U(t,0)g(0,\phi) + \int_0^t U(t,s)A(s)g(s, \bar{z}_s + x_s)ds \\ &\quad + \int_0^t U(t,s)\bar{f}(s)ds + \sum_{0<t_k<t} U(t,t_k)I_k(\bar{z}(t_k^-) + x(t_k^-)) \end{aligned}$$

Then, for each $t \in [0, n]$ and $n \in \mathbb{N}$ and using (H1) and (H3)-(H6) and (H8), we get

$$\begin{aligned}
\|h(t) - \bar{h}(t)\| \leq\ & \|g(t, z_t + x_t) - g(t, \bar{z}_t + x_t)\| \\
& + \int_0^t \|U(t,s)A(s)\Big[g(s, z_s + x_s) + g(s, \bar{z}_s + x_s)\Big]\|ds \\
& + \int_0^t \|U(t,s)\Big[f(s) - \bar{f}(s)\Big]\|ds \\
& + \sum_{0<t_k<t} \|U(t,t_k)\Big[I_k(z(t_k^-) + x(t_k^-)) - I_k(\bar{z}(t_k^-) + x(t_k^-))\Big]\| \\
\leq\ & \|A^{-1}(t)\|_{B(E)}\|A(t)g(t, z_t + x_t) - A(t)g(t, \bar{z}_t + x_t)\| \\
& + \int_0^t \|U(t,s)\|_{B(E)}\|A(s)g(s, z_s + x_s) - A(s)g(s, \bar{z}_s + x_s)\|ds \\
& + \int_0^t \|U(t,s)\|_{B(E)}\|f(s) - \bar{f}(s)\|ds \\
& + \sum_{0<t_k<t} \|U(t,t_k)\|_{B(E)}\|I_k(z(t_k^-) + x(t_k^-)) - I_k(\bar{z}(t_k^-) + x(t_k^-))\| \\
\leq\ & M_0 L_* \|z_t - \bar{z}_t\|_{\mathcal{B}_h} + \int_0^t M_1 L_* \|z_s - \bar{z}_s\|_{\mathcal{B}_h} ds \\
& + \int_0^t M_1 \|f(s) - \bar{f}(s)\|ds \\
& + M_1 \sum_{k=1}^{m} d_k \|(z(t_k^-) - \bar{z}(t_k^-))\|
\end{aligned}$$

Using (A1) and (7), we obtain

$$\begin{aligned}
\|h(t)-\bar{h}(t)\| &\leq M_0 L_* K(t)\|z(t)-\bar{z}(t)\| + \int_0^t M_1 L_* K(s)\|z(s)-\bar{z}(s)\|ds \\
&\quad + \int_0^t M_1 l_n(s) K_n \|z(s)-\bar{z}(s)\|ds \\
&\quad + M_1 \sum_{k=1}^{m} d_k \sup_{s\in[0,t_k]} \|z(s)-\bar{z}(s)\| \\
&\leq M_0 L_* K_n \|z(t)-\bar{z}(t)\| + \int_0^t M_1 K_n \left[L_* + l_n(s)\right] \|z(s)-\bar{z}(s)\|ds \\
&\quad + M_1 \sum_{k=1}^{m} d_k \sup_{s\in[0,t_k]} \|z(s)-\bar{z}(s)\| \\
&\leq M_0 L_* K_n \|z(t)-\bar{z}(t)\| + \int_0^t \bar{l}_n(s)\|z(s)-\bar{z}(s)\|ds \\
&\quad + M_1 \sum_{k=1}^{m} d_k \sup_{s\in[0,t_k]} \|z(s)-\bar{z}(s)\| \\
&\leq M_0 L_* K_n \left[e^{\tau L_n^*(t)}\right]\left[e^{-\tau L_n^*(t)}\|z(t)-\bar{z}(t)\|\right] \\
&\quad + \int_0^t \left[l_n(s) e^{\tau L_n^*(s)}\right]\left[e^{-\tau L_n^*(s)}\|z(s)-\bar{z}(s)\|\right] ds \\
&\quad + M_1 \sum_{k=1}^{m} d_k e^{\tau L_n^*(s)} e^{-\tau L_n^*(s)} \sup_{s\in[0,t_k]} \|z(s)-\bar{z}(s)\| \\
&\leq M_0 L_* K_n e^{\tau L_n^*(t)} \|z-\bar{z}\|_{B_*^k} + \int_0^t \left[\frac{e^{\tau L_n^*(s)}}{\tau}\right]' ds \|z-\bar{z}\|_{B_*^k} \\
&\quad + M_1 \sum_{k=1}^{m} d_k e^{\tau L_n^*(s)} \|z-\bar{z}\|_{B_*^k} \\
&\leq M_0 L_* K_n e^{\tau L_n^*(t)} \|z-\bar{z}\|_{B_*^k} + \frac{1}{\tau} e^{\tau L_n^*(t)} \|z-\bar{z}\|_{B_*^k} \\
&\quad + M_1 \sum_{k=1}^{m} d_k e^{\tau L_n^*(t)} \|z-\bar{z}\|_{B_*^k} \\
&\leq \left[M_0 L_* K_n + \frac{1}{\tau} + M_1 \sum_{k=1}^{m} d_k\right] e^{\tau L_n^*(t)} \|z-\bar{z}\|_{B_*^k}
\end{aligned}$$

Therefore,

$$\|h-\bar{h}\|_{B_*^k} \leq \left[M_0 L_* K_n + \frac{1}{\tau} + M_1 \sum_{k=1}^{m} d_k\right] \|z-\bar{z}\|_{B_*^k}$$

By an analogous relation, obtained by interchanging the roles of z and $\bar{z}$, it follows that

$$H_d(\mathcal{F}(z)-\mathcal{F}(\bar{z})) \leq \left[M_0 L_* K_n + \frac{1}{\tau} + M_1 \sum_{k=1}^{m} d_k\right] \|z-\bar{z}\|_{\bar{B}}$$

Thus, the operator $\mathcal{F}$ is a contraction for all $n \in \mathbb{N}$.

Now, we shall show that $\mathcal{F}$ is an admissible operator. Let $z \in B_*^k$. Set, for every $n \in \mathbb{N}$, the space,

$$B_{**}^k = \{y : (-\infty, n] \to E : y|_{[0,n]} \in PC([0,n], E), y_0 \in \mathcal{B}_h\}$$

and let us consider the multivalued operator $\mathcal{F} : B^k_{**} \to \mathcal{P}_{cl}(B^k_{**})$ defined by

$$\begin{aligned}\mathcal{F}(z) = h \in B^k_{**} : h(t) &= g(t, z_t + x_t) - U(t,0)g(0,\phi) + \int_0^t U(t,s)A(s)g(s, z_s + x_s)ds \\ &+ \int_0^t U(t,s)f(s)ds + \sum_{0<t_k<t} U(t,t_k)I_k(z(t_k^-) + x(t_k^-)), \quad t \in [0,n]\end{aligned}$$

where $f \in S^n_{F,y} = \{v \in L^1([0,n], E) : v(t) \in F(t, y_t) \quad \text{for a.e} \quad t \in [0,n]\}$.
From (H1) to (H7), and since F is multivalued map with compact values, we can prove that for every $z \in B^k_{**}, \quad \mathcal{F}(z) \in \mathcal{P}_{cl}(B^k_{**})$ and there exists $z_* \in B^k_{**}$ such that $z_* \in \mathcal{F}(z_*)$. Let $h \in B^k_{**}, \quad \bar{y} \in \bar{\mu}$ and $\epsilon > 0$. Assume that $z_* \in \mathcal{F}(\bar{z})$, then we have

$$\begin{aligned}\|\bar{z}(t) - z_*(t)\| &\leq \|\bar{z}(t) - h(t)\| + \|z_*(t) - h(t)\| \\ &\leq e^{\tau L_n^*(t)}\|\bar{z}(t) - \mathcal{F}(\bar{z})\|_{B^k_*} + \|z_* - h\|\end{aligned}$$

Since h is arbitrary, we may suppose that

$$h \in B(z_*, \epsilon) = \{h \in B^k_{**} : \|h - z_*\|_{B^k_*} \leq \epsilon\}$$

Therefore,

$$\|\bar{z} - z_*\|_n \leq \|\bar{z} - \mathcal{F}(\bar{z})\|_{B^k_*} + \epsilon$$

If z is not in $\mathcal{F}(\bar{z})$, then $\|z_* - \mathcal{F}(\bar{z})\| \neq 0$. Since $\mathcal{F}(\bar{z})$ is compact, there exists $x \in \mathcal{F}(\bar{z})$ such that

$$\|z_* - \mathcal{F}(\bar{z})\| = \|z_* - x\|$$

Then, we have

$$\begin{aligned}\|\bar{z}(t) - z_*(t)\| &< \|\bar{z}(t) - h(t)\| + \|x(t) - h(t)\| \\ &\leq e^{\tau L_n^*}\|\bar{z} - \mathcal{F}(\bar{z})\|_{B^k_*} + \|x(t) - h(t)\|\end{aligned}$$

Thus, $\|\bar{z} - x\|_{B^k_*} \leq \|\bar{z} - \mathcal{F}(\bar{z})\|_{B^k_*} + \epsilon$.

Therefore, $\mathcal{F}$ is an admissible operator contraction.

From the choice of Ω, there is no $z \in \partial\Omega$ such that $z = \lambda\mathcal{F}(z)$ for some $\lambda \in (0,1)$. Then, the statement (C2) in Theorem (2.1) does not hold. This implies that the operator $\mathcal{F}$ has a fixed point z^*. Then $y^*(t) = z^*(t) + x(t), \quad t \in (-\infty, +\infty)$ is a fixed point of the operator $\mathbb{N}$, which is a mild solution of the Problem $(1) - (3)$.
Hence the proof. □

4. Example

As an application of Theorem (3.1), we study the following impulsive neutral differential system:

$$\begin{aligned}\frac{\partial}{\partial t}\left[v(t,\xi) - \int_{-\infty}^0 T(\theta)u(t, v(t+\theta,\xi))d\theta\right] &\in a(t,\xi)\frac{\partial^2 v}{\partial \xi^2}(t,\xi) \\ &\quad + \int_{-\infty}^0 P(\theta)Q(t, v(t+\theta,\xi))d\theta \\ t \in [0,+\infty), \quad \xi \in [0,\pi] & \\ v(t,0) = v(t,\pi) = 0 & \\ z(t_k^+) - z(t_k^-) = I_k(z(t_k^-)), \quad k = 1,2,\ldots & \\ v(\theta,\xi) = v_0(\theta,\xi) \qquad -\infty < \theta \leq 0, \xi \in [0,\pi] &\end{aligned} \tag{8}$$

where $a(t,\xi)$ is a continuous function and is uniformly Holder continuous in t; $T, P : (-\infty, 0] \to R$; $u : (-\infty, 0] \times R \to R$ and $v_0 : (-\infty, 0] \times [0, \pi] \to R$ are continuous functions and $Q : [0, +\infty) \times R \to \mathcal{P}(R)$ is a multivalued map with compact convex values.

Consider $E = L^2([0,\pi], R)$ and define $A(t)$ by $A(t)w = a(t,\xi)w''$ with domain $D(A) = \{w \in E : w, w'$ are absolutely continuous, $w'' \in E, w(0) = w(\pi) = 0\}$.

Then, $A(t)$ generates an evolution system $U(t,s)$ satisfying assumption (H1)(see [41]). We can define respectively that

$$g(t,\phi)(\xi) = \int_{-\infty}^{0} T(\theta)u(t,\phi(\theta)(\xi)d\theta, \qquad -\infty < \theta \le 0, \quad \xi \in [0,\pi]$$

and

$$F(t,\phi)(\xi) = \int_{-\infty}^{0} P(\theta)Q(t,\phi(\theta)(\xi)d\theta, \qquad -\infty < \theta \le 0, \quad \xi \in [0,\pi]$$

Then, in order to prove the existence of mild solutions of the System (8), we suppose the following assumptions:

(i) u is Lipschitz with respect to its second argument. Let lip(u) denotes the Lipschitz constant of u.
(ii) There exist $p \in L^1(J, R^+)$ and a nondecreasing continuous function $\psi : [0, +\infty) \to (0, +\infty)$ such that

$$|Q(t,x)| \le p(t)\psi(|x|), \quad \text{for} \quad t \in J, \quad \text{and} \quad x \in R$$

(iii) T, P are integrable on $(-\infty, 0]$.
(iv) There exist positive constants c_k and $d_k, k = 1, 2, \ldots$ such that

$$\|I_k(x) - I_k(\bar{x})\| \le d_k\|x - \bar{x}\|,$$
$$\|I_k(z)\| \le c_k, \quad \text{for all} \quad x, \bar{x}, z \in E$$

By the dominated convergence theorem, one can show that $f \in S_{F,y}$ is a continuous function from $\mathcal{B}_h$ to E. Moreover, the mapping g is Lipschitz continuous in its argument. In fact, we have

$$\|g(t,\phi_1) - g(t,\phi_2)\| \le M_0 L_* lip(u) \int_{-\infty}^{0} \|T(\theta)\|d\theta\|\phi_1 - \phi_2\|, \quad \text{for} \quad \phi_1, \phi_2 \in \mathcal{B}_h$$

On the other hand, we have for $\phi \in \mathcal{B}_h$ and $\xi \in [0,\pi]$

$$\|F(t,\phi)(\xi)\| \le \int_{-\infty}^{0} \|p(t)P(\theta)\|\xi(\|(\phi(\theta))(\xi)\|)d\theta$$

Since the function ξ is nondecreasing, it follows that

$$\|F(t,\phi)\|_{\mathcal{P}(E)} \le p(t) \int_{-\infty}^{0} \|P(\theta)\|d\theta\xi(\|\phi\|), \quad \text{for} \quad \phi \in \mathcal{B}_h$$

Proposition: Under the above assumptions, if we assume that condition (5) in Theorem (3.1) is true, $\phi \in \mathcal{B}_h$, then the System (8) has a mild solution which is defined in $(-\infty, +\infty)$.

5. Conclusions

In this manuscript, we have proved the existence result of first order impulsive neutral evolution inclusion with infinite delay in Frechet spaces. Here, we defined a new notion of phase space and proved the result without compactness of an evolution operator, using a recently developed nonlinear alternative for contractive multivalued maps due to Frigon. The same result can be generalized for controllability of an impulsive neutral evolution inclusion with infinite delay of the form [19,21]

$$\begin{aligned} \frac{d}{dt}\left[y(t) - g(t, y_t)\right] &\in A(t)y(t) + Bu(t) + F(t, y_t) \\ t \in J = [0, +\infty), \quad t &\neq t_k, \quad k = 1, 2, \ldots \\ \Delta y|_{t=t_k} &= I_k(y(t_k^-)), \quad k = 1, 2, \ldots \\ y_0 &= \phi \in \mathcal{B}_h \end{aligned}$$

Acknowledgments: Authors wish to express their gratitude to the anonymous referees for their valuable suggestions and comments for improving this manuscript. This work is supported by Grant In Aid research fund of Virginia Military Instittue, USA.

Author Contributions: Dimplekumar N. Chalishajar defined the new phase space and compared with the phase space given by Hall and Kato. Dimplekumar N. Chalishajar and Kulandhivel Karthikeyan conceived and designed the control problem experiment and study the existence result of first order impulsive neutral evolution inclusion with infinite delay in Frechet spaces without compactness of an evolution operator; Kulandhivel Karthikeyan performed the analysis and wrote the paper; Annamalai Anguraj analyzed and contributed materials/analysis tools.

Conflicts of Interest: The authors declare no conflict of interest.

References

1. Bainov, D.D.; Simeonov, P.S. *Systems with Impulsive Effect*; Ellis Horwood: Chichester, UK, 1989.
2. Lakshimikantham, V.; Bainov, D.D.; Simeonov, P.S. *Theory of Impulsive Differential Equations*; World Science: Singapore, 1989.
3. Samoilenko, A.M.; Perestyuk, N.A. *Impulsive Differential Equations*; World Scientific Publishing: Singapore, 1995; Volume 14.
4. Anguraj, A.; Arjunan, M.M.; Hernández, M.E. Existence results for an impulsive neutral functional differential equation with state-dependent delay. *Appl. Anal.* **2007**, *86*, 861–872.
5. Chang, Y.K.; Anguraj, A.; Karthikeyan, K. Existence for impulsive neutral integrodifferential inclusions with nonlocal initial conditions via fractional operators. *Nonlin. Anal. TMA* **2009**, *71*, 4377–4386.
6. Benchohra, M.; Henderson, J.; Ntouyas, S.K. *Impulsive Differential Equations and Inclusions*; Hindawi Publishing Corporation: New York, NY, USA, 2006; Volume 2.
7. Nieto, J.J.; O'Regan, D. Variational approach to impulsive differential equations. *Nonlinear Anal. Real World Appl.* **2009**, *10*, 680–690.
8. Zavalishchin, S.T.; Sesekin, A.N. *Dynamic Impulse Systems: Theory and Applications*; Kluwer Academic Publishers: New York, NY, USA, 1997.
9. Wang, L.; Chen, L.; Nieto, J.J. The dynamics of an epidemic model for pest control with impulsive effect. *Nonlinear Anal. Real World Appl.* **2010**, *11*, 1374–1386.
10. Tai , Z.; Lun, S. On controllability of fractional impulsive neutral infinite delay evolution integrodifferential systems in Banach spaces. *Appl. Math. Lett.* **2012**, *25*, 104–110.
11. Gao, S.; Chen, L.; Nieto, J.J.; Torres, A. Analysis of a delayed epidemic model with pulse vaccination and saturation incidence. *Vaccine* **2006**, *24*, 6037–6045.
12. Hale, J.; Kato, J. Phase space for retarded equations with infinite delay. *Funkc. Ekvacio.* **1978**, *21*, 11–41.
13. Corduneanu, C.; Lakshmikantham, V. Equations with unbounded delay. *Nonlinear Anal.* **1980**, *4*, 831–877.
14. Graef, J.R.; Ouahab, A. Some existence and uniqueness results for first order boundary value problems for impulsive functional differential equations with infinite delay in Frechet spaces. *Int. J. Math Math Sci.* **2006**, *2006*, 1–16.
15. Baghli, S.; Benchohra, M. Uniqueness results for partial functional differential equations in Frechet spaces. *Fixed Point Theory* **2008**, *9*, 395–406.
16. Baghli, S.; Benchohra, M. Multivalued evolution equations with infinite delay in Frechet spaces. *Electron. J. Qual. Theory Differ. Equ.* **2008**, *33*, 1–24.
17. Henderson, J.; Ouahab, A. Existence results for nondensely defined semilinear functional differential inclusions in Frechet spaces. *Electron. J. Qual. Theory Differ. Equ.* **2005**, *17*, 1–17.

18. Hernández, E.M.; Rabello, M.; Henriquez, H. Existence of solution of impulsive partial neutral functional differential equations. *J. Math. Anal. Appl.* **2007**, *331*, 1135–1158.
19. Arthi, G.; Balachandran, K. Controllability of second order impulsive functional differential equations with state dependent delay. *Bull. Korean Math. Soc.* **2011**, *48*, 1271–1290.
20. Hino, Y.; Murakami, S.; Naito, T. *Functional Differential Equations with Infinite Delay*; Springer-Verlag: Berlin, Germany, 1991; Volume 1473.
21. Chalishajar, D.N. Controllability of second order impulsive neutral functional differential inclusions with infinite delay. *J. Optim. Theory. Appl.* **2012**, doi:10.1007/s10957-012-0025-6.
22. Chalishajar, D.N.; Acharya, F.S. Controllability of second order semilinear neutral functional impulsive differential inclusion with infinite delay in Banach spaces. *Bull. Korean Math. Soc.* **2011**, *48*, 813–838.
23. Triggiani, R. Addendum : A note on lack of exact controllability for mild solution in Banach spaces. *SIAM J. Control Optim.* **1977**, *15*, 407–411.
24. Ntouyas, S.; O'Regan, D. Some remarks on controllability of evolution equations in Banach spaces. *Electron. J. Differ. Equ.* **2009**, *2009*, 1–6.
25. Fečkan, M.; Wang, J.-R.; Zhou, Y. On the new concept of solutions and existence results for impulsive fractional evolution equations. *Dyn. Part. Differ. Equ.* **2011**, *8*, 345–361.
26. Zhou, Y. *Basic Theory of Fractional Differential Equations*; World Scientific: Singapore, 2014.
27. Fu, X.; Liu, X.; Lu, B. On a new class of impulsive fractional evolution equations. *Adv. Differ. Equ.* **2015**, *2015*, 1–16.
28. Yosida, K. *Functional Analysis*, 6th ed.; Springer-Verlag: Berlin, Dermany, 1980.
29. Kisielewicz, M. *Differential Inclusions and Optimal Control*; Kluwer: Dordrecht, The Netherlands, 1991.
30. Ahmed, N.U. *Semigroup Theory with Applications to Systems and Control*; John Wiley & Sons: New York, NY, USA, 1991.
31. Engel, K.J.; Nagel, R. *One-parameter Semigroups for Linear Evolution Equations*. Springer-Verlag: New York, NY, USA, 2000.
32. Pazy, A. *Semigroups of Linear Operators and Applications to Partial Differential Equations*; Springer-Verlag: New York, NY, USA, 1988.
33. Aubin, J.P.; Cellina, A. *Differential Inclusions*; Berlin Heidelberg: Tokyo, Japan, 1984.
34. Deimling, K. *Multivalued Differential Equations*; Walter de Gruyter: Berlin, Germany, 1992.
35. Gorniewicz, L. *Topological Fixed Point Theory of Multivalued Mappings*; Kluwer Academic Publishers: Dordrecht, The Netherlands, 1999; Volume 495.
36. Hu, S.; Papageorgiou, N. *Handbook of Multivalued Analysis*; Wolters Kluwer: Dordrecht, The Netherlands, 1997.
37. Tolstonogov, A.A. *Differential Inclusions in a Banach Space*; Kluwer Academic Publishers: Dordrecht, The Netherlands, 2000.
38. Frigon, M. Fixed point results for multivalued contractions on gauge spaces. In *Set Valued Mapping Applications in Nonlinear Analysis*; Taylor & Francis: London, UK, 2002; Volume 4, pp. 175–181.
39. Frigon, M. Fixed point and continuation results for contractions in metric and gauge spaces. In *Fixed Point Theory & Applications*; Banach Center Publications: Warsaw, Poland, 2007; Volume 77, pp. 89–114.
40. Castaing, C.; Valadier, M. *Convex Analysis and Measurable Multifunctions*; Springer-Verlag: New York, NY, 1977; Volume 580.
41. Freidman, A. *Partial Differential Equations*; Holt McDougal: New York, NY, USA, 1969.

5

SIC-POVMs and Compatibility among Quantum States

Blake C. Stacey

Physics Department, University of Massachusetts Boston, 100 Morrissey Boulevard, Boston, MA 02125, USA; blake.stacey@umb.edu

Academic Editors: Paul Busch, Takayuki Miyadera and Teiko Heinosaari

Abstract: An unexpected connection exists between compatibility criteria for quantum states and Symmetric Informationally Complete quantum measurements (SIC-POVMs). Beginning with Caves, Fuchs and Schack's "Conditions for compatibility of quantum state assignments", I show that a qutrit SIC-POVM studied in other contexts enjoys additional interesting properties. Compatibility criteria provide a new way to understand the relationship between SIC-POVMs and mutually unbiased bases, as calculations in the SIC representation of quantum states make clear. This, in turn, illuminates the resources necessary for magic-state quantum computation, and why hidden-variable models fail to capture the vitality of quantum mechanics.

Keywords: SIC-POVM; qutrit; post-Peierls compatibility; Quantum Bayesian; QBism

PACS: 03.65.Aa, 03.65.Ta, 03.67.-a

---

1. A Compatibility Criterion for Quantum States

This article presents an unforeseen connection between two subjects originally studied for separate reasons by the Quantum Bayesians, or to use the more recent and specific term, QBists [1,2]. One of these topics originates in the paper "Conditions for compatibility of quantum state assignments" [3] by Caves, Fuchs and Schack (CFS). Refining CFS's treatment reveals an unexpected link between the concept of compatibility and other constructions of quantum information theory.

We begin by taking up the question of what it means for probability distributions to be compatible with one another. Consider a thoroughly classical scenario, in which Alice and Bob are gambling on the outcome of a coin toss. Both Alice and Bob are certain the toss is rigged, but Alice is convinced that the outcome will be heads, while Bob is equally steadfast in maintaining that it will be tails. That is, $p_A(H) = 1$ and $p_A(T) = 0$, while $p_B(H) = 0$ and $p_B(T) = 1$. No matter which way the coin lands, one party will be disappointed—or, depending on the stakes, bankrupt.

We can equally well present a single-user version of this scenario. Imagine that Alice, and only Alice, is gambling on the coin flip, and that conditional on some other information, she will choose to do so in accordance with either the probability distribution

$$p_A(H) = 1, \qquad p_A(T) = 0 \tag{1}$$

or the alternative,

$$p'_A(H) = 0, \qquad p'_A(T) = 1 \tag{2}$$

For example, Alice may be confident that the toss will be rigged and that she can deduce which way it will be rigged once she can observe the handedness of the coin-tosser. The incompatibility between p_A and p'_A is then relevant to Alice, regardless of the presence or absence of other players.

Having formulated the scenario in single-user terms, we can develop a quantum analogue; a single-user statement avoids the conceptual problem of whether an event occurring for an agent Alice could, in the quantum setting, itself be an event for any other agent [2]. The quantum version of this kind of incompatibility is a condition on pairs of quantum states. Take two density matrices ρ_A and ρ'_A. If there exists a measurement $\{E_i\}$ such that

$$\sum_i \mathrm{tr}(\rho_A E_i)\mathrm{tr}(\rho'_A E_i) = 0 \tag{3}$$

then ρ_A and ρ'_A are called *post-Peierls incompatible* [3]. If two states are post-Peierls (PP) compatible, then for *all* experiments, there is at least one outcome for which both states yield nonzero probability via the Born rule. We can naturally extend this criterion to sets of more than two quantum states. In general, a set of N states is PP incompatible when, for some experiment $\{E_i\}$,

$$\sum_i \prod_{a=1}^{N} \mathrm{tr}(\rho^{(a)} E_i) = 0 \tag{4}$$

A von Neumann measurement is a Positive Operator Valued Measure (POVM) with d possible outcomes, specified by a set of d orthonormal vectors in Hilbert space. In terms of projection operators, each von Neumann measurement comprises d one-dimensional orthogonal projectors. We abbreviate this phrase as ODOP. Therefore, compatibility with respect to von Neumann measurements is known as *PP-ODOP* compatibility [3]. When we formulate an exact criterion for PP-ODOP compatibility of *qutrit* pure states, *i.e.*, pure states in $d = 3$, we find something interesting.

In the next two sections, we will examine PP-ODOP compatibility in more detail and find a connection to another topic of much interest to the QBist research program. For completeness, we note that compatibility criteria have also recently entered the quantum foundations discourse through a different route. They play a key role in discussions of whether quantum states can be treated as encoding information about the values of hidden variables [4,5]. In this paper, we disregard the issue of hidden variables and treat quantum states as directly specifying the probabilities of experiment outcomes.

2. Three Pure States in Dimension Three

It is possible to have triplets of states which are PP-ODOP incompatible when taken all together, even though they are compatible when taken in pairs. CFS provide a specific example, which for convenience we reproduce here. Pick an orthonormal basis $\{|0\rangle, |1\rangle, |2\rangle\}$, and consider the following three possible states which can be ascribed to a qutrit:

$$\begin{aligned} |\psi\rangle &= \frac{1}{\sqrt{2}}(|1\rangle + |2\rangle) \\ |\psi'\rangle &= \frac{1}{\sqrt{2}}(|2\rangle + |0\rangle) \\ |\psi''\rangle &= \frac{1}{\sqrt{2}}(|0\rangle + |1\rangle) \end{aligned} \tag{5}$$

Now, Alice the graduate student performs a von Neumann experiment in the computational basis of $\{|0\rangle, |1\rangle, |2\rangle\}$. Whatever result she experiences, there is a state assignment in the set $\{|\psi\rangle, |\psi'\rangle, |\psi''\rangle\}$ according to which that experience is an event of probability zero.

By pursuing this argument, we will have the opportunity, incidentally, to correct two mathematical errors in CFS. Both are slight, but one is a missed opportunity, as it introduced an inconsistency into CFS's calculations and obscured the connection which we shall examine here.

A general set of three pure states $\{|\psi\rangle, |\psi'\rangle, |\psi''\rangle\}$ is PP-ODOP incompatible at the tertiary level if some orthonormal basis $\{|0\rangle, |1\rangle, |2\rangle\}$ exists such that

$$\begin{aligned} |\psi\rangle &= e^{i\chi}\left(\cos\theta|1\rangle + e^{i\phi}\sin\theta|2\rangle\right) \\ |\psi'\rangle &= e^{i\chi'}\left(\cos\theta'|2\rangle + e^{i\phi'}\sin\theta'|0\rangle\right) \\ |\psi''\rangle &= e^{i\chi''}\left(\cos\theta''|0\rangle + e^{i\phi''}\sin\theta''|1\rangle\right) \end{aligned} \tag{6}$$

The three θ angles are restricted to the interval $(0, \pi/2)$, while the χ and ϕ angles must all lie in the interval $[0, 2\pi)$. Taking the inner products picks out one basis vector per pair:

$$\begin{aligned} \langle\psi|\psi'\rangle &= e^{i(\chi'-\chi)}e^{-i\phi}\sin\theta\cos\theta' \\ \langle\psi'|\psi''\rangle &= e^{i(\chi''-\chi')}e^{-i\phi'}\sin\theta'\cos\theta'' \\ \langle\psi''|\psi\rangle &= e^{i(\chi-\chi'')}e^{-i\phi''}\sin\theta''\cos\theta \end{aligned} \tag{7}$$

This is the first of the two errata mentioned above: because the ϕ-dependent phase factors come from the bra vectors rather than the ket vectors, the sign in the exponential should be negative, instead of positive as written in CFS's Equation (19). The sign error cancels in the next step, which is to multiply these quantities by their complex conjugates, yielding

$$\begin{aligned} |\langle\psi|\psi'\rangle|^2 &= \sin^2\theta\cos^2\theta' \\ |\langle\psi'|\psi''\rangle|^2 &= \sin^2\theta'\cos^2\theta'' \\ |\langle\psi''|\psi\rangle|^2 &= \sin^2\theta''\cos^2\theta \end{aligned} \tag{8}$$

After some algebra, one can prove from Equation (8) that three pure states in $d = 3$ are PP-ODOP incompatible if and only if

$$|\langle\psi|\psi'\rangle|^2 + |\langle\psi'|\psi''\rangle|^2 + |\langle\psi''|\psi\rangle|^2 < 1 \tag{9}$$

$$\left(|\langle\psi|\psi'\rangle|^2 + |\langle\psi'|\psi''\rangle|^2 + |\langle\psi''|\psi\rangle|^2 - 1\right)^2 \geq 4|\langle\psi|\psi'\rangle|^2|\langle\psi'|\psi''\rangle|^2|\langle\psi''|\psi\rangle|^2 \tag{10}$$

This is the second glitch in CFS: the latter inequality should be $\geq$ rather than $>$, as written in CFS's Formula (28). This is the more consequential error, since if the inequality is strict, then the example of three-party PP-incompatible states which CFS provide—our triplet Equation (5)—is not consistent with CFS's compatibility criterion.

3. Qutrit SIC POVMs

The second inequality has a more intricate structure than the first. What happens when we try to *saturate* it? Suppose we require that the three squared overlaps all have the same value, x. Then, saturating the second inequality implies that x satisfies the cubic equation

$$4x^3 - 9x^2 + 6x - 1 = 0 \tag{11}$$

This cubic polynomial has a zero at $x = \frac{1}{4}$, and a double zero at $x = 1$, which is disallowed by the requirement that the states are nonidentical.

How many states can we push simultaneously to the edge of PP-ODOP incompatibility in this way? That is, how many states in qutrit state space can we find such that for any two of them,

$$|\langle\psi_i|\psi_j\rangle|^2 = \frac{1 + 3\delta_{ij}}{4} \tag{12}$$

This is the problem of finding the *maximal set of equiangular lines* in three-dimensional complex vector space.

Beginning with the question of compatibility among probability distributions, we have arrived at *Symmetric Informationally Complete POVMs* [2,6–9]. This term is abbreviated to SIC-POVM, or just to SIC (pronounced like "seek"). A SIC for a d-dimensional Hilbert space is a set of d^2 operators $\{E_i = \frac{1}{d}\Pi_i\}$ where the rank-one projection operators $\{\Pi_i\}$ satisfy

$$\mathrm{tr}(\Pi_k \Pi_l) = \frac{d\delta_{kl} + 1}{d+1} \tag{13}$$

It is known that SICs are maximal in this regard, *i.e.*, no more than d^2 operators can simultaneously satisfy Equation (13). For qutrits, this means that a set of states such that any three saturate the edge of PP-ODOP incompatibility can contain at most *nine* states.

We have shown that any triple of states chosen from a qutrit SIC will be PP-incompatible. One might expect that a large number of different von Neumann measurements would be required to cover all the possible choices of triples, perhaps comparable to the number of triples themselves. Surprisingly, this is not the case; our toolbox can be much more economical. Take $\omega = e^{2\pi i/3}$, and construct the set of states $\{|\psi_j\rangle\}$ given by the columns of the following matrix:

$$\frac{1}{\sqrt{2}}\begin{pmatrix} 0 & -1 & 1 & 0 & -1 & 1 & 0 & -1 & 1 \\ 1 & 0 & -1 & \omega & 0 & -\omega & \omega^2 & 0 & -\omega^2 \\ -1 & 1 & 0 & -\omega^2 & \omega^2 & 0 & -\omega & \omega & 0 \end{pmatrix} \tag{14}$$

The set of nine states $\{|\psi_i\rangle\}$ forms a SIC known as the *Hesse SIC* [10]. For the Hesse SIC, a set of *four* orthogonal bases is sufficient to reveal the PP-incompatibility of all possible triples. Moreover, the requisite bases have an interesting property: they are *mutually unbiased* with respect to each other. In general, two bases are mutually unbiased if, for any vector $|\psi\rangle$ in one basis and any vector $|\phi\rangle$ in the other, $|\langle\psi|\phi\rangle|^2 = 1/d$. Any set of three states drawn from the SIC will be revealed as PP-incompatible by a measurement in one or more of the Mutually Unbiased Bases (MUB). We construct each basis vector by finding a state orthogonal to three of the SIC states. Specifically, each basis vector corresponds to an element in a Steiner triple system [11] of order 9, which we build by cyclically tracing all the horizontal, vertical and diagonal lines in the array

$$\begin{matrix} 0 & 1 & 2 \\ 3 & 4 & 5 \\ 6 & 7 & 8 \end{matrix}\;; \qquad \text{that is to say,} \qquad S(9) = \begin{matrix} (012) & (345) & (678) \\ (036) & (147) & (258) \\ (048) & (156) & (237) \\ (057) & (138) & (246) \end{matrix} \tag{15}$$

Each possible value of the index i occurs in exactly four entries of $S(9)$, and each possible *pair* of index values occurs exactly once. It is easiest to see the meaning of this construction using the SIC representation of quantum states. Any quantum state, pure or mixed, is equivalent to a probability distribution over the outcomes of an informationally complete measurement, and the qutrit SIC of Equation (14) furnishes such a measurement. An arbitrary qutrit density matrix ρ can be decomposed as

$$\rho = \sum_{i=0}^{8} \left(4p(i) - \frac{1}{3}\right)\Pi_i = 4\sum_{i=0}^{8} p(i)\Pi_i - I \tag{16}$$

where $\Pi_i = |\psi_i\rangle\langle\psi_i|$ and $p(i)$ is the Born-rule probability

$$p(i) - \frac{1}{3}\mathrm{tr}(\rho\Pi_i) \tag{17}$$

To construct the state orthogonal to SIC vectors $|\psi_i\rangle$, $|\psi_j\rangle$ and $|\psi_k\rangle$, we simply write a probability vector $\mathbf{p}$ which is zero in entries i, j and k, and $\frac{1}{6}$ everywhere else.

To see why this construction yields a complete set of MUB, we use the fact that the Hilbert–Schmidt inner product of density matrices is just an affine transformation of the Euclidean inner product of the corresponding probability distributions [2,12]. For qutrit states,

$$\operatorname{tr} \rho\sigma = 12\mathbf{p} \cdot \mathbf{q} - 1 \tag{18}$$

This means that if ρ and σ are orthogonal, then $\mathbf{p} \cdot \mathbf{q} = 1/12$.

With this, we can see that the vectors orthogonal to (012) and (345), for example, must be orthogonal to each other, because when we take the dot product of their SIC representations, we only have three nonzero contributions. If instead we take the vectors orthogonal to (012) and (036), say, a zero in one vector coincides with a zero in the other, the dot product can come out larger. These within-row and between-rows relationships hold generally. Each row corresponds to a set of three mutually orthogonal vectors, and when we take vectors from two different rows, we always get the same nonzero overlap: the Hilbert space inner product of their density matrices is always $\frac{1}{3}$.

We have fashioned a *complete set of mutually unbiased bases.* Starting with the SIC Equation (14), we constructed 12 pure states which fall into four sets of three. Each set of three, corresponding to a row in our table, is an orthonormal basis. When we take the Hilbert–Schmidt inner product of a state from one basis with a state from another, we get $1/3$ every time. This is the requirement for two bases to be mutually unbiased in $d = 3$ (in older language, the observables associated with any two such bases are complementary [13]). Furthermore, the largest number of MUB that can exist in d dimensions is $d + 1$ [14], and we constructed four. From now on, we will refer to these $d(d + 1) = 12$ vectors as *MUB states* for short. While the relation between qutrit SIC and MUB states has been known for some time [10,15], the convenience of the SIC representation has so far not been appreciated.

Given any three distinct elements from the SIC set, a measurement in at least one of the MUB will reveal PP-ODOP incompatibility among those three states. For example, say we pick the SIC elements $|\psi_0\rangle$, $|\psi_1\rangle$ and $|\psi_4\rangle$. Then, we measure in the basis given by the vectors orthogonal to the fourth row: (057), (138), (246). Each possible outcome of the experiment conflicts with one of the three given states: the first with the state ascription $|\psi_0\rangle$, the second with the ascription $|\psi_1\rangle$ and the third with $|\psi_4\rangle$. Therefore, we have PP-ODOP incompatibility at the ternary level, while of course any two distinct states in the SIC are *pairwise* compatible, having an inner-product-squared of $1/(d+1) = 1/4$.

Note that the second error in CFS, writing $>$ instead of $\geq$ in the condition Equation (10), mistakenly implies that triplets of SIC states are PP-ODOP compatible, though barely so. This is clearly incorrect, as we can see by testing $|\psi_0\rangle$, $|\psi_1\rangle$ and $|\psi_2\rangle$ in the computational basis.

4. Additional Properties of the Hesse SIC and Associated MUB

In order to be a SIC representation of a *pure* quantum state, a probability distribution $p(i)$ must satisfy two conditions [2]. The first is quadratic:

$$\sum_i p(i)^2 = \frac{2}{d(d+1)} \tag{19}$$

and the second is cubic, or "QBic":

$$\sum_{ijk} \operatorname{Re}[\operatorname{tr}(\Pi_i \Pi_j \Pi_k)] p(i) p(j) p(k) = \frac{d+7}{(d+1)^3} \tag{20}$$

The quadratic condition has an interesting interpretation that provides a handy mnemonic for it [16].

Whenever we have a probability distribution, we can compute indices that summarize its properties. One well-known example is the Shannon entropy, also designated as the Shannon information or Shannon index, which reflects the extent to which a probability distribution is "spread out." The Shannon index is maximized for a uniform distribution, and it attains its minimum value of zero when the distribution is a delta function. Another way to quantify the spread of a probability distribution is an *effective number.*

Imagine that we have an urn full of marbles, and we presume that when we draw a marble from the urn, no choice is preferred over any other. If the urn contains N marbles, our probability of obtaining any individual one of them is $1/N$. However, what if our probability distribution is not uniform, as it would be if we thought the drawing was rigged in some way? In that case, we can label the marbles with the integers from 1 to N, and we say that our probability for obtaining the one labeled i is $p(i)$.

We draw one marble, replace it and repeat the drawing. What is the probability that we will draw the same marble both times? Let the result of the first drawing be j. Then, our probability for obtaining that marble again is $p(j)$, and to find the overall probability for drawing doubles, we average over all the choices of j:

$$\text{Prob(doubles)} = \sum_j p^2(j) \tag{21}$$

For a uniform distribution, this is

$$\sum_j p^2(j) = \sum_j \left(\frac{1}{N}\right)^2 = N\left(\frac{1}{N}\right)^2 = \frac{1}{N} \tag{22}$$

That is, if all draws are equally probable, then the probability of a coincidence is the reciprocal of the population size. Turning this around, we can say that whatever our probabilities for the different draws, the effective size of the population is

$$N_{\text{eff}} = \left[\sum_j p^2(j)\right]^{-1} \tag{23}$$

Amusingly, assigning a pure state to a quantum system means that the *effective number* of possible outcomes for a SIC experiment that one is willing to contemplate is a simple combinatorial quantity:

$$N_{\text{eff}} = \binom{d+1}{2} \tag{24}$$

This provides a way to remember the value on the right-hand side of the quadratic constraint, Equation (19). Another way to think of this is that when all SIC outcomes are judged as equiprobable, that is to say $p(i) = \frac{1}{d^2}$, the effective number of experimental outcomes is the total number which comprise the SIC: $N_{\text{eff}} = d^2$. Thus, if we focus on the quadratic constraint, ascribing a pure state means neglecting $\binom{d}{2}$ possible outcomes of a SIC experiment. Entertainingly, this is also the best known upper bound on the number of entries which can be zero in a quantum-state assignment $\mathbf{p}$ [12]. This is not a coincidence: we can deduce that bound by starting with the normalization of $\mathbf{p}$ and squaring to find

$$\left(\sum_i p(i)\right)^2 = 1 \tag{25}$$

We then apply the Cauchy–Schwarz inequality to find, writing n_0 for the number of zero-valued entries in $\mathbf{p}$,

$$(d^2 - n_0) \sum_{\text{nonzero}} p(i)^2 \geq \left(\sum_{\text{nonzero}} p(i)\right)^2 = 1 \tag{26}$$

We see the inverse of the effective number appearing on the left-hand side. Consequently,

$$n_0 \leq d^2 - N_{\text{eff}} \tag{27}$$

and from Equation (19) we know the right-hand side equals $d(d-1)/2$, as advertised. In earlier work [17], it was conjectured that this bound might be improved, and that the true upper bound on the number of zeros was actually d. Note that in $d = 3$, this is equivalent to the bound in Equation (27). However, using the so-called *Hoggar SIC* in dimension $d = 8$, we can construct states that saturate the bound in Equation (27), containing exactly 28 zeros. This follows readily from the recent results of Szymusiak and Słomczyński [18]. Therefore, Equation (27) is actually the tightest bound possible in general.

Using the Hesse SIC to define a representation for qutrit state space, the QBic condition can be simplified to

$$\sum_i p(i)^3 - 3 \sum_{(ijk)\in S(9)} p(i)p(j)p(k) = 0 \tag{28}$$

Here, $S(9)$ is the Steiner triple system defined above. This is a consequence [19,20] of the fact that for the Hesse SIC, the triple products

$$C_{jkl} = \operatorname{Re}\operatorname{tr}(\Pi_j \Pi_k \Pi_l) \tag{29}$$

take a particularly simple form. Note that if all three indices are equal, then $\operatorname{tr}(\Pi_j \Pi_k \Pi_l)$ reduces to $\operatorname{tr}(\Pi_j)$, which is unity. Likewise, if two of the three indices are equal, then the value of the triple product follows from the definition of a SIC, Equation (13). The nontrivial case is when all three indices differ.

All known SICs have a *group covariance* property: they can be generated by starting with a single vector (the so-called "fiducial"), and applying the elements of a group to that vector to create all the others. In all cases but one, that group is the *Weyl-Heisenberg group* [21], which is defined from the two generators

$$X|j\rangle = |j+1\rangle \text{ (modulo } d\text{)},\; Z|j\rangle = \omega^j |j\rangle, \text{ where } \omega = e^{2i\pi/d} \tag{30}$$

Combinations of X and Z, together with phase factors, yield the Weyl–Heisenberg displacement operators:

$$\sigma_{a,b} = \left(-e^{i\pi/d}\right)^{ab} X^a Z^b \tag{31}$$

Starting with the fiducial vector $|\psi_0\rangle$, we create the other vectors in the SIC by applying $\sigma_{a,b}$, with $a, b \in \{0, \ldots, d-1\}$. Because $X^d = Z^d = I$, it is convenient to visualize the Weyl–Heisenberg displacement operators as living at the points of a $d \times d$ grid. In $d = 3$, this grid is just the square array from Equation (15).

Group covariance alone tells us something about the triple products: if acting with a group element g on the projectors transforms them as

$$\Pi_i \to g\Pi_i g^\dagger = \Pi_{i'} \tag{32}$$

then the cyclic property of the trace implies

$$C_{ijk} = C_{i'j'k'} \tag{33}$$

In dimension $d = 3$, we have a grid of nine points that we can carve up into four different "striations" (horizontal, vertical and two diagonal). Each striation is a set of three parallel lines, corresponding to three vectors in an orthonormal basis. The Weyl-Heisenberg operators are horizontal and vertical shifts of this grid. These shifts map one line in a striation into another. Any triple product corresponds to a set of three points in the grid. Therefore, if a triple product belongs to one of the four striations, we can transform it into any other triple product in that striation, by applying a

Weyl-Heisenberg operator and possibly permuting indices. Consequently, triple products are constant on striations. The Hesse SIC has the additional nice property that triple products are constant from one set of parallel lines to another. The upshot of this is that for the Hesse SIC, we can find all the nontrivial triple products entirely geometrically. If j, k and l are three *collinear* points, then

$$C_{jkl} = -\frac{1}{8} \tag{34}$$

Otherwise, if j, k and l are distinct but noncollinear,

$$C_{jkl} = \frac{1}{16} \tag{35}$$

The fact that the triple products follow this geometrical rule is what allows us to reduce the QBic Equation (20) to the simpler form of Equation (28).

If we have a probability distribution, we can compute the Shannon entropy of it. We can, therefore, ask which pure states maximize or minimize the Shannon entropy of their SIC representations. In particular, if we try to minimize the Shannon entropy of $\mathbf{p}$ under the constraint that its "effective number" is

$$\left[\sum_i p(i)^2\right]^{-1} = \frac{d(d+1)}{2} \tag{36}$$

then we find that $p(i)$ must be 0 in exactly three entries, and uniformly $1/6$ elsewhere (this is pointed out, in slightly different language, by Szymusiak and Słomczyński [18]).

How many such vectors are valid quantum states? We must check them against the QBic Equation (28). For any vector of this form,

$$\sum_i p(i)^3 = 6\left(\frac{1}{6}\right)^3 = \frac{1}{36} \tag{37}$$

Therefore, we must have

$$\sum_{(ijk)\in S(9)} p(i)p(j)p(k) = \frac{1}{108} \tag{38}$$

Suppose that we fill in one line of our 3×3 grid with zeros. If i, j and k are the points on this line, or on any line that intersects with it, then $p(i)p(j)p(k)$ will evaluate to zero. Exactly two lines will correspond to nonzero products, namely, the two lines parallel to the one we filled with zeros. Therefore,

$$\sum_{(ijk)\in S(9)} p(i)p(j)p(k) = 2\left(\frac{1}{6}\right)^3 = \frac{2}{216} = \frac{1}{108} \tag{39}$$

It follows that the states we seek are the twelve states made by filling one line in the 3×3 grid with zeros and inserting $1/6$ elsewhere. These twelve states fall naturally into four sets of three, corresponding to the four rows of our table. Each row is derived from one way of carving the grid into three parallel lines.

Having a complete set of MUB, we can define a discrete Wigner function [22]. Like a SIC representation, a Wigner representation is a way of writing a quantum state as a list of real numbers. Unlike the SIC representation, Wigner functions for quantum states can have negative values, and thus are called "quasi-probability distributions" [23]. It is easiest to define a Wigner quasi-probability function when the Hilbert-space dimension is a prime number or a power of a prime.

Wootters [22] showed that one can construct a set of *phase-space point operators* that live in a $d \times d$ grid and enjoy the following properties. First, each of the d^2 operators is Hermitian and has unit trace:

$$\mathrm{tr} A_j = 1 \tag{40}$$

Second, they are orthogonal to one another:

$$\mathrm{tr} A_j A_k = d\delta_{jk} \tag{41}$$

Third, if we carve the grid into a set of parallel lines $\{\lambda\}$, then

$$P_\lambda = \frac{1}{d}\sum_{j\in\lambda} A_j \tag{42}$$

defines a set $\{P_\lambda\}$ of mutually orthogonal projection operators, the sum of which is the identity. For any density matrix ρ, we have

$$\rho = \sum_j W(j)A_j,\ W(j) = \frac{1}{d}\mathrm{tr}(\rho A_j) \tag{43}$$

Wootters' discrete Wigner function is closely related to MUB [24]. Each of the $d+1$ sets of parallel lines corresponds to a basis, and these $d+1$ bases (containing d states each) are mutually unbiased with respect to one another. Summing the Wigner function along a line yields the probability of obtaining the outcome corresponding to that state when performing a measurement in the basis to which that line belongs.

For qutrit states, the Wigner functions will be quasi-probability distributions over nine points,

$$\sum_{i=0}^{8} W(i) = 1 \tag{44}$$

where the individual $W(i)$ can go negative. By summing $W(i)$ over a line (jkl), we get the probability q_{jkl} of obtaining the outcome corresponding to that vector if we perform a measurement on that basis. Alternatively, if we know these probabilities, we can solve for the Wigner function at any point:

$$W(i) = \frac{1}{3}\left[\sum_{\lambda: i\in\lambda} q_\lambda - 1\right] \tag{45}$$

Call $\mathbf{s}_{ijk}$ the SIC representation of the MUB vector that has zeroes in positions i, j and k. Then, for example,

$$q_{012} = 12\mathbf{p}\cdot\mathbf{s}_{012} - 1 \tag{46}$$

$$= 1 - 2(p(0) + p(1) + p(2)) \tag{47}$$

If we add up all the probabilities involving point 0, we find that $p(0)$ occurs four times in the sum, and all the other SIC probabilities $p(j)$ occur once:

$$q_{012} + q_{036} + q_{045} + q_{057} = 4 - 2(3p(0) + p(0) + p(1) + \cdots + p(8)) = 2 - 6p(0) \tag{48}$$

Therefore,

$$W(0) = \frac{1}{3}(2 - 6p(0) - 1) = \frac{1}{3} - 2p(0) \tag{49}$$

The argument works analogously for all the points in the grid, and so we arrive at the relation

$$W(i) = \frac{1}{3} - 2p(i) \tag{50}$$

The MUB states affiliated with the Hesse SIC have simple SIC representations, as we have seen. Their Wigner representations are easily found using Equation (50). Let W_{jkl} be the quasi-probability function for the MUB state that is orthogonal to the SIC vectors $|\psi_j\rangle$, $|\psi_k\rangle$ and $|\psi_l\rangle$. Then,

$$W_{jkl}(i) = \begin{cases} \frac{1}{3}, & i \in \{j,k,l\} \\ 0, & \text{otherwise} \end{cases} \tag{51}$$

Any two MUB vectors belonging to different bases intersect in one point on the 3×3 grid. It is instructive to compare the Wigner functions for these MUB vectors to the twelve states of maximal knowledge in Spekkens' quasi-classical model of a three-level system [25].

Moreover, it also follows from Equation (50) that no entry in the Wigner quasi-probability can go more negative than it does for the Hesse SIC states themselves. Generally, if $|\psi_k\rangle$ is a SIC state used to define a SIC representation of quantum state space, and if we turn that representation upon $|\psi_k\rangle$ itself, then we find it corresponds to the probability distribution:

$$e_k(i) = \frac{1}{d(d+1)} + \frac{1}{d+1}\delta_{ik} \tag{52}$$

In dimension $d = 3$, this is a probability vector that contains $1/3$ in the kth position and $1/12$ everywhere else. Using Equation (50) to turn this into a Wigner quasi-probability, we find that

$$W_k(i) = \begin{cases} -\frac{1}{3}, & i = k \\ \frac{1}{6}, & i \neq k \end{cases} \tag{53}$$

No entry in a SIC representation of a quantum state can exceed $1/d$. This follows [17] from the quadratic condition, Equation (19). Therefore,

$$W(i) \geq -\frac{1}{3} \tag{54}$$

In fact, it is known [26] that the sum of *all* the negative entries in a qutrit Wigner quasi-probability function cannot exceed $1/3$ in magnitude. The SIC states themselves pack as much negativity into a single entry as a state can have. This is why the Hesse SIC states are among the *maximally magic resources* for quantum computation [26].

The Hesse SIC states are nine in number, and in the terminology of [26] are the "Strange states." The other maximally magic states—that is, the other states for which the sum of the negative entries has maximal magnitude—are designated the "Norrell states." They are 36 in number, and they spread an equal share of negativity across two entries of the Wigner representation. To illustrate, we write the Wigner representation of a Hesse SIC state as a 3×3 grid:

$$\begin{pmatrix} -\frac{1}{3} & \frac{1}{6} & \frac{1}{6} \\ \frac{1}{6} & \frac{1}{6} & \frac{1}{6} \\ \frac{1}{6} & \frac{1}{6} & \frac{1}{6} \end{pmatrix} \tag{55}$$

There are obviously nine ways to position the $-1/3$. If we instead pick two elements to be equal negative values, then we can form a state like

$$\begin{pmatrix} -\frac{1}{6} & -\frac{1}{6} & \frac{1}{3} \\ \frac{1}{6} & \frac{1}{6} & \frac{1}{6} \\ \frac{1}{6} & \frac{1}{6} & \frac{1}{6} \end{pmatrix} \tag{56}$$

There are $\binom{9}{2} = 36$ ways to do this: First, we pick a striation (horizontal, vertical, left diagonal or right diagonal). Then, we pick a line within that striation (for which we have three choices). Finally, we select which element in that line we will set to 1/3 (again, three choices). Each set of nine states derived from a striation is, in fact, a SIC. Thus, the 36 Norrell states comprise four separate SICs. In the language of group theory, they are a *Clifford orbit*, where the *Clifford group* is defined as the stabilizer of the Weyl-Heisenberg group.

We have seen how we can start with the Hesse SIC and construct four MUB by minimizing the Shannon entropy of the SIC representation, subject to the pure-state constraints. Now, take the nine SIC vectors and twelve MUB vectors, and represent each vector by a vertex of a graph. Connect two vertices with an edge if the corresponding states are orthogonal. The resulting graph has *chromatic number* 4. That is, one needs at least four different colors of paint in order to color all the vertices in such a way that no two adjacent vertices share the same color. We illustrate this in Figure 1. Because the chromatic number exceeds the dimension of the Hilbert space, this set of 21 vectors meets Cabello's necessary condition for demonstrating "state-independent contextuality" [27,28].

We now unpack the meaning of this statement.

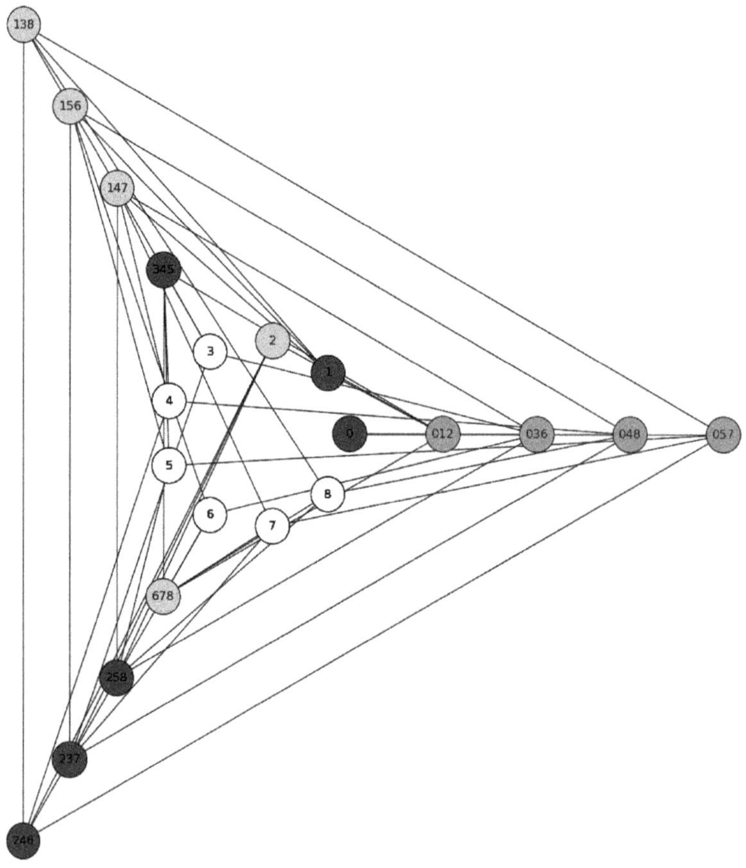

Figure 1. One possible presentation of the orthogonality graph for the Hesse SIC states and their associated MUB. Each vertex stands for a quantum state, and vertices are linked by an edge if those states are orthogonal. The nine circles near the center, labeled by single-digit numbers, denote the SIC states. The other vertices, labeled by three-digit sequences, stand for the MUB states. Each of the four nested triangles is an orthonormal basis, e.g., $\{\mathbf{s}_{012}, \mathbf{s}_{345}, \mathbf{s}_{678}\}$. The graph is colored such that no two adjacent vertices share the same color.

The issue at stake is *whether quantum mechanics could be an approximation to some deeper theory of physics that is classical in character.* Could it be that the randomness we find in quantum phenomena might be explained away as due to our ignorance of more fundamental degrees of freedom? A saga of theorems, conceived by Bell, Kochen, Specker and others, argues against this [29]. Quantum theory, they tell us, is incompatible with the idea that quantum uncertainty is a result of our ignorance about "hidden variables" contained within the systems we study (so-called "noncontextual hidden variables").

It is not *a priori* obvious exactly which phenomena truly tap into this failure of classicality. Many "quantum" effects can be emulated in models that are essentially classical. The list is remarkably long, in fact, and includes teleportation, key distribution, the no-cloning and no-broadcasting theorems, coherent superpositions turning to incoherent mixtures by becoming entangled with the environment, "quantum discord" and many more [25,30–33]. However, the Bell-Kochen-Specker results take us out of that regime.

We can think of the Kochen-Specker theorem (and its more modern descendents) in the following way. Suppose that we have some physical system, and we list a series of "questions" that we might ask it. Each question is some physical experiment that yields a quantitative result. For simplicity, we can assume that all these experimental tests are binary, yielding either 0 or 1. Prior to choosing a test and carrying it out, one can have expectations about what will transpire should one choose a particular test and perform it. If we assume that the behavior of the system is governed by some hidden internal degrees of freedom that are independent of the test one might elect to make, then this assumption will constrain the expectations that one might have for the experimental outcomes. The predictions for different experiments will be tied together in a certain way—one which quantum phenomena can violate.

When can a set of questions demonstrate this effect? In quantum theory, we can represent a binary question by a projection operator. A set of projection operators defines an orthogonality graph. A necessary condition for a set of projectors to be able to reveal the failure of the hidden-variable hypothesis is that the chromatic number of their orthogonality graph exceed the dimension of the system's Hilbert space [27,28]. As explained above, our set of SIC and MUB vectors meets this criterion.

Cabello's criterion is necessary but not sufficient. However, the Hesse SIC states and the MUB vectors we derived from them are, in fact, sufficient to demonstrate nonclassicality. A Kochen-Specker theorem that demonstrates this explicitly has been worked out [15].

When taken together, the Hesse SIC and its affiliated MUB comprise a set of questions, for which the statistics of the answers mesh together in a way that lies beyond the classical worldview.

5. Conclusions

We began with the issue of compatibility between probability assignments. Extending these considerations from the classical realm to the quantum, we found that the problem of PP-ODOP compatibility for three pure states in three-dimensional Hilbert space leads naturally to SICs and MUB. There are still open questions regarding PP compatibility in higher dimensions [34], and SIC solutions in higher dimensions grow much more complicated than Equation (14). However, the patterns of linear dependencies observed in higher-dimensional SICs [10] suggest that SIC states may have interesting compatibility properties there as well. Likewise, the strategy of finding the pure states whose SIC representations have minimal Shannon entropy yields an intriguing result in dimension 8 [18], and perhaps such states merit attention more generally.

Acknowledgments: I thank Marcus Appleby, Chris Fuchs and Rüdiger Schack for illuminating discussions. I also thank the anonymous referees for helpful feedback. At the time this paper was first drafted, I was supported by the Brandeis Geometry and Dynamics IGERT grant (NSF-DGE1068620).

Conflicts of Interest: The author declares no conflict of interests.

Abbreviations

The following abbreviations are used in this manuscript:

CFS: Caves, Fuchs and Schack
MUB: Mutually Unbiased Bases
ODOP: One-Dimensional Orthogonal Projectors
POVM: Positive Operator-Valued Measure
PP: post-Peierls
SIC: Symmetric Informationally Complete

References

1. Fuchs, C.A. QBism: The perimeter of Quantum Bayesianism. 2010, arxiv:1003.5209. arXiv.org e-Print archive. http://arxiv.org/abs/1003.5209 (accessed 19 May 2016).
2. Fuchs, C.A.; Schack, R. Quantum-Bayesian coherence. *Rev. Mod. Phys.* **2013**, *85*, 1693–1715, doi:10.1103/RevModPhys.85.1693.
3. Caves, C.M.; Fuchs, C.A.; Schack, R. Conditions for compatibility of quantum state assignments. *Phys. Rev. A* **2002**, *66*, 062111, doi:10.1103/PhysRevA.66.062111.
4. Pusey, M.F.; Barrett, J.; Rudolph, T. On the reality of the quantum state. *Nat. Phys.* **2012**, *8*, 475–478, doi:10.1038/nphys2309.
5. Schlosshauer, M.; Fine, A. No-go theorem for the composition of quantum systems. *Phys. Rev. Lett.* **2014**, *112*, 070407, doi:10.1103/PhysRevLett.112.070407.
6. Zauner, G. Quantum Designs-Foundations of a Noncommutative Theory of Designs. Ph.D. Thesis, University of Vienna, Vienna, Austria, 1999.
7. Renes, J.M.; Blume-Kohout, R.; Scott, A.J.; Caves, C.M. Symmetric informationally complete quantum measurements. *J. Math. Phys.* **2004**, *45*, 2171, doi:10.1063/1.1737053.
8. Scott, A.J.; Grassl, M. SIC-POVMs: A new computer study. *J. Math. Phys.* **2010**, *51*, 042203, doi:10.1063/1.3374022.
9. Graydon, M.A.; Appleby, D.M. Quantum conical designs. *J. Phys. A* **2016**, *49*, 085301, doi:10.1088/1751-8113/49/8/085301.
10. Dang, H.B.; Blanchfield, K.; Bengtsson, I.; Appleby, D.M. Linear dependencies in Weyl-Heisenberg orbits. *Quant. Inf. Proc.* **2013**, *12*, 3449–3475, doi:10.1007/s11128-013-0609-6.
11. Cameron, P.J. Steiner Triple Systems. Encyclopaedia of Design Theory, 2002. Available online: http://designtheory.org/library/encyc/sts/g/ (accessed on 19 May 2016).
12. Fuchs, C.A.; Stacey, B.C. Some negative remarks on operational approaches to quantum theory. In *Quantum Theory: Informational Foundations and Foils*; Springer: Dordrecht, Germany, 2016; pp. 283–305.
13. Schwinger, J. Unitary operator bases. *Proc. Natl. Acad. Sci. USA* **1960**, *46*, 570–579. Available online: http://www.pnas.org/content/46/4/570 (accessed on 19 May 2016).
14. Appleby, D.M. SIC-POVMs and MUBs: Geometrical relationships in prime dimension. 2009, arxiv:0905.1428. arXiv.org e-Print archive. http://arxiv.org/abs/0905.1428 (accessed 19 May 2016).
15. Bengtsson, I.; Blanchfield, K.; Cabello, A. A Kochen-Specker inequality from a SIC. *Phys. Lett. A* **2012**, *376*, 374–376, doi:10.1016/j.physleta.2011.12.011.
16. Stacey, B.C. Multiscale Structure in Eco-Evolutionary Dynamics. Ph.D. Thesis, Brandeis University, Waltham, MA, USA, 2015. arxiv:1509.02958. arXiv.org e-Print archive. http://arxiv.org/abs/1509.02958 (accessed 19 May 2016).
17. Appleby, D.M.; Ericsson, Å.; Fuchs, C.A. Properties of QBist state spaces. *Found. Phys.* **2011**, *41*, 564–579, doi:10.1007/s10701-010-9458-7.
18. Szymusiak, A.; Słomczyński, W. Informational power of the Hoggar SIC-POVM. 2015, arxiv:1512.01735. arXiv.org e-Print archive. http://arxiv.org/abs/1512.01735 (accessed 19 May 2016).
19. Tabia, G.N.M. Experimental scheme for qubit and qutrit symmetric informationally complete positive operator-valued measurements using multiport devices. *Phys. Rev. A* **2012**, *86*, 062107, doi:10.1103/PhysRevA.86.062107.

20. Tabia, G.N.M.; Appleby, D.M. Exploring the geometry of qutrit state space using symmetric informationally complete probabilities. *Phys. Rev. A* **2013**, *88*, 012131, doi:10.1103/PhysRevA.88.012131.
21. Zhu, H. Super-symmetric informationally complete measurements. *Ann. Phys.* **2015**, *362*, 311–326, doi:10.1016/j.aop.2015.08.005.
22. Wootters, W.K. A Wigner-function formulation of finite-state quantum mechanics. *Ann. Phys.* **1987**, *176*, 1–21, doi:10.1016/0003-4916(87)90176-X.
23. Zhu, H. Quasiprobability representations of quantum mechanics with minimal negativity. 2016, arxiv:1604.06974. arXiv.org e-Print archive. http://arxiv.org/abs/1604.06974 (accessed 19 May 2016).
24. Gibbons, K.; Hoffman, M. J.; Wootters, W. K. Discrete phase space based on finite fields. *Phys. Rev. A* **2004**, *70*, 062101, doi:10.1103/PhysRevA.70.062101.
25. Spekkens, R.W. Quasi-quantization: Classical statistical theories with an epistemic restriction. In *Quantum Theory: Informational Foundations and Foils*; Springer: Dordrecht, Germany, 2016; pp. 83–135.
26. Veitch, V.; Mousavian, S.A.H.; Gottesman, D.; Emerson, J. The resource theory of stabilizer computation. *New J. Phys.* **2014**, *16*, 013009, doi:10.1088/1367-2630/16/1/013009.
27. Cabello, A. State-independent quantum contextuality and maximum nonlocality. 2011, arxiv:1112.5149. arXiv.org e-Print archive. http://arxiv.org/abs/1112.5149 (accessed 19 May 2016).
28. Cabello, A.; Kleinmann, M.; Budroni, C. Necessary and sufficient condition for quantum state-independent contextuality. *Phys. Rev. Lett.* **2015**, *114*, 250402, doi:10.1103/PhysRevLett.114.250402.
29. Mermin, N.D. Hidden variables and the two theorems of John Bell. *Rev. Mod. Phys.* **1993**, *65*, 803, doi:10.1103/RevModPhys.65.803.
30. Spekkens, R.W. Evidence for the epistemic view of quantum states: A toy theory. *Phys. Rev. A* **2007**, *75*, 032110, doi:10.1103/PhysRevA.75.032110.
31. Van Enk, S.J. A toy model for quantum mechanics. *Found. Phys.* **2007**, *37*, 1447–1460, doi:10.1007/s10701-007-9171-3.
32. Bartlett, S.D.; Rudolph, T.; Spekkens, R.W. Reconstruction of Gaussian quantum mechanics from Liouville mechanics with an epistemic restriction. *Phys. Rev. A* **2012**, *86*, 012103, doi:10.1103/PhysRevA.86.012103.
33. Combes, J.; Ferrie, C.; Leifer, M.S.; Pusey, M.F. Why protective measurement does not establish the reality of the quantum state. 2015, arxiv:1509.08893. arXiv.org e-Print archive. http://arxiv.org/abs/1509.08893 (accessed 19 May 2016).
34. Brun, T.A.; Hsieh, M.-H.; Perry, C. Compatibility of state assignments and pooling of information. *Phys. Rev. A* **2015**, *92*, 012107, doi:10.1103/PhysRevA.92.012107.

6

Solution of Excited Non-Linear Oscillators under Damping Effects Using the Modified Differential Transform Method

H. M. Abdelhafez

Department of Physics and Engineering Mathematics, Faculty of Electronic Engineering, Menoufia University, Menouf 32952, Egypt; hassanma0@yahoo.com

Academic Editor: Reza Abedi

Abstract: The modified differential transform method (MDTM), Laplace transform and Padé approximants are used to investigate a semi-analytic form of solutions of nonlinear oscillators in a large time domain. Forced Duffing and forced van der Pol oscillators under damping effect are studied to investigate semi-analytic forms of solutions. Moreover, solutions of the suggested nonlinear oscillators are obtained using the fourth-order Runge-Kutta numerical solution method. A comparison of the result by the numerical Runge-Kutta fourth-order accuracy method is compared with the result by the MDTM and plotted in a long time domain.

Keywords: forced duffing oscillator; forced van der Pol Oscillator; Padé approximant; laplace transform; semi-analytical solution

1. Introduction

Our concern in this work is to give semi-analytic solutions of excited Duffing and excited van der Pol oscillators under damping effect which are given in the forms

Duffing equation $$\frac{d^2x}{dt^2} + \eta\frac{dx}{dt} + \omega^2 x + \alpha x^3 = A\sin(\Omega\, t) \tag{1a}$$

van der Pol equation $$\frac{d^2x}{dt^2} - \epsilon(1 - x^2)\frac{dx}{dt} + \omega^2 x = A\sin(\Omega\, t) \tag{1b}$$

where x is the position coordinate which is a function of the time t, ω is the system's natural frequency, η is a scalar parameter indicating the damping factor in Duffing equation, ϵ the nonlinearity and strength of the damping in van der Pol equation, respectively. α is a nonlinear parameter factor, A and Ω are the forcing amplitude and frequency, respectively.

Those two considered nonlinear oscillators have received remarkable attention in recent decades due to the variety of their engineering applications. For example, Duffing Equation (1a) used in studying the magneto-elastic mechanical systems [1], nonlinear vibrations of beams and plates [2,3] and vibrations induced by fluid flow [4] which are modeled by the nonlinear Duffing equation.

On the other hand, during the first half of the twentieth century, Balthazar van der Pol pioneered the fields of radio and telecommunications [5–10]. In an era when these areas were much less advanced than they are today, vacuum tubes were used to control the flow of electricity of transmitters and receivers. Simultaneously with Lorenz, Thompson, and Appleton, van der Pol experimented with oscillations in a vacuum tube triode circuit and concluded that all initial conditions converged to the same orbit of a finite amplitude. Since this behavior is different from the behavior of solutions of linear equations, van der Pol proposed a nonlinear differential Equation (1b) without excitation force, i.e., $A = 0$, commonly referred to as the (unforced) van der Pol equation [8], as a model for the behavior observed in the experiment. In studying the case $\eta << 1$, van der Pol discovered the importance of what has become known as relaxation oscillations [8].

The most common methods for constructing approximate analytical solutions to the nonlinear oscillator's equations are the perturbation methods [11]. These methods include the harmonic balance method, the elliptic Lindstedt-Poincaré method [11–13]. The Krylov Bogoliubov Mitropolsky method [14,15], the averaging [11–16] and multiple scales method [12] are widely used to obtain approximate solutions of nonlinear oscillators. A general common factor to all of these methods is that they solve weakly the nonlinear systems by using perturbation techniques to reduce the system into simpler equations which transform the physical problem into a purely mathematical one, for which a solution is readily available.

This work is the derivation to obtain approximate analytical oscillatory solutions for the nonlinear oscillator Equations (1a) and (1b) with initial conditions $x(0) = a$ and $\dot{x}(0) = b$ using the modified differential transform method. This is a powerful method for solving linear and nonlinear differential equations. This method was at first used as differential transform method in the engineering domain by Zhou [17] and in fluid flow problems [18–25]. The differential transform solution diverges by using finite number of terms. To solve this problem the modified differential transform method [26–30] was developed by combining the differential transform method (DTM) with the Laplace transform and Padé approximant [31] which can successfully predict the solution of differential equations with finite numbers of terms [32,33].

2. Differential Transform Method

A brief explanation of the differential transform method (DTM) is given in [27–31]; for an analytic function $x(t)$ in domain G, which can be represented by a power series around any arbitrary point in this domain. The differential transform of $x(t)$ is defined as follows:

$$X(k) = \frac{1}{k!}\left[\frac{d^k x(t)}{dt^k}\right]_{t=0} \tag{2}$$

In Equation (2), $x(t)$ is the original function and $X(k)$ is the transformed corresponding function. The inverse transform of $X(k)$ is defined as

$$x(t) = \sum_{k=0}^{\infty} X(k) t^k \tag{3}$$

Combining equations (2) and (3), we obtain the following equation

$$x(t) = \sum_{k=0}^{\infty} \frac{t^k}{k!}\left[\frac{d^k x(t)}{dt^k}\right]_{t=0} \tag{4}$$

In applications, $x(t)$ takes finite number of terms and Equation (4) can be written as

$$x(t) = \sum_{k=0}^{N} \frac{t^k}{k!}\left[\frac{d^k x(t)}{dt^k}\right]_{t=0} \tag{5}$$

Some basic transformation rules of the differential transform method which are used in this work are tabulated in Table 1.

Table 1. The fundamental operations of the differential transform method (DTM).

Original Function	Transformed Function
$\alpha u(t) \pm \beta v(t)$	$\alpha U(k) \pm \beta V(t)$
$u(t)\, v(t)$	$\sum_{l=0}^{k} U(l)\, V(k-l)$
$u(t)\, v(t)\, w(t)$	$\sum_{s=0}^{k} \sum_{m=0}^{k-s} U(s)\, V(m)\, W(k-s-m)$
$\dfrac{d^m u(t)}{dt^m}$	$\dfrac{(k+m)!}{k!} U(k+m)$
$\exp(t)$	$\dfrac{1}{k!}$
$\sin(\omega t + \alpha)$	$\dfrac{\omega^k}{k!} \sin(k\pi/2 + \alpha)$
$\cos(\omega t + \alpha)$	$\dfrac{\omega^k}{k!} \cos(k\pi/2 + \alpha)$

Now, we will apply the MDTM to obtain semi-analytic solutions for forced nonlinear Duffing and van der Pol differential equations under damping effect.

3. Forced Duffing Oscillator under Damping Effect

Consider a nonlinear differential Equation (1a) which describes the forced Duffing oscillator with damping effect and initial conditions $x(0) = a$ and $\dot{x}(0) = b$. The differential transform of this equation gives the recurrence relation

$$\begin{aligned} &(k+2)(k+1)X(k+2) + \omega^2 X(k) + \eta(k+1)X(k+1) \\ &+\alpha \sum_{n=0}^{k} \sum_{m=0}^{n} X(m)X(n-m)X(k-n) - \frac{A\Omega^k}{k!} \sin k\pi/2 = 0 \end{aligned} \tag{6}$$

$$X(0) = a, \qquad X(1) = b \tag{7}$$

The recursive equations deduced from Equations (6) and (7) for different values of k is obtained as follows:

$$k=0: \quad 2X(2) + \eta\, b + \alpha\, a^3 + \omega^2 a = 0 \tag{8}$$

$$k=1: \quad 6X(3) + 2\eta\, X(2) + (\omega^2 + 3\alpha\, a^2)\, b - A\,\Omega = 0 \tag{9}$$

$$k=2: \quad 12X(4) + \omega^2 X(2) + 3\eta\, X(3) + 3\,\alpha\, a(b^2 + aX(2)) = 0 \tag{10}$$

and so on.

The recursive relation in Equations (8)–(10) can be solved successively and then by taking the inverse differential transform $x(t)$ is obtained.

3.1. *Example 1: Free Duffing Oscillator under Damping Effect*

Let

$$a = 1.0, \quad b = 0.0, \quad \omega = 1, \quad \eta = 0.05, \quad \alpha = 0.15, \quad A = 0 \tag{11}$$

The analytic expansion $x(t)$ of Equation (1a) for the given values in (11) is given as follows:

$$\begin{aligned} x(t) = 1.0 - 0.575\, t^2 + 0.00958333\, t^3 + 0.0693594\, t^4 \\ -0.00138839\, t^5 - 0.00830017\, t^6 \\ +0.0002253\, t^7 + 0.00136295\, t^8 + \cdots \end{aligned} \tag{12}$$

Figure 1 shows the comparison between the results obtained using the DTM, Equation (12), and the numerical results obtained by Runge-Kutta fourth-order accuracy method. It is clear that the results using the DTM have a reasonable agreement with the results obtained using only the fourth-order Runge-Kutta numerical method in a small range of the solution domain.

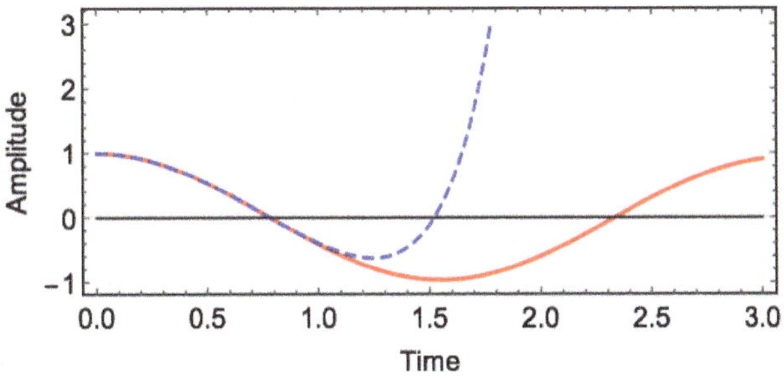

Figure 1. The red curve is the numerical solution and the dashed blue curve is the solution by the DTM.

Now we are improving the accuracy of the differential transform solution using the MDTM [22]. We first apply Laplace transform to the series solution given by Equation (12) to obtain

$$L[x(t)] = \frac{54.954}{s^9} + \frac{1.1355}{s^8} - \frac{5.9761}{s^7} - \frac{0.1666}{s^6} + \frac{1.6646}{s^5} + \frac{0.0575}{s^4} - \frac{1.15}{s^3} + \frac{1}{s} \tag{13}$$

As the first step of the procedure of the MDTM [22] replacing s by $1/t$, calculating the Padé approximant of $[4/4]$ and letting $t = 1/s$ gives the following:

$$\left[\frac{4}{4}\right] = \frac{0.5525 + 10.6558\,s + 0.16667\,s^2 + s^3}{11.9025 + 0.6867\,s + 11.8058\,s^2 + 0.16667\,s^3 + s^4} \tag{14}$$

Taking the inverse Laplace transform to the Padé approximant $[4/4]$, Equation (14), to obtain the solution by the MDTM as follows:

$$x(t) = 0.9962\,e^{-0.0262\,t}\cos(1.0551\,t) + 0.0038\,e^{-0.0572\,t}\cos(3.2685\,t) \tag{15}$$

Figure 2 depicts the comparison between the MDTM results obtained by the Padé approximant of $[4/4]$ and the results obtained using the fourth-order Runge-Kutta numerical method. It is clear that the MDTM result obtained by the real part of Padé approximate gives an excellent agreement with the result obtained using the fourth-order Runge-Kutta numerical method.

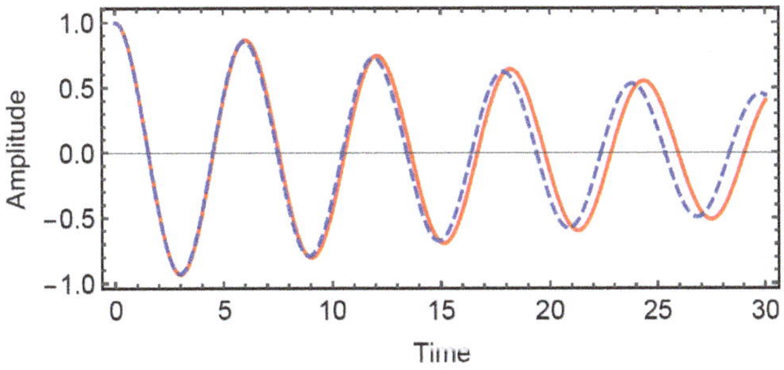

Figure 2. The red curve is the numerical solution and the dashed blue curve is the solution by the modified differential to transform method (MDTM) in Equation (15).

3.2. *Example 2: Forced Nonlinear Duffing Oscillator with Damping Effect*

In Equation (1a) let

$$a = 1.0,\ b = 0.0,\ \eta = 0.03,\ \alpha = 0.03,\ A = 0.15,\ \omega = 1.0, \text{ and } \Omega = 0.8 \tag{16}$$

In this case the analytic expansion takes the form:

$$x(t) = 1.0 - 0.515\,t^2 + 0.02515\,t^3 + 0.0440155\,t^4 - 0.00219932\,t^5 \\ -0.0015002\,t^6 + 0.0000701175\,t^7 + 0.0000273301\,t^8 + \cdots \tag{17}$$

Figure 3 illustrates the comparison of results obtained by the differential transform method which given by Equation (17) and the fourth-order Runge-Kutta numerical method. Clearly, the weakness of accuracy of the DTM result even for short time domain.

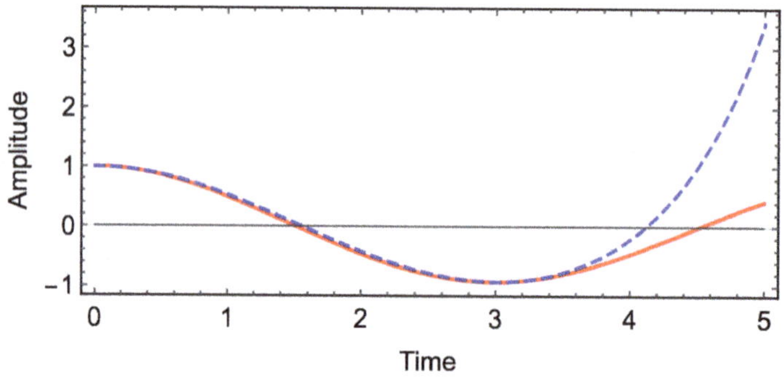

Figure 3. The red curve is the numerical solution and the dashed blue curve is the solution by the DTM in Equation (17).

To apply the modified differential transform method, we first get the Laplace transform to the time series solution which is given by Equation (17), yields

$$L[x(t)] = \frac{1}{s} - \frac{1.03}{s^3} + \frac{0.1509}{s^4} + \frac{1.0564}{s^5} - \frac{0.2639}{s^6} \\ - \frac{1.08015}{s^7} + \frac{0.3534}{s^8} - \frac{1.10195}{s^9} \tag{18}$$

Using the concept of the MDTM [22]. Replace s by $1/t$ in Equation (18), calculate the Padé approximant of $[4/4]$ and after that let $t = 1/s$ which gives the following:

$$\left[\frac{4}{4}\right] = \frac{0.1392 + 0.64\,s + 0.03\,s^2 + s^3}{0.6592 + 0.0192\,s + 1.67\,s^2 + 0.03s^3 + s^4} \tag{19}$$

Applying the inverse Laplace transform to the Padé approximant of $[4/4]$ in Equation (19) to obtain the solution by the MDTM in the form:

$$x(t) = -0.0236\cos(0.8\,t) + 0.3832\sin(0.8\,t) \\ +e^{-0.015\,t}(1.0236\cos(1.0148\,t) - 0.287\sin(1.0148\,t)) \tag{20}$$

Figure 4 shows the comparison between the MDTM result obtained by the Padé approximant of $[4/4]$ and the result obtained using the fourth-order Runge-Kutta numerical method. The MDTM results obtained by the real part of the Padé approximant of $[4/4]$ are clearly in excellent agreement with the result obtained using Runge-Kutta fourth-order accuracy numerical method.

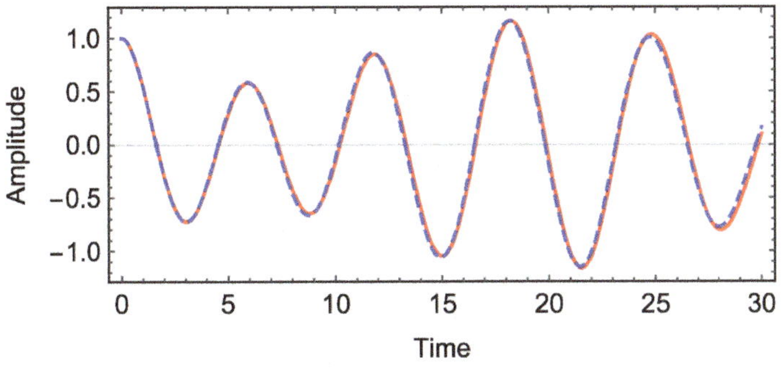

Figure 4. The red curve is the numerical result and the dashed blue curve is the solution by the DTM given in Equation (20).

4. Forced van der Pol Oscillator under Damping Effect

Nayfeh and Mook from [12] derived the ***Rayleigh equation*** in the form

$$y'' + \epsilon\Big[\frac{1}{3}(y')^2 - 1\Big]y' + y = 0, \quad \text{with} \quad y' = dy/d\tau \tag{21}$$

Setting $x = y'$ we find the ***van der Pol equation***

$$x'' + \epsilon\big[x^2 - 1\big]x + x = 0 \quad \text{with} \quad x = dx/d\tau \tag{22}$$

which arises in the study of circuits containing vacuum tubes and given by Equation (1b) with $A = 0$.

Electrical circuit involving a triode results a forced van der Pol oscillator Equation (1b) where $A \neq 0$, see Figure 5. The circuit contains: a triode, a resistor R, a coupled inductor L and mutual inductance M. In the serial RCL circuit there is a current i, and towards the triode anode (plate) a current i_a, while there is a voltage u_g on the triode control grid, see Figure 5. The van der Pol oscillator is forced by an AC voltage source E_s.

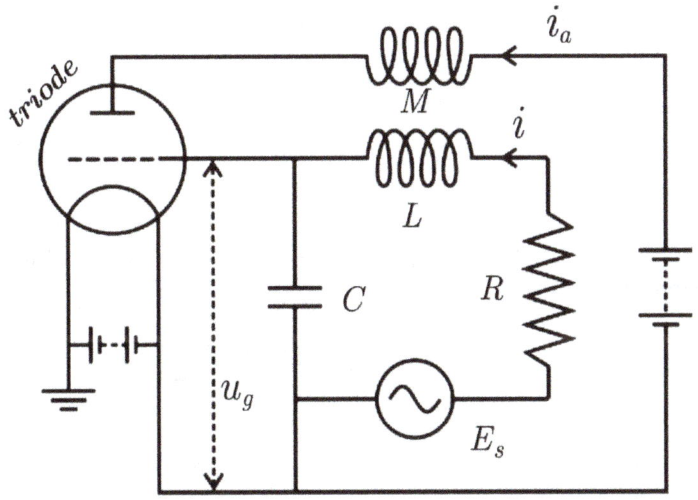

Figure 5. Electrical circuit involving a triode, resulting in a forced van der Pol oscillator.

To investigate a semi-analytic solution of forced van der Pol oscillator under damping effect we apply the MDTM to Equation (1b) with initial conditions $x(0) = a$ and $\dot{x} = b$. The differential transform of this equation has the recurrence relation

$$
\begin{aligned}
&(k+2)(k+1)X(k+2)+\omega^2X(k)+\epsilon\Big[(k+1)X(k+1)\\
&\quad-\sum_{n=0}^{k}\sum_{m=0}^{n}(k-n+1)X(m)X(n-m)X(k-n+1)\Big]\\
&\quad-\frac{A\,\Omega^k}{k!}\sin\left(\frac{k\pi}{2}\right)=0 \qquad (23)\\
&X(0)=a,\quad X(1)=b \qquad (24)
\end{aligned}
$$

The recursive equations deduced from Equations (23) and (24) for $k=0,1,2$ are obtained as follows:

$$k=0:\qquad 2X(2)+\omega^2a+\epsilon\,(a^2-1)b=0 \qquad (25)$$

$$k=1:\qquad 6X(3)+\omega^2b+2\epsilon\,X(2)(a^2-1)+2\epsilon\,a\,b^2-A\,\Omega=0 \qquad (26)$$

$$k=2:\qquad 12X(4)+\omega^2X(2)+\epsilon[3(a^2-1)X(3)+6\,a\,b\,X(2)+b^3]=0 \qquad (27)$$

and so on.

4.1. *Example 3:*

Let

$$a=1.0,\ b=0.0,\ \omega=1.0,\ \epsilon=0.04,\ A=0.04,\ \text{and}\ \Omega=1.4 \qquad (28)$$

For the values given in Equation (28) the analytic expansion of the nonlinear van der Pol Equation (1b) is given by:

$$
\begin{aligned}
x(t)&=1.0-0.5\,t^2+0.009333\,t^3+0.041667\,t^4t^4-0.003381\,t^5\\
&\quad-0.001327\,t^6+0.000599\,t^7-.000009\,t^8+\cdots \qquad (29)
\end{aligned}
$$

To use the modified differential transform method we take the Laplace transform of the series solution in Equation (29), yields:

$$
\begin{aligned}
L[x(t)]&=\frac{1}{s}-\frac{1}{s^3}+\frac{0.056}{s^4}+\frac{1}{s^5}-\frac{0.40576}{s^6}-\frac{0.9552}{s^7}+\frac{3.01838}{s^8}\\
&\quad-\frac{0.353677}{s^9}-\frac{25.0076}{s^{10}}+\frac{31.4072}{s^{11}} \qquad (30)
\end{aligned}
$$

Forthwith as in [22] we replace s by $1/t$, calculating the Padé approximant of [3/3] and [4/4] and letting $t=1/s$ gives us the following:

$$\left[\frac{3}{3}\right]=\frac{5.24571+93.6735\,s+s^2}{93.6175+6.24571\,s+93.6735\,s^2+s^3} \qquad (31)$$

$$\left[\frac{4}{4}\right]=\frac{0.13895+7.72697\,s+0.168318\,s^2+s^3}{7.71754+0.251268\,s+8.72697\,s^2+0.168318\,s^3+s^4} \qquad (32)$$

Taking the inverse Laplace transform of the Padé approximant of [3/3] and [4/4] in Equations (31) and (32) we obtain the semi-analytic solutions respectively, as follows:

$$x(t) \approx e^{-0.02802\,t}(\cos t + 0.02803\,\sin t), \tag{33}$$

$$x(t) \approx e^{-0.0062\,t}(\cos t + 0.01264\,\sin t) + e^{-0.07798\,t}(0.0003\cos(2.7785\,t) - 0.0023\sin(2.7785\,t)) \tag{34}$$

Figure 6a illustrated to show a comparison of the solution by the differential transform method, Equation (29), and the numerical solution by the fourth-order Runge-Kutta method. It is clear that, the result obtained by DTM have not reasonable agreement with the numerical result by Runge-Kutta for a long time domain.

Figure 6b shows the comparison between the MDTM results obtained by the real part of Padé approximant of orders [3/3] and [4/4] whose given by Equations (33) and (34), respectively, and the result obtained by the fourth-order Runge-Kutta numerical method. The MDTM result obtained by the real part of Padé approximant [3/3] in Equation (33) shows some discrepancies in comparison to the result obtained using the fourth-order Runge-Kuta numerical method. However, it is clear that the result of the result of the MDTM by the real part of Padé approximant [4/4] in Equation (34) has excellent agreement and seems to coincide with the numerical result.

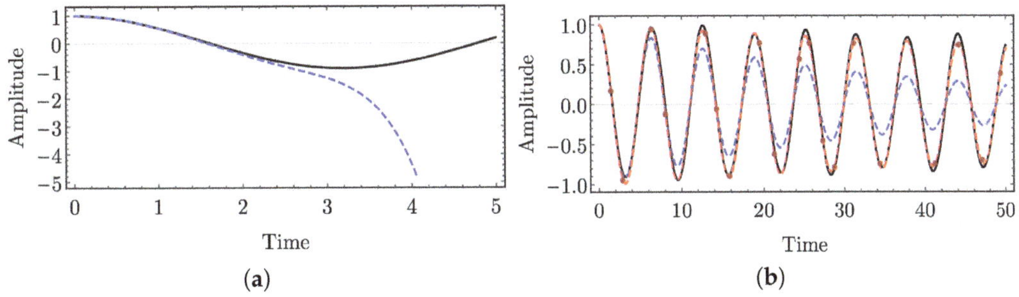

Figure 6. Example 3. (**a**) The black curve is the numerical solution and the dashed blue curve is the solution by the DTM given by Equation (29); (**b**) The dashed blue curve is the solution by the MDTM and Padé [3/3] approximant solution. The dashed red curve is the solution by the MDTM and Padé [4/4] approximant. The black curve is the numerical solution.

4.2. *Example 4:*

Let

$$a = 1.0,\ b = 0.0,\ \epsilon = 0.004,\ A = 0.9,\ \omega = 2.0 \text{ and } \Omega = 1.4 \tag{35}$$

The nonlinear differential Equation (1b) in this case has the analytic solution in the form:

$$x(t) = 1 - 2\,t^2 + 0.21\,t^3 + 0.666667\,t^4 - 0.06578\,t^5 - 0.088329'\,t^6 + 0.010248\,t^7 + 0.005618\,t^8 - 0.002084\,t^9 + 0.000145\,t^{10} + 0.000545\,t^{11} - 0.000143\,t^{12} + \cdots \tag{36}$$

To apply the MDTM we first take Laplace transform of the DTM solution given in Equation (36), yields:

$$L[x(t)] = \frac{1}{s} - \frac{4}{s^2} + \frac{1.26}{s^4} + \frac{16}{s^5} - \frac{7.8936}{s^6} - \frac{63.5968}{s^7} + \frac{51.647808}{s^8} + \frac{226.503501}{s^9} - \frac{756.140505}{s^{10}} + \frac{524.935235}{s^{11}} + \frac{21755.582199}{s^{12}} - \frac{68682.662616}{s^{13}} \quad (37)$$

Replace s by $1/t$ in Equation (37) and calculate the Padé approximants [4/4] and [6/6] and after that let $t = 1/s$ to obtain

$$\left[\frac{4}{4}\right] = \frac{0.554007 + 3.493991\,s - 0.860981\,s^2 - s}{15.0608 - 1.629917\,s - 0.506009\,s^2 - 0.860981\,s^3 - s^4} \quad (38)$$

$$\left[\frac{6}{6}\right] = \frac{45.6864 + 71.9781\,s + 3.2114\,s^2 + 38.5489\,s^3 + 0.7508\,s^4 + s^5}{283.4087 + 7.7739s + 225.2278s^2 + 4.9547s^3 + 42.5489s^4 + 0.7508s^5 + s^6} \quad (39)$$

The inverse Laplace transform of (38) and (39), respectively, gives the semi-analytical solutions as follows:

$$x(t) = -0.0058\,e^{-2.2603\,t} + 0.01583\,e^{1.6398\,t} + e^{-0.12024\,t}(0.9899\cos(2.0122\,t) + 0.0394\sin(2.0122\,t)) \quad (40)$$

$$x(t) = e^{-0.00698\,t}(0.0091\ \cos(1.3895\,t) + 0.4314\ \sin(1.3895\,t)) + e^{-0.00015\,t}(\cos(2.00267\,t) - 0.29905\ \sin 92.00267\,t)) \quad (41)$$

Figure 7 illustrates the comparison of the fourth-order Runge-Kutta numerical solution and the solution of the MDTM by the real part of the Padé approximant of order [4/4]. In Figure 8, it is clear that the solution by the MDTM is unstable. For this reason we have another attempt to obtain a more accurate and stable semi-analytical solution. To resolve this problem we use again the MDTM but with the real part of the Padé approximant of order [6/6] given by Equation (41).

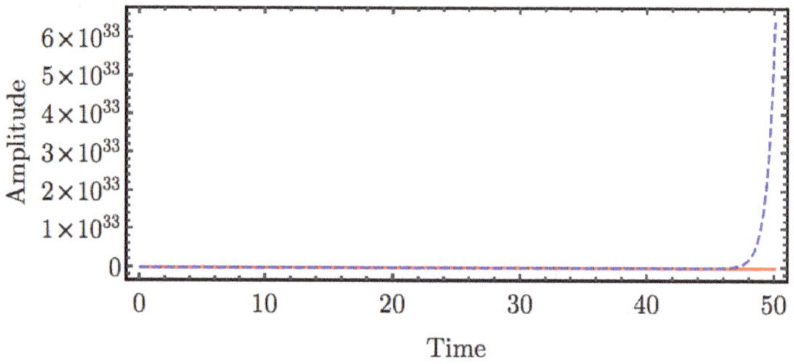

Figure 7. The red curve is the numerical solution and the dashed blue curve is the DTM solution in Equation (36).

Example 4 shows that the solution by the MDTM with the real part of the Padé approximant [6/6] not only matches perfectly with the numerical solution for a long time domain, as in Figure 8, but also shows that the phase plane given by the two methods seems to be identical, show Figure 9.

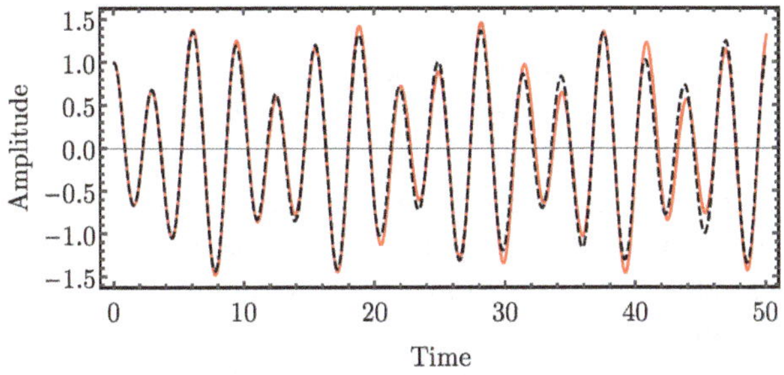

Figure 8. The red curve is the numerical solution and the dashed blue curve is the MDTM solution by Padé approximant [6/6] in Equation (41).

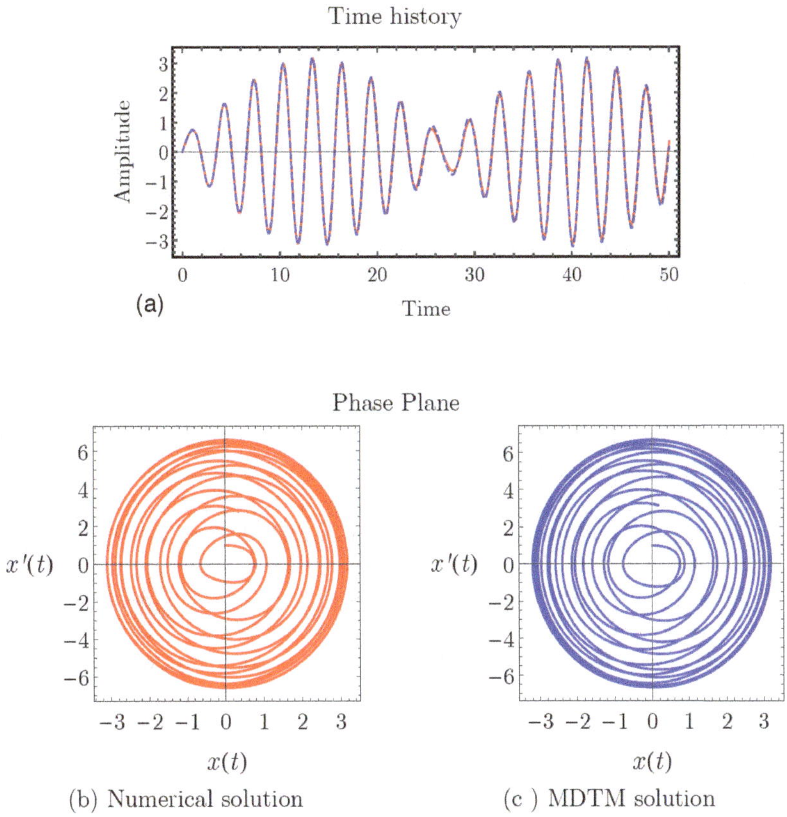

Figure 9. Red plot is the numerical solution and the blue plot is the Padé approximant solution of order [6/6] given by Equation (41).

5. Conclusions

The main goal of researchers who are interested in solving nonlinear differential equations is to obtain analytical solutions along with numerical solutions. These researchers have relied on some methods such as the multiple time scales method and the harmonic balance method and others. Other researchers have taken another turn and used the modified differential to transform method (MDTM) to obtain semi-analytic solutions of free non-linear oscillation by adding Laplace transform and Padé approximant [4/4]. Here, we extend their studies and provided semi-analytical solutions of forced oscillations of Duffing and van der Pol under damping effects.

Through some applications, we have provided records to indicate the success of the semi-analytical solution by the MDTM of nonlinear differential equations. The study focused on van der Pol and Duffing nonlinear oscillators under damping effects, due to their importance in science and engineering.

The numerical results demonstrate the validity and applicability of analytic solutions that we obtained by using the MDTM with appropriate values of parameters. This assures us the extent of success of using the modified differential to transform method to obtain analytical solutions of non-linear differential equations.

To get the analytical solution using the MDTM method one can obtain the result of the real part of Padé approximant of any order like [3/3], [4/4], [5/5] and higher where at least one of these approximant gives a satisfactory accurate analytic solution.

Conflicts of Interest: The authors declare no conflict of interest.

References

1. Guckenheimer, J.; Holmes, P. *Nonlinear Oscillations, Dynamical Systems and Bifurcations of Vector Fields*; Springer-Verlag: New York, NY, USA, 1983.
2. Ahmadian, M.T.; Mojahedi, M.; Moeenfard, H. Free vibration analysis of a nonlinear beam using homotopy and modified LindstedtPoincaré methods. *J. Solid Mech.* **2009**, *1*, 29–36.
3. Bakhtiari-Nejad, F.; Nazari, M. Nonlinear vibration analysis of isotropic cantilever plate with viscoelastic laminate. *Nonlinear Dyn.* **2009**, *56*, 325–356.
4. Srinil, N.; Zanganeh, H. Modeling of coupled cross-flow/in-line vortex-induced vibrations using double Duffing and van der Pol oscillators. *Ocean Eng.* **2012**, *53*, 83–97.
5. Cartwright, M. L. Balthazar van der Pol. *J. Lond. Math. Soc.* **1960**, *35*, 367–376.
6. Cartwright, M. L. *Van der Pols Equation for Relaxation Oscillations, in Contributions to the Theory of Nonlinear Oscillations II*; Princeton Annals of Mathematics Studies 2; Princeton University Press: Princeton, NJ, USA, 1952; p. 318.
7. Stumpers, F.L.H.M. Balth. van der Pols work on nonlinear circuits. *IRE Trans. Circuit Theory* **1960**, *7*, 366–367.
8. Van der Pol, B. A theory of the amplitude of free and forced triode vibrations. *Radio Rev.* **1920**, *1*, 701–710.
9. Van der Pol, B. Relaxation Oscillations I. *Phil. Mag.* **1926**, *2*, 978–992.
10. Van der Pol, B. The nonlinear theory of electric oscillations. *Proc. IRE* **1934**, *22*, 1051–1086.
11. Nayfeh, A.H. *Perturbation Methods*; John Wiley: New York, NY, USA, 1973.
12. Nayfeh, A.H.; Mook, D.T. *Nonlinear Oscillations*; John Wiley: New York, NY, USA, 1979.
13. Nayfeh, A.H. *Introduction to Perturbation Methods*; Jhon Wiley: New York, NY, USA, 1981.
14. Krylov, N.; Bogolioubov, N. *Introduction to Nonlinear Mechanics*; Princeton University Press: Princeton, NJ, USA, 1943.
15. Bogolioubov, N.N.; Mitropolsky, Y.A. *Asymptotic Methods in the Theory of Nonlinear Oscillations*; Gordon and Breach: New York, NY, USA, 1961.
16. Sanders, J.A.; Verhulst, F. *Averaging Methods in Nonlinear Dynamical Systems*; Springer-Verlag: New York, NY, USA, 1985.
17. Zhou, J.K. *Differential Transformation and Its Application for Electrical Circuits*; Huazhong University Press: Wuhan, China, 1986.
18. Mao, Q. Design of shaped piezoelectric modal sensors for cantilever beams with intermediate support by using differential transformation method. *Appl. Acoust.* **2012**, *73*, 144–149.
19. Rashidi, M.M.; Erfani, E. Analytical method for solving steady MHD convective and slip flow due to a rotating disk with viscous dissipation and Ohmic heating. *Eng. Comput.* **2012**, *29*, 562–579.
20. Rashidi, M.M.; Domairry, G. New analytical solution of the three dimensional Navier Stokes equations. *Mod. Phys. Lett. B* **2009**, *23*, 3147–3155.
21. Su, X.H.; Zheng, L.C. Approximate solutions to MHD Falkner Skan flow over permeable wall. *Appl. Math. Mech. (Engl. Ed.)* **2011**, *32*, 401–408.
22. Momani, S.; Odibat, Z.; Erturk, V.S. Generalized differential transform method for solving a space- and time-fractional diffusion-wave equation. *Phys. Lett. A* **2007**, *370*, 379–387.

23. Kangalgil, F.; Ayaz, F. Solitary wave solutions for the KdV and mKdV equations by differential transform method. *Chaos Solitons Fractals* **2009**, *41*, 464–472.
24. Biazar, J.; Eslami, M. Analytic solution for Telegraph equation by differential transform method. *Phys. Lett. A* **2010**, *374*, 2904–2906.
25. Hesam, S.; Nazemi, A.R.; Haghbin, A. Analytical solution for the Fokker Planck equation by differential transform method. *Sci. Iran.* **2012**, *19*, 1140–1145.
26. El-Shahed, M. Application of differential transform method to non-linear oscillatory systems. *Commun. Nonlinear Sci. Numer. Simul.* **2008**, *13*, 1714–1720.
27. Momani, S.; Erturk, V.S. Solutions of non-linear oscillators by the modified differential transform method. *Comput. Math. Appl.* **2008**, *55*, 833–842.
28. Yildirim, A.; Gokdogan, A.; Merdan, M. Chaotic systems via multistep differential transformation method. *Can. J. Phys.* **2012**, *90*, 391–406.
29. Mirzabeigy, A.; Yildirim, A. Approximate periodic solution for nonlinear jerk equation as a third-order nonlinear equation via modified differential transform method. *Eng. Comput.* **2014**, *31*, 622–633.
30. Erturk, V.S.; Yildirim, A.; Momani, S.; Khan, Y. The differential transform method and Padé approximants for a fractional population growth model. *Int. J. Numer. Methods Heat Fluid Flow* **2012**, *22*, 791–802.
31. Baker, G.A. *Essentials of Padé Approximants*; Academic Press: London, UK, 1975.
32. Nourazar, S.; Mirzabeigy, A. Approximate solution for nonlinear Duffing oscillator with damping effect using the modified differential transform method. *Sci. Iran. B* **2013**, *20*, 364–368.
33. Momani, S.; Ertürk, V.S. Solutions of non-linear oscillators by the modified differential transform method. *Comput. Math. Appl.* **2008**, *55*, 833–842.

7

Exponential Energy Decay of Solutions for a Transmission Problem With Viscoelastic Term and Delay

Danhua Wang *, Gang Li and Biqing Zhu

College of Mathematics and Statistics, Nanjing University of Information Science and Technology, Nanjing 210044, China; ligang@nuist.edu.cn (G.L.); brucechu@163.com (B.Z.)
* Correspondence: matdhwang@yeah.net

Academic Editors: Reza Abedi and Indranil SenGupta

Abstract: In our previous work (Journal of Nonlinear Science and Applications 9: 1202–1215, 2016), we studied the well-posedness and general decay rate for a transmission problem in a bounded domain with a viscoelastic term and a delay term. In this paper, we continue to study the similar problem but without the frictional damping term. The main difficulty arises since we have no frictional damping term to control the delay term in the estimate of the energy decay. By introducing suitable energy and Lyapunov functionals, we establish an exponential decay result for the energy.

Keywords: wave equation; transmission problem; exponential decay; viscoelastic; delay

MSC: AMS Subject Classification (2000): 35B37, 35L55, 93D15, 93D20

---

1. Introduction

In our previous work [1], we considered the following transmission system with a viscoelastic term and a delay term:

$$\begin{cases} u_{tt}(x,t) - au_{xx}(x,t) + \int_0^t g(t-s)u_{xx}(x,s)ds \\ \qquad\qquad + \mu_1 u_t(x,t) + \mu_2 u_t(x,t-\tau) = 0, \quad (x,t) \in \Omega \times (0,+\infty) \\ v_{tt}(x,t) - bv_{xx}(x,t) = 0, \qquad\qquad (x,t) \in (L_1, L_2) \times (0,+\infty) \end{cases} \tag{1}$$

where $0 < L_1 < L_2 < L_3$, $\Omega = (0, L_1) \cup (L_2, L_3)$, a, b, μ_1, μ_2 are positive constants, and $\tau > 0$ is the delay. In that work, we first proved the well-posedness by using the Faedo–Galerkin approximations together with some energy estimates when $\mu_2 \leq \mu_1$. Then, a general decay rate result was established under the hypothesis that $\mu_2 < \mu_1$. As for the previous results and developments of transmission problems, and the research of wave equations with viscoelastic damping or time delay effects, we have stated and summarized in great detail in our previous work [1], thus we just omit it here. The readers, for a better understanding of present work, are strongly recommended to [1] and the reference therein (see [2–33]).

It is worth pointing out that, in our previous work, the assumption "$\mu_2 < \mu_1$" plays an important role in the proof of the above-mentioned general decay result. In this paper, we intend to investigate system (1) with $\mu_1 = 0$. That is, we study the exponential decay rate of the solutions for the following transmission system with a viscoelastic term and a delay term but without the frictional damping:

$$\begin{cases} u_{tt}(x,t) - au_{xx}(x,t) + \int_0^t g(t-s)u_{xx}(x,s)\mathrm{d}s \\ \qquad\qquad + \mu_2 u_t(x,t-\tau) = 0, & (x,t) \in \Omega \times (0,+\infty) \\ v_{tt}(x,t) - bv_{xx}(x,t) = 0, & (x,t) \in (L_1,L_2) \times (0,+\infty) \end{cases} \tag{2}$$

under the boundary and transmission conditions

$$\begin{cases} u(0,t) = u(L_3,t) = 0, \\ u(L_i,t) = v(L_i,t), & i = 1,2 \\ \left(a - \int_0^t g(s)\mathrm{d}s\right) u_x(L_i,t) = bv_x(L_i,t), & i = 1,2 \end{cases} \tag{3}$$

and the initial conditions

$$\begin{cases} u(x,0) = u_0(x), \quad u_t(x,0) = u_1(x), & x \in \Omega \\ u_t(x,t-\tau) = f_0(x,t-\tau), & x \in \Omega, \quad t \in [0,\tau] \\ v(x,0) = v_0(x), \quad v_t(x,0) = v_1(x), & x \in (L_1,L_2) \end{cases} \tag{4}$$

where μ_2 is a real number, a, b are positive constants and $u_1(x) = f_0(x,0)$.

The main difficulty in dealing with this problem is that in the first equation of system (2), we have no frictional damping term to control the delay term in the estimate of the energy decay. To overcome this difficulty, our basic idea is to control the delay term by making use of the viscoelastic term. In order to achieve this goal, a restriction of the size between the parameter μ_2 and the relaxation function g and a suitable energy is needed. This is motivated by Dai and Yang's work [34], in which the viscoelastic wave equation with delay term but without a frictional damping term was studied and an exponential decay result was established. In the work here, we will establish an exponential decay rate result for the energy.

The remaining part of this paper is organized as follows. In Section 2, we give some notations and hypotheses needed for our work and state the main results. In Section 3, under some restrictions of μ_2 (see (35) below), we prove the exponential decay of the solutions for the relaxation function satisfying assumption (H_1) and (H_2).

2. Preliminaries and Main Results

In this section, we present some materials that shall be used in order to prove our main result. Let us first introduce the following notations:

$$(g * h)(t) := \int_0^t g(t-s)h(s)\mathrm{d}s$$

$$(g \diamond h)(t) := \int_0^t g(t-s)(h(t) - h(s))\mathrm{d}s$$

$$(g \Box h)(t) := \int_0^t g(t-s)|h(t) - h(s)|^2\mathrm{d}s$$

We easily see that the above operators satisfy

$$(g * h)(t) = \left(\int_0^t g(s)\mathrm{d}s\right) h(t) - (g \diamond h)(t)$$
$$|(g \diamond h)(t)|^2 \leq \left(\int_0^t |g(s)|\mathrm{d}s\right) (|g|\Box h)(t)$$

Lemma 1. *For any $g, h \in C^1(\mathbb{R})$, the following equation holds*

$$2[g * h]h' = g'\Box h - g(t)|h|^2 - \frac{d}{dt}\left\{g\Box h - \left(\int_0^t g(s)\mathrm{d}s\right)|h|^2\right\}$$

For the relaxation function g, we assume
(H1) g: $\mathbb{R}_+ \to \mathbb{R}_+$ is a C^1 function satisfying

$$g(0) > 0, \quad \beta_0 := a - \int_0^\infty g(s)\mathrm{d}s = a - \bar{g} > 0$$

(H2) There exists a positive constant ξ satisfying $\xi > \xi_0 > 0$ (ξ_0 defined by (39) below) and

$$g'(t) \leq -\xi g(t), \quad \forall t \geq 0$$

According to previous results in the literature (see [1]), we state the following well-posedness result, which can be proved by using the Faedo–Galerkin method.

Theorem 2. *Assume that (H1) and (H2) hold. Then, given $(u_0, v_0) \in \mathcal{V}$, $(u_1, v_1) \in \mathcal{L}^2$, and $f_0 \in L^2((0,1), H^1(\Omega))$, Equations (2)–(4) have a unique weak solution in the following class:*

$$(u, v) \in C(0, \infty; \mathcal{V}) \cap C^1(0, \infty; \mathcal{L}^2)$$

where

$$\mathcal{V} = \Big\{(u, v) \in H^1(\Omega) \cap H^1(L_1, L_2) : u(0, t) = u(L_3, 0) = 0, u(L_i, t) = v(L_i, t)$$
$$\left(a - \int_0^t g(s)\mathrm{d}s\right) u_x(L_i, t) = bv_x(L_i, t), i = 1, 2\Big\}$$

and

$$\mathcal{L}^2 = L^2(\Omega) \times L^2(L_1, L_2)$$

To state our decay result, we introduce the following energy functional:

$$\begin{aligned} E(t) =& \frac{1}{2}\int_\Omega u_t^2(x,t)\mathrm{d}x + \frac{1}{2}\left(a - \int_0^t g(s)\mathrm{d}s\right)\int_\Omega u_x^2(x,t)\mathrm{d}x + \frac{1}{2}\int_\Omega (g\Box u_x)\mathrm{d}x \\ &+ \frac{1}{2}\int_{L_1}^{L_2}\left[v_t^2(x,t) + bv_x^2(x,t)\right]\mathrm{d}x + \frac{\zeta}{2}\int_{t-\tau}^t\int_\Omega e^{\sigma(s-t)}u_s^2((x,s)\mathrm{d}x\mathrm{d}s \end{aligned} \tag{5}$$

where σ and ζ are positive constants to be determined later.

Remark 1. We note that the energy functional defined here is different from that of [1] in the construction of the last term. This is motivated by the idea of [28], in which wave equations with time dependent delay was studied.

Our decay results read as follows:

Theorem 3. *Let (u, v) be the solution of Equations* (2)–(4). *Assume that* $(H1)$, $(H2)$ *and*

$$a > \frac{8(L_2 - L_1)}{L_1 + L_3 - L_2}\beta_0, \quad b > \frac{8(L_2 - L_1)}{L_1 + L_3 - L_2}\beta_0 \tag{6}$$

hold. Let a_0 be the constants defined by (35) *below. If $|\mu_2| < a_0$, then there exists constants $\gamma_1, \gamma_2 > 0$ such that*

$$E(t) \leq \gamma_2 e^{-\gamma_1 \xi(t-t_0)}, \quad t \geq t_0 \tag{7}$$

3. Proof of Theorem 3

For the proof of Theorem 3, we use the following lemmas.

Lemma 4. *Let (u, v) be the solution of Equations* (2)–(4). *Then, we have the inequality*

$$\begin{aligned}\frac{d}{dt}E(t) \leq & \frac{1}{2}\int_\Omega (g'\Box u_x)(t)\mathrm{d}x + \left(\frac{|\mu_2|}{2} + \frac{\zeta}{2}\right)\int_\Omega u_t^2(x,t)\mathrm{d}x - \left(\frac{\zeta}{2}e^{-\sigma\tau} - \frac{|\mu_2|}{2}\right)\int_\Omega u_t^2(x,t-\tau)\mathrm{d}x \\ & -\frac{1}{2}g(t)\int_\Omega u_x^2(x,t)\mathrm{d}x - \frac{\sigma\zeta}{2}\int_{t-\tau}^t\int_\Omega e^{-\sigma(t-s)}u_s^2((x,s)\mathrm{d}x\mathrm{d}s\end{aligned} \tag{8}$$

Proof. Differentiating (5) and using (2), we have

$$\begin{aligned}\frac{d}{dt}E(t) = & \int_\Omega \left[u_t u_{tt} + \left(a - \int_0^t g(s)\mathrm{d}s\right)u_x u_{xt} - \frac{1}{2}g(t)u_x^2\right]\mathrm{d}x + \int_{L_1}^{L_2}[v_t v_{tt} + b v_x v_{xt}]\,\mathrm{d}x \\ & + \int_0^t g(t-s)\int_\Omega u_{xt}(u_x(t) - u_x(s))\mathrm{d}x\mathrm{d}s + \frac{1}{2}\int_\Omega g'\Box u_x\mathrm{d}x + \frac{\zeta}{2}\int_\Omega u_t^2(x,t)\mathrm{d}x \\ & - \frac{\zeta}{2}\int_\Omega e^{-\sigma\tau}u_t^2(x,t-\tau)\mathrm{d}x - \frac{\sigma\zeta}{2}\int_{t-\tau}^t\int_\Omega e^{-\sigma(t-s)}u_s^2(x,s)\mathrm{d}x\mathrm{d}s \\ = & \frac{1}{2}\int_\Omega g'\Box u_x\mathrm{d}x - \frac{1}{2}g(t)\int_\Omega u_x^2\mathrm{d}x - |\mu_2|\int_\Omega u_t(t)u_t(t-\tau)\mathrm{d}x + \frac{\zeta}{2}\int_\Omega u_t^2(x,t)\mathrm{d}x \\ & - \frac{\zeta}{2}\int_\Omega e^{-\sigma\tau}u_t^2(x,t-\tau)\mathrm{d}x - \frac{\sigma\zeta}{2}\int_{t-\tau}^t\int_\Omega e^{-\sigma(t-s)}u_s^2(x,s)\mathrm{d}x\mathrm{d}s\end{aligned} \tag{9}$$

By Cauchy inequalities, we get

$$\begin{aligned}\frac{d}{dt}E(t) \leq & \frac{1}{2}\int_\Omega (g'\Box u_x)(t)\mathrm{d}x + \left(\frac{|\mu_2|}{2} + \frac{\zeta}{2}\right)\int_\Omega u_t^2(x,t)\mathrm{d}x + \left(\frac{|\mu_2|}{2} - \frac{\zeta}{2}e^{-\sigma\tau}\right)\int_\Omega u_t^2(x,t-\tau)\mathrm{d}x \\ & -\frac{1}{2}g(t)\int_\Omega u_x^2(x,t)\mathrm{d}x - \frac{\sigma\zeta}{2}\int_{t-\tau}^t\int_\Omega e^{-\sigma(t-s)}u_s^2((x,s)\mathrm{d}x\mathrm{d}s\end{aligned}$$

The proof is complete. □

Remark 2. In ([1] Lemma 4.1), we proved that the energy functional defined in [1] is non-increasing. However, since $\left(\frac{|\mu_2|}{2} + \frac{\zeta}{2}\right)\int_\Omega u_t^2\mathrm{d}x \geq 0$, $E(t)$ may not be non-increasing here.

Now, we define the functional $\mathscr{D}(t)$ as follows:

$$\mathscr{D}(t) = \int_\Omega u u_t\mathrm{d}x + \int_{L_1}^{L_2} v v_t\mathrm{d}x$$

Then, we have the following estimate:

Lemma 5. *The functional $\mathscr{D}(t)$ satisfies*

$$\begin{aligned}\frac{d}{dt}\mathscr{D}(t) \le \int_\Omega u_t^2 \mathrm{d}x + \int_{L_1}^{L_2} v_t^2 \mathrm{d}x + \left(\delta_1|\mu_2|L^2 + \delta_1 - \left(a - \int_0^t g(s)\mathrm{d}s\right)\right)\int_\Omega u_x^2 \mathrm{d}x \\ + \frac{1}{4\delta_1}\int_0^t g(s)\mathrm{d}s \int_\Omega (g\Box u_x)\mathrm{d}x + \frac{|\mu_2|}{4\delta_1}\int_\Omega u_t^2(x,t-\tau)\mathrm{d}x - \int_{L_1}^{L_2} b v_x^2 \mathrm{d}x\end{aligned} \tag{10}$$

Proof. Taking the derivative of $\mathscr{D}(t)$ with respect to t and using (2), we have

$$\begin{aligned}\frac{d}{dt}\mathscr{D}(t) &= \int_\Omega u_t^2 \mathrm{d}x - \int_\Omega (au_x - g * u_x)u_x \mathrm{d}x - |\mu_2|\int_\Omega u_t(x,t-\tau)u\mathrm{d}x + \int_{L_1}^{L_2} v_t^2 \mathrm{d}x - \int_{L_1}^{L_2} b v_x^2 \mathrm{d}x \\ &= \int_\Omega u_t^2 \mathrm{d}x - \left(a - \int_0^t g(s)\mathrm{d}s\right)\int_\Omega u_x^2 \mathrm{d}x - \int_\Omega (g \diamond u_x)u_x \mathrm{d}x - |\mu_2|\int_\Omega u_t(x,t-\tau)u\mathrm{d}x \\ &+ \int_{L_1}^{L_2} v_t^2 \mathrm{d}x - \int_{L_1}^{L_2} b v_x^2 \mathrm{d}x\end{aligned} \tag{11}$$

By the boundary condition (3), we have

$$u^2(x,t) = \left(\int_0^x u_x(x,t)\mathrm{d}x\right)^2 \le L_1 \int_0^{L_1} u_x^2(x,t)\mathrm{d}x, \quad x \in [0, L_1]$$

$$u^2(x,t) \le (L_3 - L_2)\int_{L_2}^{L_3} u_x^2(x,t)\mathrm{d}x, \quad x \in [L_2, L_3]$$

which implies

$$\int_\Omega u^2(x,t)\mathrm{d}x \le L^2 \int_\Omega u_x^2 \mathrm{d}x, \quad x \in \Omega \tag{12}$$

where $L = \max\{L_1, L_3 - L_2\}$. By exploiting Young's inequality and (12), we get for any $\delta_1 > 0$

$$-|\mu_2|\int_\Omega u_t(x,t-\tau)u\mathrm{d}x \le \frac{|\mu_2|}{4\delta_1}\int_\Omega u_t^2(x,t-\tau)\mathrm{d}x + \delta_1|\mu_2|L^2\int_\Omega u_x \mathrm{d}x \tag{13}$$

Young's inequality implies that

$$\int_\Omega (g \diamond u_x)u_x \mathrm{d}x \le \delta_1 \int_\Omega u_x^2 \mathrm{d}x + \frac{1}{4\delta_1}\int_0^t g(s)\mathrm{d}s \int_\Omega (g\Box u_x)\mathrm{d}x \tag{14}$$

Inserting the estimates (13) and (14) into (11), then (10) is fulfilled. The proof is complete. □

Now, as in Lemma 4.5 of [24], we introduce the function

$$q(x) = \begin{cases} x - \dfrac{L_1}{2}, & x \in [0, L_1] \\ \dfrac{L_1}{2} - \dfrac{L_1 + L_3 - L_2}{2(L_2 - L_1)}(x - L_1), & x \in (L_1, L_2) \\ x - \dfrac{L_2 + L_3}{2}, & x \in [L_2, L_3] \end{cases} \tag{15}$$

It is easy to see that $q(x)$ is bounded, that is, $|q(x)| \le M$, where $M = \max\left\{\frac{L_1}{2}, \frac{L_3 - L_2}{2}\right\}$ is a positive constant. In addition, we define the functionals:

$$\mathscr{F}_1(t) = -\int_\Omega q(x)u_t(au_x - g * u_x)\mathrm{d}x, \quad \mathscr{F}_2(t) = -\int_{L_1}^{L_2} q(x)v_x v_t \mathrm{d}x \tag{16}$$

Then, we have the following estimates.

Lemma 6. *The functionals $\mathscr{F}_1(t)$ and $\mathscr{F}_2(t)$ satisfy*

$$\begin{aligned}\frac{d}{dt}\mathscr{F}_1(t) \leq & \left[-\frac{q(x)}{2}(au_x - g * u_x)^2\right]_{\partial\Omega} - \left[\frac{a}{2}q(x)u_t^2\right]_{\partial\Omega} + \left[\frac{a}{2} + \frac{M^2}{4\delta_2}\right]\int_\Omega u_t^2 \mathrm{d}x \\ & + \left[2a^2 + \delta_2 M^2 a^2|\mu_2| + g^2(0)\delta_2 + (4+\delta_2|\mu_2|)\left(\int_0^t g(s)\mathrm{d}s\right)^2\right]\int_\Omega u_x^2 \mathrm{d}x \\ & + \left[\delta_2|\mu_2|M^2 + \frac{|\mu_2|}{4\delta_2} + \frac{|\mu_2|M^2}{4\delta_2}\right]\int_\Omega u_t^2(x,t-\tau)\mathrm{d}x \\ & + \left(4 + \frac{|\mu_2|}{4\delta_2}\right)\left(\int_0^t g(s)\mathrm{d}s\right)\int_\Omega (g\Box u_x)\mathrm{d}x - g(0)\delta_2\int_\Omega (g'\Box u_x)\mathrm{d}x\end{aligned} \tag{17}$$

and

$$\begin{aligned}\frac{d}{dt}\mathscr{F}_2(t) \leq & -\frac{L_1+L_3-L_2}{4(L_2-L_1)}\left(\int_{L_1}^{L_2} v_t^2\mathrm{d}x + \int_{L_1}^{L_2} bv_x^2\mathrm{d}x\right) + \frac{L_1}{4}v_t^2(L_1) + \frac{L_3-L_2}{4}v_t^2(L_2) \\ & + \frac{b}{4}\left((L_3-L_2)v_x^2(L_2,t) + L_1 v_x^2(L_1,t)\right)\end{aligned} \tag{18}$$

Proof. Taking the derivative of $\mathscr{F}_1(t)$ with respect to t and using (2), we get

$$\begin{aligned}\frac{d}{dt}\mathscr{F}_1(t) = & -\int_\Omega q(x)u_{tt}(au_x - g*u_x)\mathrm{d}x - \int_\Omega q(x)u_t\left(au_{xt} - g(t)u_x(t) + (g'\diamond u_x)(t)\right)\mathrm{d}x \\ = & \left[-\frac{q(x)}{2}(au_x - g*u_x)^2\right]_{\partial\Omega} + \frac{1}{2}\int_\Omega q'(x)(au_x - g*u_x)^2\mathrm{d}x - \left[\frac{a}{2}q(x)u_t^2\right]_{\partial\Omega} \\ & + \frac{a}{2}\int_\Omega q'(x)u_t^2\mathrm{d}x - \int_\Omega q(x)|\mu_2|u_t(x,t-\tau)(g*u_x)\mathrm{d}x \\ & + \int_\Omega q(x)au_x|\mu_2|u_t(x,t-\tau)\mathrm{d}x - \int_\Omega q(x)u_t[(g'\diamond u_x)(t) - g(t)u_x]\mathrm{d}x\end{aligned} \tag{19}$$

We note that

$$\begin{aligned} & \frac{1}{2}\int_\Omega q'(x)(au_x - g*u_x)^2\mathrm{d}x \\ \leq & 2\int_\Omega a^2u_x^2\mathrm{d}x + 2\int_\Omega (g*u_x)^2\mathrm{d}x \\ \leq & 2\int_\Omega a^2u_x^2\mathrm{d}x + 2\int_\Omega\left(\int_0^t g(t-s)(u_x(s) - u_x(t) + u_x(t))\mathrm{d}s\right)^2\mathrm{d}x \\ \leq & 2a^2\int_\Omega u_x^2\mathrm{d}x + 4\left(\int_0^t g(s)\mathrm{d}s\right)^2\int_\Omega u_x^2\mathrm{d}x + 4\left(\int_0^t g(s)\mathrm{d}s\right)\int_\Omega (g\Box u_x)\mathrm{d}x\end{aligned} \tag{20}$$

Young's inequality gives us for any $\delta_2 > 0$,

$$\int_\Omega q(x)au_x|\mu_2|u_t(x,t-\tau)\mathrm{d}x \leq \delta_2 M^2 a^2|\mu_2|\int_\Omega u_x^2\mathrm{d}x + \frac{|\mu_2|}{4\delta_2}\int_\Omega u_t^2(x,t-\tau)\mathrm{d}x \tag{21}$$

$$\begin{aligned} & \int_\Omega q(x)|\mu_2|u_t(x,t-\tau)(g*u_x)\mathrm{d}x \\ = & |\mu_2|\int_\Omega (g\diamond u_x)q(x)u_t(x,t-\tau)\mathrm{d}x + |\mu_2|\int_0^t g(s)\mathrm{d}s\int_\Omega q(x)u_t(x,t-\tau)u_x\mathrm{d}x \\ \leq & \delta_2 M^2|\mu_2|\int_\Omega u_t^2(x,t-\tau)\mathrm{d}x + \frac{|\mu_2|}{4\delta_2}\int_0^t g(s)\mathrm{d}s\int_\Omega (g\Box u_x)\mathrm{d}x + \delta_2|\mu_2|\left(\int_0^t g(s)\mathrm{d}s\right)^2\int_\Omega u_x^2\mathrm{d}x \\ & + \frac{|\mu_2|M^2}{4\delta_2}\int_\Omega u_t^2(x,t-\tau)\mathrm{d}x\end{aligned} \tag{22}$$

and

$$
\begin{aligned}
&-\int_\Omega q(x)u_t[(g'\diamond u_x)(t)-g(t)u_x]\mathrm{d}x\\
\leq&\frac{M^2}{4\delta_2}\int_\Omega u_t^2\mathrm{d}x+g^2(0)\delta_2\int_\Omega u_x^2\mathrm{d}x-g(0)\delta_2\int_\Omega(g'\Box u_x)\mathrm{d}x
\end{aligned}
\tag{23}
$$

Inserting (20)–(23) into (19), we get (17).

By the same method, taking the derivative of $\mathscr{F}_1(t)$ with respect to t, we obtain

$$
\begin{aligned}
\frac{d}{dt}\mathscr{F}_2(t)=&-\int_{L_1}^{L_2}q(x)v_{xt}v_t\mathrm{d}x-\int_{L_1}^{L_2}q(x)v_xv_{tt}\mathrm{d}x\\
=&\left[-\frac{1}{2}q(x)v_t^2\right]_{L_1}^{L_2}+\frac{1}{2}\int_{L_1}^{L_2}q'(x)v_t^2\mathrm{d}x+\frac{1}{2}\int_{L_1}^{L_2}bq'(x)v_x^2\mathrm{d}x+\left[-\frac{b}{2}q(x)v_x^2\right]_{L_1}^{L_2}\\
\leq&-\frac{L_1+L_3-L_2}{4(L_2-L_1)}\left(\int_{L_1}^{L_2}v_t^2\mathrm{d}x+\int_{L_1}^{L_2}bv_x^2\mathrm{d}x\right)+\frac{L_1}{4}v_t^2(L_1)+\frac{L_3-L_2}{4}v_t^2(L_2)\\
&+\frac{b}{4}\left((L_3-L_2)v_x^2(L_2,t)+L_1v_x^2(L_1,t)\right)
\end{aligned}
$$

Thus, the proof of Lemma 6 is finished. □

In [1], the authors pointed out that if $\mu_2<\mu_1$, then the energy is non-increasing. Thus, the negative term $-\int_\Omega u_t^2\mathrm{d}x$ appeared in the derivative energy can be used to stabilize the system. However, in this paper, the energy is not non-increasing. In this case, we need some additional negative term $-\int_\Omega u_t^2\mathrm{d}x$. For this purpose, let us introduce the functional

$$
\mathscr{F}_3(t)=-\int_\Omega u_t(g\diamond u)\mathrm{d}x
$$

Then, we have the following estimate.

Lemma 7. *The functionals $\mathscr{F}_3(t)$ satisfies*

$$
\begin{aligned}
\frac{d}{dt}\mathscr{F}_3(t)\leq&-\left(\int_0^t g(s)\mathrm{d}s-\frac{\alpha_4}{2}\right)\int_\Omega u_t^2(x,t)\mathrm{d}x+\left[\delta_4+\delta_4\left(\int_0^t g(s)\mathrm{d}s\right)^2\right]\int_\Omega u_x^2(x,t)\mathrm{d}x\\
&+\delta_4|\mu_2|\int_\Omega u_t^2(x,t-\tau)\mathrm{d}x+\left[\left(\delta_4+\frac{1}{2\delta_4}+\frac{a^2}{4\delta_4}+\frac{|\mu_2|L^2}{4\delta_4}\right)\int_0^t g(s)\mathrm{d}s\right]\int_\Omega(g\Box u_x)\mathrm{d}x\\
&-\frac{g(0)L^2}{2\alpha_4}\int_\Omega(g'\Box u_x)\mathrm{d}x
\end{aligned}
\tag{24}
$$

Proof. Taking the derivative of $\mathscr{F}_3(t)$ with respect to t and using (2), we get

$$
\begin{aligned}
\frac{d}{dt}\mathscr{F}_3(t)=&-\int_\Omega u_{tt}(g\diamond u)\mathrm{d}x-\int_0^t g(s)\mathrm{d}s\int_\Omega u_t^2(x,t)\mathrm{d}x-\int_\Omega u_t(g'\diamond u)\mathrm{d}x\\
=&-\int_\Omega\left(au_{xx}(x,t)-\int_0^t g(t-s)u_{xx}(x,s)\mathrm{d}s-|\mu_2|u_t(x,t-\tau)\right)(g\diamond u)\mathrm{d}x\\
&-\int_0^t g(s)\mathrm{d}s\int_\Omega u_t^2(x,t)\mathrm{d}x-\int_\Omega u_t(g'\diamond u)\mathrm{d}x\\
=&-\int_\Omega\left(\int_0^t g(t-s)u_x(x,s)\mathrm{d}s\right)(g\diamond u_x)\mathrm{d}x+a\int_\Omega u_x(g\diamond u_x)\mathrm{d}x\\
&+|\mu_2|\int_\Omega u_t(x,t-\tau)(g\diamond u)\mathrm{d}x-\int_0^t g(s)\mathrm{d}s\int_\Omega u_t^2(x,t)\mathrm{d}x-\int_\Omega u_t(g'\diamond u)\mathrm{d}x
\end{aligned}
\tag{25}
$$

Young's inequality implies that for any $\delta_4 > 0$:

$$\begin{aligned}
&-\int_\Omega \left(\int_0^t g(t-s)u_x(x,s)\mathrm{d}s\right)(g\diamond u_x)\mathrm{d}x \\
&\leq \frac{\delta_4}{2}\int_\Omega \left(\int_0^t g(t-s)(u_x(t)-u_x(s)-u_x(t))\mathrm{d}s\right)^2 \mathrm{d}x + \frac{1}{2\delta_4}\int_\Omega (g\diamond u_x)^2\mathrm{d}x \\
&\leq \delta_4\left(\int_0^t g(s)\mathrm{d}s\right)^2\int_\Omega u_x^2(x,t)\mathrm{d}x + \left(\delta_4 + \frac{1}{2\delta_4}\right)\int_\Omega (g\diamond u_x)^2\mathrm{d}x \\
&\leq \delta_4\left(\int_0^t g(s)\mathrm{d}s\right)^2\int_\Omega u_x^2(x,t)\mathrm{d}x + \left(\delta_4 + \frac{1}{2\delta_4}\right)\int_0^t g(s)\mathrm{d}s\int_\Omega (g\Box u_x)\mathrm{d}x
\end{aligned} \tag{26}$$

and

$$a\int_\Omega u_x(g\diamond u_x)\mathrm{d}x \leq \delta_4\int_\Omega u_x^2(x,t)\mathrm{d}x + \frac{a^2}{4\delta_4}\int_0^t g(s)\mathrm{d}s\int_\Omega (g\Box u_x)\mathrm{d}x \tag{27}$$

By Young's inequality and (12), we get for any $\delta_4 > 0, \alpha_1 > 0$

$$|\mu_2|\int_\Omega u_t(x,t-\tau)(g\diamond u)\mathrm{d}x \leq \delta_4|\mu_2|\int_\Omega u_t^2(x,t-\tau)\mathrm{d}x + \frac{|\mu_2|L^2}{4\delta_4}\int_0^t g(s)\mathrm{d}s\int_\Omega (g\Box u_x)\mathrm{d}x \tag{28}$$

and

$$\begin{aligned}
-\int_\Omega u_t(g'\diamond u)\mathrm{d}x &\leq \frac{\alpha_4}{2}\int_\Omega u_x^2(x,t)\mathrm{d}x + \frac{1}{2\alpha_4}\int_0^t (-g'(s))\mathrm{d}s\int_\Omega (-g'\Box u)\mathrm{d}x \\
&\leq \frac{\alpha_4}{2}\int_\Omega u_x^2(x,t)\mathrm{d}x - \frac{g(0)L^2}{2\alpha_4}\int_\Omega (g'\Box u_x)\mathrm{d}x
\end{aligned} \tag{29}$$

Inserting (26)–(29) into (25), we get (24). □

Now, we are ready to prove Theorem 3.

Proof. We define the Lyapunov functional:

$$L(t) = N_1E(t) + N_2\mathscr{D}(t) + \mathscr{F}_1(t) + N_4\mathscr{F}_2(t) + N_5\mathscr{F}_3(t) \tag{30}$$

where N_1, N_2, N_4 and N_5 are positive constants that will be fixed later.

Since g is continuous and $g(0) > 0$, then for any $t \geq t_0 > 0$, we obtain

$$\int_0^t g(s)\mathrm{d}s \geq \int_0^{t_0} g(s)\mathrm{d}s = g_0 \tag{31}$$

Taking the derivative of (30) with respect to t and making the use of the above lemmas, we have

$$\begin{aligned}
\frac{d}{dt}L(t) \le & -\left\{N_5\left(g_0-\frac{\alpha_4}{2}\right)-N_1\left(\frac{|\mu_2|}{2}+\frac{\zeta}{2}\right)-N_2-\left(\frac{a}{2}+\frac{M^2}{4\delta_2}\right)\right\}\int_\Omega u_t^2 dx \\
& -\{N_2\beta_0-N_2(\delta_1+L^2\delta_1|\mu_2|)-(2a^2+4\bar{g}^2+\delta_2|\mu_2|\bar{g}^2+\delta_2M^2a^2|\mu_2|+g^2(0)\delta_2) \\
& -N_5(\delta_4+\delta_4\bar{g}^2)\}\int_\Omega u_x^2 dx \\
& +\left\{N_1\left(\frac{|\mu_2|}{2}-\frac{\zeta}{2}e^{-\sigma\tau}\right)+\frac{N_2|\mu_2|}{4\delta_1}+\left(\delta_2|\mu_2|M^2+\frac{|\mu_2|(1+M^2)}{4\delta_2}\right)+N_5\delta_4|\mu_2|\right\}\int_\Omega u_t^2(x,t-\tau)dx \\
& -\left\{\frac{b(L_1+L_3-L_2)}{4(L_2-L_1)}N_4+N_2b\right\}\int_{L_1}^{L_2} v_x^2 dx-\left\{\frac{L_1+L_3-L_2}{4(L_2-L_1)}N_4-N_2\right\}\int_{L_1}^{L_2} v_t^2 dx \\
& -(b-N_4)\frac{b}{4}\left((L_3-L_2)v_x^2(L_2,t)+L_1v_x^2(L_1,t)\right)-(a-N_4)\left[\frac{L_1}{4}v_t^2(L_1,t)+\frac{L_3-L_2}{4}v_t^2(L_2,t)\right] \\
& +\left[\frac{N_2\bar{g}}{4\delta_1}+\left(4\bar{g}+\frac{|\mu_2|\bar{g}}{4\delta_2}\right)+N_5\left(\delta_4+\frac{1}{2\delta_4}+\frac{(a^2+|\mu_2|L^2)}{4\delta_4}\right)\bar{g}\right]\int_\Omega(g\Box u_x)dx \\
& +\left[\frac{N_1}{2}-g(0)\delta_2-\frac{N_5g(0)L^2}{2\alpha_4}\right]\int_\Omega(g'\Box u_x)dx
\end{aligned} \tag{32}$$

At this moment, we wish all coefficients except the last two in (32) will be negative. We want to choose N_2 and N_4 to ensure that

$$\begin{cases} a-N_4\ge 0 \\ b-N_4\ge 0 \\ \dfrac{L_1+L_3-L_2}{4(L_2-L_1)}N_4-N_2>0 \end{cases} \tag{33}$$

For this purpose, since $\dfrac{8l(L_2-L_1)}{L_1+L_3-L_2}<\min\{a,b\}$, we first choose N_4 satisfying

$$\frac{8l(L_2-L_1)}{L_1+L_3-L_2}<N_4\le\min\{a,b\}$$

Then, we pick

$$\alpha_4=g_0,\quad \delta_1<\frac{\beta_0}{8}\quad\text{and}\quad \delta_2<\frac{1}{g^2(0)}$$

such that

$$N_5\left(g_0-\frac{\alpha_4}{2}\right)=\frac{N_5g_0}{2},\quad N_2\beta_0-N_2\delta_1>\frac{7N_2\beta_0}{8},\quad\text{and}\quad g^2(0)\delta_2<1$$

Once δ_2 is fixed, we take N_2 satisfying

$$N_2>\frac{8(2a^2+4\bar{g}+g^2(0)\delta_2)}{\beta_0}$$

such that

$$(2a^2+4\bar{g}+g^2(0)\delta_2)<\frac{N_2\beta_0}{8}$$

Furthermore, we choose N_5 satisfying

$$\frac{N_5g_0}{8}>N_2+\frac{a}{2}+\frac{M^2}{4\delta_2}$$

such that

$$N_2<\frac{N_5\beta_0}{8},\quad \frac{a}{2}<\frac{N_5\beta_0}{8},\quad \frac{M^2}{4\delta_2}<\frac{N_5\beta_0}{8}\quad\text{and}\quad \frac{N_5g_0}{8}-N_2-\left(\frac{a}{2}+\frac{M^2}{4\delta_2}\right)>0$$

Then, we pick δ_4 satisfying

$$\delta_4 < \frac{\beta_0 N_2}{8N_5(1+\bar{g}^2)}$$

such that

$$N_5(\delta_4 + \delta_4\bar{g}^2) < \frac{N_2\beta_0}{8}$$

Once the above constants are fixed, we choose N_1 satisfying

$$\frac{N_1}{2} > g(0)\delta_2 + \frac{N_5 g(0)L^2}{2\alpha_4}$$

Now, we need to choose suitable $|\mu_2|$ and ζ such that

$$\begin{cases} K_1 - N_1\left(\frac{|\mu_2|}{2} + \frac{\zeta}{2}\right) > 0 \\ \frac{5\beta_0 N_2}{8} - K_2|\mu_2| > 0 \\ K_3\frac{|\mu_2|}{2} - N_1\frac{\zeta}{2}e^{-\sigma\tau} < 0 \end{cases} \tag{34}$$

where

$$K_1 = \frac{N_5 g_0}{8} - N_2 - \left(\frac{a}{2} + \frac{M^2}{4\delta_2}\right), \quad K_2 = N_2\delta_1 L^2 + \delta_2\bar{g}^2 + \delta_2 M^2 a^2$$

$$K_3 = N_1 + \frac{N_2}{2\delta_1} + 2\delta_2 M^2 + \frac{1}{2\delta_2} + \frac{M^2}{2\delta_2} + 2N_5\delta_4$$

We first choose ζ satisfying

$$\frac{2K_1}{N_1} - \zeta > 0$$

Then, we pick $|\mu_2|$ satisfying

$$|\mu_2| < \min\left\{\frac{5\beta_0 N_2}{8K_2}, \frac{N_1\zeta}{K_3 e^{\sigma\tau}}, \frac{2K_1}{N_1} - \zeta\right\} := a_0 \tag{35}$$

From the above, we deduce that there exist two positive constants α_5 and α_6 such that (32) becomes

$$\frac{d}{dt}L(t) \leq -\alpha_5 E(t) + \alpha_6 \int_\Omega (g\Box u_x)\mathrm{d}x \tag{36}$$

Multiplying (36) by ξ, we have

$$\xi\frac{d}{dt}L(t) \leq -\alpha_5\xi E(t) + \alpha_6\xi \int_\Omega (g\Box u_x)\mathrm{d}x$$

On the other hand, by the definition of the functionals $\mathscr{D}(t)$, $\mathscr{F}_1(t)$, $\mathscr{F}_2(t)$, $\mathscr{F}_3(t)$ and $E(t)$, for N_1 large enough, there exists a positive constant α_3 satisfying

$$|N_2\mathscr{D}(t) + N_3\mathscr{F}_1(t) + N_4\mathscr{F}_2(t) + \mathscr{F}_3(t)| \leq \eta_1 E(t)$$

which implies that

$$(N_1 - \eta_1)E(t) \leq L(t) \leq (N_1 + \eta_1)E(t)$$

Exploiting (H2) and (8), we have

$$\xi\int_{\Omega}(g\Box u_x)\mathrm{d}x \le -\int_{\Omega}(g'\Box u_x)\mathrm{d}x \le -2\frac{d}{dt}E(t) + \frac{2K_1}{N_1}\int_{\Omega}u_t^2\mathrm{d}x \tag{37}$$

Thus, (36) becomes

$$\xi\frac{d}{dt}L(t) \le -\alpha_5\xi E(t) - 2\alpha_6\frac{d}{dt}E(t) + \frac{4K_1\alpha_6}{N_1}E(t) \tag{38}$$

We add a restriction condition on ξ, that is, we suppose that

$$\xi > \frac{4K_1\alpha_6}{N_1} := \xi_0 \tag{39}$$

Then, (38) becomes, for some positive constants

$$\xi\frac{d}{dt}L(t) \le -\alpha_7\xi E(t) - \alpha_8\frac{d}{dt}E(t)$$

Now, we define functionals $\mathscr{L}(t)$ as

$$\mathscr{L}(t) = \xi L(t) + \alpha_8 E(t)$$

It is clear that

$$\mathscr{L}(t) \sim E(t) \tag{40}$$

Then, we have

$$\frac{d}{dt}\mathscr{L}(t) \le -\alpha_7\xi E(t) \tag{41}$$

A simple integration of (41) over (t_0, t) leads to

$$\mathscr{L}(t) \le \mathscr{L}(t_0)e^{-c\xi(t-t_0)}, \quad \forall t \ge t_0 \tag{42}$$

Recalling (40), Equation (42) yields the desired result (7). This completes the proof of Theorem 3. □

4. Conclusions

The main purpose of present work is to investigate decay rate for a transmission problem with a viscoelastic term and a delay term but without the frictional damping term. It is based upon our previous work ([1]), in which we studied the well-posedness and general decay rate for a transmission problem in a bounded domain with a viscoelastic term and a delay term. The main difficulty in dealing with the problem here is that in the first equation of system (2), we have no frictional damping term to control the delay term in the estimate of the energy decay. To overcome this difficulty, our basic idea is to control the delay term by making use of the viscoelastic term. In order to achieve this target, a restriction of the size between the parameter μ_2 and the relaxation function g and a suitable energy is needed. This is motivated by Dai and Yang's work [34], in which the viscoelastic wave equation with delay term but without a frictional damping term was considered and an exponential decay result was established. In Section 2, we give some notations and hypotheses needed for our work and state the main results. In Section 3, because the energy is not non-increasing, we introduce the additional functional to produce negative term $-\int_{\Omega}u_t^2\mathrm{d}x$. Then by introducing suitable Lyaponov functionals, we prove the exponential decay of the solutions for the relaxation function satisfying assumption (H_1) and (H_2).

Acknowledgments: This work was supported by the project of the Chinese Ministry of Finance (Grant No. GYHY200906006), the National Natural Science Foundation of China (Grant No. 11301277), and the Natural Science Foundation of Jiangsu Province (Grant No. BK20151523).

Author Contributions: There was equal contribution by the authors. Danhua Wang, Gang Li, Biqing Zhu read and approved the final manuscript.

Conflicts of Interest: The authors declare no conflict of interest.

References

1. Wang, D.H.; Li, G.; Zhu, B.Q. Well-posedness and general decay of solution for a transmission problem with viscoelastic term and delay. *J. Nonlinear Sci. Appl.* **2016**, *9*, 1202–1215.
2. Alabau-Boussouira, F.; Nicaise, S.; Pignotti, C. Exponential stability of the wave equation with memory and time delay. In *New Prospects in Direct, Inverse and Control Problems for Evolution Equations*; Springer: Cham, Switzerland, 2014; Volume 10, pp. 1–22.
3. Ammari, K.; Nicaise, S.; Pignotti, C. Feedback boundary stabilization of wave equations with interior delay. *Syst. Control Lett.* **2010**, *59*, 623–628.
4. Apalara, T.A.; Messaoudi, S.A.; Mustafa, M.I. Energy decay in thermoelasticity type III with viscoelastic damping and delay term. *Electron. J. Differ. Equ.* **2012**, *2012*, 1–15.
5. Bae, J.J. Nonlinear transmission problem for wave equation with boundary condition of memory type. *Acta Appl. Math.* **2010**, *110*, 907–919.
6. Bastos, W.D.; Raposo, C.A. Transmission problem for waves with frictional damping. *Electron. J. Differ. Equ.* **2007**, *2007*, 1–10.
7. Berrimi, S.; Messaoudi, S.A. Existence and decay of solutions of a viscoelastic equation with a nonlinear source. *Nonlinear Anal.* **2006**, *64*, 2314–2331.
8. Cavalcanti, M.M.; Cavalcanti, V.N.D.; Filho, J.S.P.; Soriano, J.A. Existence and uniform decay rates for viscoelastic problems with nonlinear boundary damping. *Differ. Integral Equ.* **2001**, *14*, 85–116.
9. Cavalcanti, M.M.; Cavalcanti, V.N.D.; Soriano, J.A. Exponential decay for the solution of semilinear viscoelastic wave equations with localized damping. *Electron. J. Differ. Equ.* **2002**, *2002*, 1–14.
10. Datko, R.; Lagnese, J.; Polis, M.P. An example on the effect of time delays in boundary feedback stabilization of wave equations. *SIAM J. Control Optim.* **1986**, *24*, 152–156.
11. Guesmia, A. Well-posedness and exponential stability of an abstract evolution equation with infinite memory and time delay. *IMA J. Math. Control Inf.* **2013**, *30*, 507–526.
12. Guesmia, A. Some well-posedness and general stability results in Timoshenko systems with infinite memory and distributed time delay. *J. Math. Phys.* **2014**, *55*, 081503.
13. Guesmia, A.; Tatar, N. Some well-posedness and stability results for abstract hyperbolic equations with infinite memory and distributed time delay. *Commun. Pure Appl. Anal.* **2015**, *14*, 457–491.
14. Kirane, M.; Said-Houari, B. Existence and asymptotic stability of a viscoelastic wave equation with a delay. *Z. Angew. Math. Phys.* **2011**, *62*, 1065–1082.
15. Li, G.; Wang, D.H.; Zhu, B.Q. Well-posedness and general decay of solution for a transmission problem with past history and delay. *Electron. J. Differ. Equ.* **2016**, *2016*, 1–21.
16. Li, G.; Wang, D.H.; Zhu, B.Q. Blow-up of solutions for a viscoelastic equation with nonlinear boundary damping and interior source. *Azerbaijan J. Math.* **2016**, *7*, in press.
17. Lions, J.L. *Quelques Méthodes de Résolution des Problèmes aux Limites Non Linéaires*; Dunod: Paris, France, 1969.
18. Liu, W.J. Arbitrary rate of decay for a viscoelastic equation with acoustic boundary conditions. *Appl. Math. Lett.* **2014**, *38*, 155–161.
19. Liu, W.J.; Chen, K.W. Existence and general decay for nondissipative distributed systems with boundary frictional and memory dampings and acoustic boundary conditions. *Z. Angew. Math. Phys.* **2015**, *66*, 1595–1614.
20. Liu, W.J.; Chen, K.W. Existence and general decay for nondissipative hyperbolic differential inclusions with acoustic/memory boundary conditions. *Math. Nachr.* **2016**, *289*, 300–320.
21. Liu, W.J.; Chen, K.W.; Yu, J. Extinction and asymptotic behavior of solutions for the ω-heat equation on graphs with source and interior absorption. *J. Math. Anal. Appl.* **2016**, *435*, 112–132.

22. Liu, W.J.; Chen, K.W.; Yu, J. Existence and general decay for the full von Kármán beam with a thermo-viscoelastic damping, frictional dampings and a delay term. *IMA J. Math. Control Inf.* **2015**, in press, doi:10.1093/imamci/dnv056.
23. Liu, W.J.; Sun, Y.; Li, G. On decay and blow-up of solutions for a singular nonlocal viscoelastic problem with a nonlinear source term. *Topol. Methods Nonlinear Anal.* **2016**, *47*, in press.
24. Marzocchi, A.; Rivera, J.E.M.; Naso, M.G. Asymptotic behaviour and exponential stability for a transmission problem in thermoelasticity. *Math. Methods Appl. Sci.* **2002**, *25*, 955–980.
25. Messaoudi, S.A. General decay of solutions of a viscoelastic equation. *J. Math. Anal. Appl.* **2008**, *341*, 1457–1467.
26. Rivera, J.E.M.; Oquendo, H.P. The transmission problem of viscoelastic waves. *Acta Appl. Math.* **2000**, *62*, 1–21.
27. Nicaise, S.; Pignotti, C. Stability and instability results of the wave equation with a delay term in the boundary or internal feedbacks. *SIAM J. Control Optim.* **2006**, *45*, 1561–1585.
28. Nicaise, S.; Pignotti, C. Interior feedback stabilization of wave equations with time dependent delay. *Electron. J. Differ. Equ.* **2011**, *2011*, 1–20.
29. Raposo, C.A. The transmission problem for Timoshenko's system of memory type. *Int. J. Mod. Math.* **2008**, *3*, 271–293.
30. Tahamtani, F.; Peyravi, A. Asymptotic behavior and blow-up of solutions for a nonlinear viscoelastic wave equation with boundary dissipation. *Taiwan. J. Math.* **2013**, *17*, 1921–1943.
31. Wu, S.-T. Asymptotic behavior for a viscoelastic wave equation with a delay term. *Taiwan. J. Math.* **2013**, *17*, 765–784.
32. Wu, S.-T. General decay of solutions for a nonlinear system of viscoelastic wave equations with degenerate damping and source terms. *J. Math. Anal. Appl.* **2013**, *406*, 34–48.
33. Yang, Z. Existence and energy decay of solutions for the Euler-Bernoulli viscoelastic equation with a delay. *Z. Angew. Math. Phys.* **2015**, *66*, 727–745.
34. Dai, Q.; Yang, Z. Global existence and exponential decay of the solution for a viscoelastic wave equation with a delay. *Z. Angew. Math. Phys.* **2014**, *65*, 885–903.

8

Lie Symmetries of (1+2) Nonautonomous Evolution Equations in Financial Mathematics

Andronikos Paliathanasis [1,*], Richard M. Morris [2,†] and Peter G. L. Leach [2,3,4,†]

1 Instituto de Ciencias Físicas y Matemáticas, Universidad Austral de Chile, Valdivia 5090000, Chile
2 Department of Mathematics, Institute of Systems Science, Durban University of Technology, PO Box 1334, Durban 4000, South Africa; rmcalc85@gmail.com (R.M.M.); leach@ucy.ac.cy (P.G.L.L.)
3 School of Mathematics, Statistics and Computer Science, University of KwaZulu-Natal, Private Bag X54001, Durban 4000, South Africa
4 Department of Mathematics and Statistics, University of Cyprus, Lefkosia 1678, Cyprus
* Correspondence: anpaliat@phys.uoa.gr
† These authors contributed equally to this work.

Academic Editor: Indranil SenGupta

Abstract: We analyse two classes of $(1+2)$ evolution equations which are of special interest in Financial Mathematics, namely the Two-dimensional Black-Scholes Equation and the equation for the Two-factor Commodities Problem. Our approach is that of Lie Symmetry Analysis. We study these equations for the case in which they are autonomous and for the case in which the parameters of the equations are unspecified functions of time. For the autonomous Black-Scholes Equation we find that the symmetry is maximal and so the equation is reducible to the $(1+2)$ Classical Heat Equation. This is not the case for the nonautonomous equation for which the number of symmetries is submaximal. In the case of the two-factor equation the number of symmetries is submaximal in both autonomous and nonautonomous cases. When the solution symmetries are used to reduce each equation to a $(1+1)$ equation, the resulting equation is of maximal symmetry and so equivalent to the $(1+1)$ Classical Heat Equation.

Keywords: lie point symmetries; financial mathematics; prices of commodities; black-scholes equation

MSC: 22E60; 35Q91

---

1. Introduction

In the early 1970s, Black and Scholes [1,2] and, independently, Merton [3] introduced a mathematical model for the pricing of European options. The Black-Scholes-Merton (BS) Model is described by an $(1+1)$ evolution equation. The mathematical expression of the BS equation is

$$\frac{1}{2}\sigma^2 S^2 u_{,SS} + rSu_{,S} - ru + u_{,t} = 0 \tag{1}$$

in which t is time, S is the current value of the underlying asset, for example a stock price, r is the rate of return on a safe investment, such as government bonds and $u = u(t,S)$ is the value of the option. The solution of Equation (1) is subject to the satisfaction of the terminal condition $u(T,S) = U$, when $t = T$.

For the prices of commodities, Schwartz [4] proposed three models which study the stochastic behaviour of the prices of commodities that take into account several aspects of possible influence on the prices. In the simplest model he assumed that the logarithm of the spot price followed a mean-reversion process of Ornstein-Uhlenbeck type. This is termed the one-factor model. The one-factor model is described by the equation:

$$\frac{1}{2}\sigma^2 S^2 F_{,SS} + \kappa\left(\mu - \lambda - \log S\right) S u_{,S} - F_{,t} = 0 \tag{2}$$

where $\kappa > 0$ measures the degree of reversion to the long-run mean log price, λ is the market price of risk, μ is the drift rate of S and $F = F(t, S)$ is the current value of the futures contract. The solution of Equation (2) satisfies the initial condition $F(0, S) = S$.

The BS Equation (1) and the one-factor Equation (2) are of the same equivalence class as the Schrödinger equation and the Heat diffusion equation. All four equations model random phenomena of different contexts. The two first are in financial mathematics, the third in quantum physics and the fourth in dispersion.

It has been proven that all four equations are maximally symmetric and invariant under the same group of invariant transformations of dimension $5 + 1 + \infty$ which span the Lie algebra $\{sl(2, R) \oplus_s W_3\} \oplus_s \infty A_1$, where W_3 is a representation of the three-dimensional Weyl–Heisenberg Group, (in the Mubarakzyanov Classification Scheme [5–8] this is $\{A_{3,8} \oplus_s A_{3,1}\} \oplus_s \infty A_1$). This means that there exists a point transformation which transforms one equation to another. The Lie symmetries of the BS Equation (1) have been found in [9], whereas the Lie symmetries of the one-factor model (2) were found in [10].

The parameters of the models (1) and (2) are generally assumed to be constant. However, in real problems they may vary with time if the time-span of the model is sufficiently long. In [11] it has been shown that, when the parameters σ, and r of the BS equation are time-dependent, *i.e.*, $\sigma = \sigma(t)$ and $r = r(t)$, the time-dependent BS equation is invariant under the same group of invariant transformations as that of the "static" BS equation. The same result has been found for the time-dependent one-factor model of commodities [12]. Hence the autonomous and the nonautonomous Equations (1) and (2) are maximally symmetric and equivalent under point transformations.

In Classical Mechanics the slowly lengthening pendulum with equation of motion in the linear approximation,

$$\ddot{x} + \omega^2(t)\, x = 0 \tag{3}$$

in which the time dependence in the "spring constant" is due to the length of the pendulum's string increasing slowly [13], admits the conservation law [14,15] (note that the case of a slowly shortening pendulum is quite different [16]),

$$I = \frac{1}{2}\left\{ \left(\rho\dot{x} - \dot{\rho}x\right) + \left(\frac{x}{\rho}\right)^2 \right\} \tag{4}$$

where $\rho = \rho(t)$, is a solution of the second-order differential equation,

$$\ddot{\rho} + \omega^2(t)\,\rho = \frac{1}{\rho^3} \tag{5}$$

This result is independent of the rate of change of the length of the pendulum.

The latter equation is the well-known Ermakov-Pinney equation [17]. The solution was given by Pinney in [18] and it is:

$$\rho(t) = \sqrt{Av_1^2 + 2Bv_1v_2 + Cv_2^2} \tag{6}$$

subject to a constraint on the three constants, A, B and C. Functions $v_1(t)$, $v_2(t)$, are two linearly independent solutions of Equation (3).

Equation (3) is invariant under the action of the group invariant transformations in which the generators of the infinitesimal transformations form the $sl(3, R)$ algebra. This is the Lie algebra admitted by the harmonic oscillator, $\omega(t) = \omega_0$, and the equation of the free particle, $\omega(t) = 0$ [19–21]. The transformation which connects the nonautonomous linear Equation (3) with the autonomous oscillator is a time-dependent linear canonical transformation of the form:

$$Q = \frac{x}{\dot{x}}\ ,\ P = \rho\dot{x} - \dot{\rho}x\ ,\ T = \int^{t} \rho^{-2}(\eta)\, d\eta \tag{7}$$

where ρ is given by Equation (6).

The connection of the number of symmetries of the corresponding Schrödinger Equation with the Noether point symmetries of the classical Lagrangian [22,23] was seen to extend to the time-dependent case [24] and, indeed, be seen to be the same as the equivalent autonomous systems [25] and in the case of maximal symmetry is $\{sl(2,R) \oplus_s W_3\} \oplus_s \infty A_1$ which is that of the $(1+1)$ classical heat equation.

In this context we wish to see what happens when we pass from an autonomous $(1+2)$ evolution equation to the corresponding nonautonomous case. For that we study the Lie symmetries of the nonautonomous models of: (a) the two-factor model of commodities and (b) the two-dimensional BS equation.

We find that, for the two-factor model, the autonomous and the nonautonomous equations are invariant under the same group of invariant transformations $\{A_1 \oplus_s W_5\} \oplus_s \infty A_1$. However, that it is not true for the two-dimensional BS equation. The reason for that is that the Lie symmetries of the two-factor model follow from the translation group of the two-dimensional Euclidian space (except the homogeneous and the infinite number of solution symmetries). The translation group generates Lie symmetries for both the autonomous system and for the nonautonomous system.

On the other hand the autonomous two-dimensional BS equation is maximally symmetric, *i.e.*, it admits nine Lie symmetries plus the infinite number of solution symmetries, which form the $\{\{sl(2,R) \oplus_s so(2)\} \oplus_s W_5\} \oplus_s \infty A_1$ Lie algebra. This result completes the analysis of [26] in which they found that the two-dimensional BS equation admits seven Lie point symmetries plus the ∞A_1.

The nonautonomous two-dimensional BS equation is invariant under the Lie algebra $\{\{A_1 \oplus_s so(2)\} \oplus_s W_5\} \oplus_s \infty A_1$, that is, the $sl(2,R)$ subalgebra is lost. The reason for that is that the Lie symmetries of the autonomous two-dimensional BS equation arise from the homothetic algebra of the two-dimensional Euclidian space which defines the Laplace operator of the evolution equation and, when the parameters in the second derivatives are not constants, the homothetic algebra of the Euclidian space does not generate Lie symmetries. Moreover, in the case for which the parameters of the second derivatives are time-indepedent, the two-dimensional BS equation is maximally symmetric, *i.e.*, it is invariant under the same group of point transformations as the (1 + 2) autonomous BS and Heat conduction equations.

The plan of the paper is as follows. In Section 2 we study the Lie symmetries of the two-factor model of commodities for the autonomous and nonautonomous cases. We show that in both cases the two-factor model is invariant under the $\{A_1 \oplus_s W_5\} \oplus_s \infty A_1$ Lie algebra. The Lie symmetries of the two-dimensional BS equation, the autonomous and the nonautonomous, are studied in Section 3. Finally in Section 4 we give some applications and we draw our conclusions.

2. The Two-Factor Model of Commodities

The two-factor model adds to the spot price, S, of Equation (2) the instantaneous convenience yield, δ, which may be interpreted as the flow of services accruing to the holder of the spot commodity but not to the owner of a futures contract. The evolution partial differential equation for this model is

$$\frac{1}{2}\sigma_1^2 S^2 F_{,SS} + \rho\sigma_1\sigma_2 F_{,S\delta} + \frac{1}{2}\sigma_2^2 F_{,\delta\delta} + (r-\delta)\, SF_{,S} + (\kappa(\alpha-\delta)-\lambda)\, F_{,\delta} - F_{,t} = 0 \tag{8}$$

for which the terminal condition is now $F(0,S,\delta) = S$.

Equation (8) is an $(1+2)$ evolution equation and under the coordinate transformation

$$S = \exp(\sigma_1 x) \ , \ \delta = \sigma_2 \left(\rho x + \sqrt{1-\rho^2 y} \right) \tag{9}$$

becomes

$$F_{,xx} + F_{,yy} - (p_1 x + p_2 y + p_3) F_{,x} - (q_1 x + q_2 y + q_3) F_{,y} - 2F_{,t} = 0 \tag{10}$$

in which the new parameters are expressed on the terms of the old ones according to

$$p_1 = 2\rho \frac{\sigma_2}{\sigma_1} \ , \ p_2 = 2\sqrt{1-\rho^2} \frac{\sigma_2}{\sigma_1} \ , \ p_3 = -2r \tag{11}$$

$$q_1 = \frac{\kappa \sigma_1 - \rho \sigma_2}{\sigma_1 \sqrt{1-\rho^2}}, \ q_2 = \frac{\kappa \sigma_1 - \rho \sigma_2}{\sigma_1} \tag{12}$$

and

$$q_3 = -\frac{\left(\sigma_1^2 \sigma_2 \rho - 2\sigma_2 \rho r + 2\sigma_1 \kappa \alpha - 2\sigma_1 \lambda\right)}{\sigma_1 \sigma_2 \sqrt{1-\rho^2}} \tag{13}$$

The Lie symmetries for the autonomous two-factor model (8) have been reported in [10]. However, for the convenience of the reader we present the results.

2.1. Lie Symmetries of the Autonomous Equation

Consider the infinitesimal one-parameter point transformation

$$t' = t + \varepsilon \xi^1 (t, x, y, F) \ , \ x' = x + \varepsilon \xi^2 (t, x, y, F) \tag{14}$$

$$y' = y + \varepsilon \xi^3 (t, x, y, F) \ , \ F' = y + \varepsilon \eta (t, x, y, F) \tag{15}$$

where ε is an infinitesimal number so that $\varepsilon^2 \to 0$. From the transformation we define the generator X, as

$$X = \frac{\partial t'}{\partial \varepsilon} \partial_t + \frac{\partial x'}{\partial \varepsilon} \partial_x + \frac{\partial y'}{\partial \varepsilon} \partial_y + \frac{\partial F'}{\partial \varepsilon} \partial_F \tag{16}$$

or, equivalently,

$$X = \xi^1 (t, x, y, F) \partial_t + \xi^2 (t, x, y, F) \partial_x + \xi^3 (t, x, y, F) \partial_y + \eta (t, x, y, F) \partial_F \tag{17}$$

The differential equation, Θ, Equation (10), is invariant under the action of the one-parameter point transformation Equations (14) and (15) if there exists a function Λ such that [27,28]

$$X^{[2]} \Theta = \Lambda \Theta \tag{18}$$

in which $X^{[2]}$ is the second prologation of X defined in the space $\{t, x, y, F, F_{,x}, F_{,y}, F_{,xx}, F_{,yy}, F_{,xy}\}$. When condition (18) holds, we say that X is a Lie (point) symmetry of Θ.

Therefore from Equation (18) we have the following Lie symmetries admitted by Equation (10)

$$X_t = \partial_t \ , \ X_F = F \partial_F \ , \ X_\infty = f(t, x, y) \partial_f \tag{19}$$

$$X_1 = e^{c_+ t} (a_1 \partial_x + a_2 \partial_y) \tag{20}$$

$$X_2 = e^{c_- t} (a_1' \partial_x + a_2' \partial_y) \tag{21}$$

$$X_3 = e^{c_+ t} (b_1 \partial_x + b_2 \partial_y + (b_3 x + b_4 x + b_5) F \partial_F) \tag{22}$$

and

$$X_4 = e^{c_- t} (b_1' \partial_x + b_2' \partial_y + (b_3' x + b_4' x + b_5') F \partial_F) \tag{23}$$

The parameters $a_{1,2}$, $a'_{1,2}$, b_{1-5}, b'_{1-5} and $c_{\pm}$ are functions of p_{1-3} and q_{1-3}. The Lie symmetries form the $\{A_1 \oplus_s W_5\} \oplus_s \infty A_1$ Lie algebra. We note that for special cases of the parameters p_{1-3}, q_{1-3}, the representation of the admitted Lie symmetries of Equation (10) can be different. For instance, when all the parameters q_{1-3} vanish, $q_{1-3} = 0$, the Lie symmetries X_{1-4} become

$$X'_1 = p_2\partial_x - p_1\partial_y \;,\; X'_2 = e^{\frac{p_1}{2}t}\partial_x \tag{24}$$

$$X'_3 = (p_1p_2t + 2p_2)\,\partial_x - tp_1^2\partial_y + p_1^2yF\partial_F \tag{25}$$

and

$$X'_4 = e^{-\frac{p_1}{2}t}\left(\left(p_1^2 - p_2^2\right)\partial_x + 2p_1p_2\partial_y + p_1^2\,(p_1x + p_2y + p_3)\,F\partial_F\right) \tag{26}$$

For the remaining cases see [10].

Below, the nonautonomous two-factor model is defined and the group invariant point transformations are derived.

2.2. Lie Symmetries of the Nonautonomous Equation

We consider that the parameters σ_I, ρ, r, κ, α and λ of Equation (8) are well-defined functions of time. Without loss of generality we can select a new time variable τ and eliminate, for instance, the function $\sigma_1\,(t)$. Therefore we select $\sigma_1 = 1$.

Under the time-depedent coordinate transformation, Equation (9), the two-factor model (8) has the following mathematical expression

$$F_{,xx} + F_{,yy} - (P_1\,(t)\,x + P_2\,(t)\,y + P_3\,(t))\,F_{,x} - (Q_1\,(t)\,x + Q_2\,(t)\,y + Q_3\,(t))\,F_{,y} - 2F_{,t} = 0 \tag{27}$$

where now the new time-depedent parameters of the model are

$$P_1\,(t) = 2\rho\sigma \;,\; P_2\,(t) = 2\sigma_2\sqrt{1-\rho^2} \;,\; P_3\,(t) = 1 - 2r\,(t) \tag{28}$$

$$Q_1\,(t) = -\frac{2\,(\rho\sigma_2)^2 + (\rho\sigma_2)_{,t} + \rho\sigma_2\kappa}{\sigma_2\sqrt{1-\rho^2}} \tag{29}$$

$$Q_2\,(t) = -\left(2\rho\sigma_2 + \kappa + 2\frac{\sigma_{2,t}}{\sigma_2}\right) + \frac{2\rho_2\rho_{2,t}}{\sqrt{1-\rho^2}} \tag{30}$$

and

$$Q_3\,(t) = -\left(\frac{\sigma_2\,(\rho - 2r\rho) - 2\kappa\alpha + 2\lambda}{\sigma_2\sqrt{1-\rho^2}}\right) \tag{31}$$

Therefore, from the symmetry condition (18) for Equation (27), we find that the generic Lie symmetry vector is

$$
\begin{aligned}
X_G \quad = \quad & a\partial_t + \left(b_1 + y\left(B_2 + \frac{1}{4}aP_2 - \frac{1}{4}aQ_1\right) + \frac{xa'}{2}\right)\partial_x + \\
& \left(g - x\left(B_2 + \frac{1}{4}aP_2 - \frac{1}{4}aQ_1\right) + \frac{ya'}{2}\right)\partial_y + \\
& \frac{1}{4}\left[4h + 2xb_1P_1 + 2xgP_2 + x^2(-P_2 - Q_1)\left(B_2 + \frac{1}{4}aP_2 - \frac{1}{4}aQ_1\right)\right]F\partial_F + \\
& \frac{1}{4}\left[2x\left(B_2 + \frac{1}{4}aP_2 - \frac{1}{4}aQ_1\right)(yP_1 - yQ_2 - Q_3) + x^2P_1a' + 2xyP_2a'\right]F\partial_F + \\
& \frac{1}{4}\left[xP_3a' - 4xb_1'^2aP_1' + 2xyaP_2' + 2xaP_3'\right]F\partial_F + \\
& \frac{1}{4}\left[-4xy\left(\frac{1}{4}P_2a' - \frac{1}{4}Q_1a' + \frac{1}{4}aP_2' - \frac{1}{4}aQ_1'\right) - x^2a''\right]F\partial_F + \\
& \frac{1}{4}\left[2yb_1Q_1 + 2yP_3\left(B_2 + \frac{1}{4}aP_2 - \frac{1}{4}aQ_1\right) + y^2(P_2 + Q_1)\left(B_2 + \frac{1}{4}aP_2 - \frac{1}{4}aQ_1\right)\right]F\partial_F + \\
& \frac{1}{4}\left[2ygQ_2 + y^2Q_2a' + yQ_3a' - 4yg'^2aQ_2' + 2yaQ_3'^2a''\right]F\partial_F
\end{aligned} \tag{32}
$$

where B_2 is constant, $a = a(t)$, $b_1 = b_1(t)$, $f = f(t)$ and $g = g(t)$, given by the system of equations of Appendix A. Furthermore, from the generic vector field (32) and the system of Appendix A, we know that the nonautonomous two-factor model of commodities is invariant under the $\{A_1 \oplus_s W_5\} \oplus_s \infty A_1$ Lie algebra, the same algebra as the autonomous model but in a different representation.

We continue our analysis with the two-dimensional Black-Scholes equation.

3. The Two-Dimensional Black-Scholes Equation

Consider a basket containing two assets the prices of which are S_1 and S_2 and that the the prices of the underlying assets obey the system of stochastic differential equations,

$$
dS_{I,t} = S_{I,t}\left(\mu_I dt + \frac{\sigma_I}{\sqrt{1+\rho^2}}\left(dW_{I,t} + \rho dW_{J,t}\right)\right) \tag{33}
$$

where $I, J = 1, 2$, $I \neq J$, and $W_{I,t}$ are two independent standard Brownian motions. Let $u = u(t, S_1, S_2)$ be the payoff function on a European option on this two-asset basket. Then the evolution equation which u satisfies is an $(1+2)$ linear evolution equation given by [29]

$$
\frac{1}{2}\sigma_1^2 u_{,11} + \rho\sigma_1\sigma_2 u_{,12} + \frac{1}{2}\sigma_2^2 u_{,22} - rS_1u_{,1} - rS_2u_{,2} - ru + u_{,t} = 0 \tag{34}
$$

with the terminal condition $u(T, S_1, S_2) = U$, when $t = T$.

Equation (34) is a generalisation of the BS equation and it is called the two-dimensional BS equation. The Lie symmetry analysis of Equation (1) has been presented in [9] and recently a Lie symmetry analysis for Equation (1), with a general potential function, was performed in [30]. The algebraic properties of the autonomous form of Equation (34) have been studied in [26] and it was found that Equation (34) is invariant under a seven-dimensional Lie algebra, plus the infinite number of solution symmetries. As we see below, the analysis of the autonomous Equation (34) in [26] is not complete. In particular we find that it is maximally symmetric, *i.e.*, invariant under a nine-dimensional Lie algebra, plus the infinite number of solution symmetries. In [26] the authors considered the following equation

$$
\frac{1}{2}\sigma_1^2 u_{,11} + \rho\sigma_1\sigma_2 u_{,12} + \frac{1}{2}\sigma_2^2 u_{,22} - \mu_1S_1u_{,1} - \mu_2S_2u_{,2} - ku + u_{,t} = 0 \tag{35}
$$

which reduces to Equation (34) when $\mu_1 = \mu_2 = k = r$.

Below we determine the Lie symmetries of Equation (35) for the autonomous and nonautonomous system.

3.1. Lie Symmetries of the Autonomous Equation

We introduce the coordinate transformation

$$S_1 = \exp(\sigma_1 x) \; , \; S_2 = \exp\left(\sigma_2 \rho x + \sigma_2 \sqrt{1-\rho^2 y}\right) \tag{36}$$

under which Equation (35) becomes

$$u_{,xx} + u_{,yy} - \phi_1 u_{,x} - \phi_2 u_{,y} - 2ku + 2u_{,t} = 0 \tag{37}$$

where now the new constants, ϕ_1 and ϕ_2, are

$$\phi_I = \frac{\sigma_1^2 + 2\mu_I}{\sigma_I} \tag{38}$$

On application of the Lie symmetry condition (18) for (37) we find that the Lie symmetry vectors are

$$X_t = \partial_t \; , \; X_u = F\partial_u \; , \; X_\infty = f(t,x,y)\,\partial_u \tag{39}$$

$$X_1 = \partial_x \; , \; X_2 = t\partial_x + \frac{1}{2}x'x\,(x + \phi_1 t)\,u\partial_u$$

$$X_3 = \partial_y \; , \; X_4 = t\partial_y + \frac{1}{2}(y + \phi_2 t)\,u\partial_u \tag{40}$$

$$X_5 = y\partial_x - x\partial_y + \frac{1}{2}(\phi_1 y - \phi_2 x)\,u\partial_u \tag{41}$$

$$X_6 = 2t\partial_t + x\partial_x + y\partial_y + \frac{1}{2}\left(\phi_1 x + \phi_2 y + t\left(\phi_1^2 + \phi_2^2 + 8k\right)\right)u\partial_u \tag{42}$$

and

$$X_7 = t^2\partial_t + tx\partial_x + ty\partial_y + \frac{1}{4}\left(x^2 + y^2 + t^2\left(\phi_1^2 + \phi_2^2 + 8k\right) + 2t\,(\phi_1 x + \phi_2 y - 2)\right)u\partial_u \tag{43}$$

which are $8 + 1 + \infty$ symmetries. This is the admitted group invariant algebra of the two-dimensional Heat Equation, that is, $\{\{sl(2,R) \oplus_s so(2)\} \oplus_s W_5\} \oplus_s \infty A_1$. Hence the two-dimensional BS Equation (35) is maximally symmetric and equivalent with the two-dimensional Heat and Schrödinger equations [31]. This result does not hold for the two-factor model of commodities. An analysis does hold when in Equation (35), $\mu_1 = \mu_2 = k = r$; that is, for Equation (34).

When we apply the transformations

$$t = -\frac{1}{2}T \; , \; x = \bar{x} - \frac{1}{2}\phi_1 t \tag{44}$$

and

$$\bar{y} = y - \frac{1}{2}\phi_2 t \; , \; u = e^{2kt} v(t,x,y) \tag{45}$$

to Equation (37), the equation becomes

$$v_{,\bar{x}\bar{x}} + v_{,\bar{y}\bar{y}} - v_{,t} = 0 \tag{46}$$

which is the two-dimensional Heat conduction equation.

We proceed to the determination of the Lie symmetries for the nonautonomous Equation (35).

3.2. Lie Symmetries of the Nonautonomous Equation

We take the parameters, σ_I, ρ, μ_I and k, of Equation (35) to be well-defined functions of time. Moreover without loss of generality we select $\sigma_1(t) = 1$.

We apply the time-dependent transformation Equation (36) to Equation (35) and we have

$$u_{,xx} + u_{,yy} - P_1(t)\,u_{,x} - (Q_1(t)\,x + Q_2(t)\,y + Q_3(t))\,u_{,y} - 2k(t)\,u + 2u_{,t} = 0 \tag{47}$$

in which

$$P_1(t) = 1 + 2\mu_1(t)\ ,\ Q_1(t) = \frac{2(\rho\sigma_2)_{,t}}{\sigma_2\sqrt{1-\rho^2}} \tag{48}$$

$$Q_2(t) = -\frac{2(\sigma_{2,t}\rho^2 + \sigma_2\rho\rho_{,t} - \sigma_{2,t})}{\sigma_2(1-\rho^2)} \tag{49}$$

and

$$Q_3(t) = \frac{\sigma_2(\sigma_2 - \rho - 2\mu_2\rho) + 2\mu_2}{\sigma_2\sqrt{1-\rho^2}} \tag{50}$$

From the symmetry condition (18) for Equation (47) we find that the generic Lie symmetry vector has the following mathematical expression

$$\begin{aligned} X_G \;=\; & a\partial_t + \left(b_1 + y\left(B_2 + \frac{1}{4}aQ_1\right) + \frac{xa'}{2}\right)\partial_x + \left(f - x\left(B_2 + \frac{1}{4}aQ_1\right) + \frac{ya'}{2}\right)\partial_y + \\ & \frac{1}{4}\left[4g + \left(-x^2Q_1\left(B_2 + \frac{1}{4}aQ_1\right) - 2x\left(B_2 + \frac{1}{4}aQ_1\right)(yQ_2 + Q_3)\right)\right]u\partial_u + \\ & \frac{1}{4}\left[xP_1a' + 4xb_1' + 2xaP_1' + x^2a'' + 4xy\left(\frac{1}{4}Q_1a' + \frac{1}{4}aQ_1'\right)\right]u\partial_u + \\ & \frac{1}{4}\left[+2yb_1Q_1 + 2yP_1\left(B_2 + \frac{1}{4}aQ_1\right) + y^2Q_1\left(B_2 + \frac{1}{4}aQ_1\right)\right]u\partial_u + \\ & \frac{1}{4}\left[2yfQ_2 + y^2Q_2a' + yQ_3a' + 4yf'^2aQ_2' + 2yaQ_3'^2a''\right]u\partial_u \end{aligned} \tag{51}$$

where B_2 is a constant, $a = a(t)$, $b_1 = b_1(t)$, $f = f(t)$ and $g = g(t)$ which given by the system of differential equations of Appendix B. Furthermore, from Equation (51) and the system of Appendix B, we observe that the nonautonomous Equation (34) is invariant under the group of transformations in which the generators form the $\{\{A_1 \oplus_s so(2)\} \oplus_s W_5\} \oplus_s \infty A_1$ Lie algebra. Below we consider a special case for which $\sigma_1(t) \simeq \sigma_2(t)$ and $\rho = const$.

Special Case: $\rho = const$ and $\sigma_1(t) \simeq \sigma_2(t)$

As a special case of the nonautonomous Equation (35) we consider $\sigma_2(t) = \sigma_0\sigma_1(t)$, where σ_0 is a constant and $\rho(t)$ is a constant. The nonautonomous two-dimensional BS equation becomes

$$\sigma_1^2(t)\left(\frac{1}{2}u_{,11} + \rho\sigma_0 u_{,12} + \frac{1}{2}\sigma_0^2 u_{,22}\right) - \mu_1(t)\,S_1u_{,1} - \mu_2(t)\,S_2u_{,2} - k(t)\,u + u_{,t} = 0 \tag{52}$$

where without loss of generality we can select $\sigma_1(t) = 1$. Under the transformation Equations (36) and (52) becomes

$$u_{,xx} + u_{,yy} - \Lambda_1(t)\,u_{,x} - \Lambda_2(t)\,u_{,y} - 2k(t)\,u + 2u_{,t} = 0 \tag{53}$$

where the new functions $\Lambda_1(t)$, $\Lambda_2(t)$ are defined as

$$\Lambda_1(t) = 1 + 2\mu_1(t) \tag{54}$$

and

$$\Lambda_2(t) = \frac{\sigma_0(\sigma_0 - \rho - 2\mu_2(t)\rho) + 2\mu_2(t)}{\sigma_0\sqrt{1-\rho^2}} \tag{55}$$

From the symmetry condition (18) for Equation (47) the following symmetry vectors arise

$$X_u = u\partial_u \ , \ X_\infty = f(t,x,y)\partial_F \tag{56}$$

$$Z_1 = \partial_x \ , \ Z_2 = t\partial_x + \left(\frac{1}{2}\int \Lambda_1 dt + x\right)u\partial_u \tag{57}$$

$$Z_3 = \partial_y \ , \ Z_4 = t\partial_y + \left(\frac{1}{2}\int \Lambda_2 dt + y\right)u\partial_u \tag{58}$$

$$Z_5 = \left(y + \frac{1}{2}\int \Lambda_2 dt\right)\partial_x - \left(x + \frac{1}{2}\int \Lambda_1 dt\right)\partial_y + \frac{1}{2}\left(\Lambda_1 y - \frac{1}{2}\Lambda_2 x\right)u\partial_u \tag{59}$$

$$Z_6 = \partial_t - \frac{1}{2}\Lambda_1\partial_x - \frac{1}{2}\Lambda_2\partial_y + ku\partial_u \tag{60}$$

$$Z_7 = 2t\partial_t + \left(x - \frac{1}{2}\int \Lambda_1 dt - \int t\Lambda_1 dt\right)\partial_x + \left(y - \frac{1}{2}\int \Lambda_2 dt - \int t\Lambda_2 dt\right)\partial_y + tku\partial_u \tag{61}$$

and

$$\begin{aligned} Z_8 \ = \ & t^2\partial_t + \left(tx - \frac{1}{2}\int\int\left(t^2\Lambda_{1,tt} + 3t\Lambda_{1,t}\right)dt\right)\partial_x + \left(ty - \frac{1}{2}\int\int t^2\Lambda_{2,tt} + 3t\Lambda_{2,t}\right)\partial_y + \\ & \left[-\frac{1}{2}x\left(\int t^2\Lambda_{1,tt}dt + 3\int t\Lambda_{1,t}dt - t^2\Lambda_{1,t} - t\Lambda_1 - x\right)\right]u\partial_u + \\ & \left[-\frac{1}{2}y\left(\int t^2\Lambda_{2,tt}dt + 3\int t\Lambda_{2,t}dt - t^2\Lambda_{2,t} - t\Lambda_2 - y\right)\right]u\partial_u + \\ & \frac{1}{4}\left[4t(t-1) - \int \Lambda_1\left(\int t^2\Lambda_{1,tt}dt\right)dt - \int \Lambda_2\left(\int t^2\Lambda_{2,tt}dt\right)dt\right]u\partial_u + \\ & \frac{1}{4}\left[-3\int \Lambda_1\int t\Lambda_{1,t}dt - 3\int \Lambda_2\int t\Lambda_{2,t}dt\right]u\partial_u \\ & \frac{1}{4}\left[\int t^2\Lambda_1\Lambda_{1,t} + \int t^2\Lambda_2\Lambda_{2,t} + \int t\left(\Lambda_1^2 + \Lambda_2^2\right)dt\right]u\partial_u \end{aligned} \tag{62}$$

Hence the nonautonomous Equation (52) is maximally symmetric, just as the autonomous two-dimensional BS equation, in contrast to the nonautonomous Equation (47) which is invariant under another group of point transformations.

Moreover Equation (53) can be written in the form of Equation (46) and the transformation which does that is

$$t = -\frac{1}{2}T \ , \ u = e^{2kt}v(t,x,y) \tag{63}$$

and

$$x = \bar{x} - \frac{1}{2}\int \Lambda_1 dt \ , \ y = \bar{y} - \frac{1}{2}\int \Lambda_2 dt \tag{64}$$

Below we discuss our results and draw our conclusions.

4. Conclusions

The purpose of this work is to study the algebraic properties of nonautonomous $(1+2)$ evolution equations in financial mathematics. Specifically, we examined the relation among the admitted group of invariant transformations between the autonomous and the nonautonomous equations of the two-factor model of commodities and of the two-dimensional BS equation was performed.

For the two-factor model of commodities we proved that the autonomous and the nonautonomous equations are invariant under the same group of point transformations in which the generators form the $\{A_1 \oplus_s W_5\} \oplus_s \infty A_1$ Lie algebra.

As far as the autonomous two-dimensional BS equation is concerned, we proved that it is maximally symmetric and admits as Lie symmetries the generators of the Lie algebra $\{\{sl(2,R) \oplus_s so(2)\} \oplus_s W_5\} \oplus_s \infty A_1$ This corrects the existing result in the literature. However, the admitted Lie symmetries of the nonautonomous two-dimensional BS equation form a different Lie algebra than that of the autonomous equation and is of lower dimension. Specifically the admitted Lie algebra is $\{\{A_1 \oplus_s so(2)\} \oplus_s W_5\} \oplus_s \infty A_1$. That result differs from that for the model of commodities for which the autonomous and the nonautonomous equations are invariant under the same group of transformations, namely $\{A_1 \oplus_s W_5\} \oplus_s \infty A_1$.

In the case for which $\rho = const$ and $\sigma_1(t) \simeq \sigma_2(t)$, the two-dimensional BS equation is maximally symmetric. In order to understand why we have this special case consider the general $(1+n)$ evolution equation (We use the Einstein summation convention).

$$A^{ij}\left(t,x^k\right)u_{ij} + B^i\left(t,x^k\right)u_{,i} + f\left(t,x^k,u\right) = u_{,t} \tag{65}$$

If $X = \xi^t \partial_t + \xi^i \partial_i + \eta \partial_u$ is the generator of a Lie symmetry vector, one of the symmetry conditions can be written as

$$\mathcal{L}_{\xi^\alpha} A^{ij} = -2\psi A^{ij} \tag{66}$$

where ψ is a function of t only, and $\alpha = 1,2,...,n,t$. Therefore from Equation (66) we know that

$$\mathcal{L}_{\xi^i} A^{ij} = -2\psi A^{ij} - A^{ij}_{,t}\xi^t \tag{67}$$

From Equation (67) we know that, when $A^{ij}_{,t} = 0$, the Lie symmetries of Equation (65) are generated by the Homothetic Algebra of A_{ij}. However, that is not true when $A^{ij}_{,t} \neq 0$ and new possible generators arise. In the $(1+1)$ equations, *i.e.*, Equations (1) and (2), when $\sigma = \sigma(t)$, as we discussed above, we can always perform a time (coordinate) transformation and cause the second derivatives to be time-independent. Therefore, in order to apply this method to the two-dimensional systems, we have to select $\rho = const$ and $\sigma_1(t) \simeq \sigma_2(t)$ so that at the end the components of the second derivatives can be seen as time-independent.

Furthermore, we remark that we performed a reduction on the two nonautonomous Equations (8) and (34) by using the Lie symmetries (32) and (51), respectively, for $a(t) = 0$. We found that the reduced equations, which are $(1+1)$ evolution equations, are maximally symmetric. This is the same result as is to be found in the case of the autonomous two-factor model [10].

As a final application consider the nonautonomous two-dimensional BS Equation (53). From the application of the invariant functions of the Lie symmetries $\{Z_1 + c_1 X_u,\ Z_3 + c_2 X_u\}$ we have the solution $u(t,x,y) = w(t)\exp(c_1 x + c_2 y)$, where

$$w(t) = \exp\left(\frac{1}{2}\int\left(2k(t) - \left(c_1^2 + c_2^2\right) + \Lambda_1(t)c_1 + \Lambda_2(t)c_2\right)dt\right) \tag{68}$$

In the case for which $\mu_1(t) = \mu_2(t) = k(t) = r(t)$ and $r(t) = r_0 + \varepsilon\sin(\omega t)$, ω ,ε and r_0 are constants, the solution of the nonautonomous two-dimensional BS equation for the "$t-x$" plane is given in Figure 1. We observe that in the $t-$direction, function $u(t,x,y)$ has periodic behavior along the line $f(t) \simeq t$ with period ω.

The implication of the results of the present analysis is that for the two-factor model of commodities, the autonomous and the nonautonomous problem share the same static solutions, that is, the differences follow only from the time-dependent terms of the solution. However, that is not true for the two-dimensional Black-Scholes Equation in which the nonautonomous equation in general is not maximally symmetric and does not share the same number of static solutions with that of the autonomous equation. On the other hand we found that if and only if the time-dependence of the two volatilities $\sigma_1(t)$, $\sigma_2(t)$are the same, *i.e.*, $\frac{\sigma_1(t)}{\sigma_2(t)} = const$, and if the correlation factor ρ is constant then the nonautonomous Black-Scholes shares the same static solutions, *i.e.*, static evolution, with the autonomous equation.

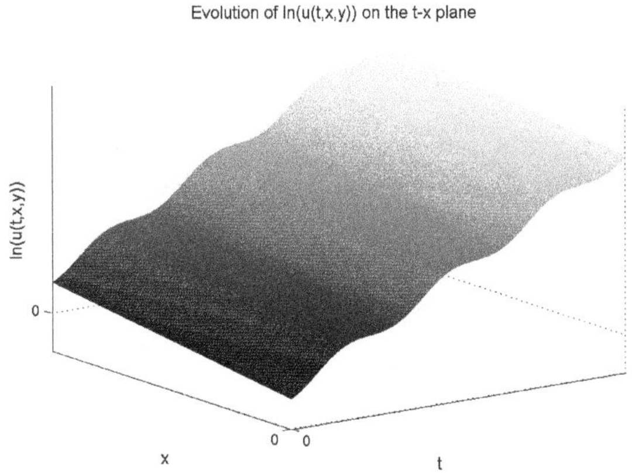

Figure 1. Qualitative evolution of the solution $u(t,x,y)$ for the nonautonomous two-dimensional Black-Scholes-Merton Equation (34) in the "t-x" plane, when σ_1, σ_2, ρ are constants and $r(t) = r_0 + \varepsilon \sin(\omega t)$.

The results of this analysis are important in the sense that by starting from the autonomous equation and with the use of coordinate transformations and only someone can analyse models with time-varying constants. On the other hand starting from real data and with the use of coordinate transformations to see if the data are well described from the autonomous system, and vice verca. The situation is not different from that which one finds on the relation between the free particle and harmonic oscillator. In order to demonstrate that, if we plot the time-position diagram of the mathematical pendulum, where we measure the distance and the time with nonlinear instruments, the graph will be a straight line, which describes the motion of the free particle.

In a forthcoming work we intend to extend our analysis to the case where the free parameters of the models are space-dependent. Such an analysis is in progress and will be published elsewhere.

Acknowledgments: The research of Andronikos Paliathanasis was supported by Fondo Nacional de Desarollo Científico y Tecnolǿgico (FONDECYT) grant No. 3160121. Richard M. Morris thanks the National Research Foundation of the Republic of South Africa for the granting of a postdoctoral fellowship with grant number 93183 while this work was being undertaken.

Author Contributions: Andronikos Paliathanasis and Peter G.L. Leach determined the problem and the method for the solution; Richard M. Morris and Peter G.L. Leach did the calculations. Andronikos Paliathanasis and Peter G.L. Leach wrote the paper and they gave interpretation to the results.

Conflicts of Interest: The authors declare no conflict of interest.

Appendix

Appendix A. Nonautonomous Two-Factor Model of Commodities

In this Appendix we give the differential equations which the functions $a(t)$, $b_1(t)$, $h(t)$ and $g(t)$ of the generic symmetry vector Equation (32) of the nonautonomous two-factor model of commodities satisfy. For the derivation of the system the symbolic package SYM of Mathematica has been used [32–34].

The system is:

$$\begin{aligned} 0 = & -\frac{1}{2}b_1P_1P_3 - \frac{1}{2}gP_2P_3 - \frac{1}{2}b_1Q_1Q_3 - \frac{1}{2}gQ_2Q_3 + \\ & \frac{1}{2}P_1a' - \frac{1}{4}P_3^2a' + \frac{1}{2}Q_2a' - \frac{1}{4}Q_3^2a' + P_3b_1' + \\ & Q_3g' - 2h' + \frac{1}{2}aP_1' - \frac{1}{2}aP_3P_3' + \frac{1}{2}aQ_2' - \frac{1}{2}aQ_3Q_3' - a'' \end{aligned} \tag{A1}$$

$$\begin{aligned} 0 = & -\frac{1}{2}b_1P_1^2 - \frac{1}{2}gP_1P_2 + \frac{1}{2}B_2P_2P_3 + \frac{1}{8}aP_2^2P_3 - \frac{1}{8}aP_2P_3Q_1 \\ & -\frac{1}{2}b_1Q_1^2 - \frac{1}{2}gQ_1Q_2 + \frac{1}{2}B_2Q_2Q_3 + \frac{1}{8}aP_2Q_2Q_3 - \frac{1}{8}aQ_1Q_2Q_3 \\ & -\frac{3}{4}P_1P_3a' - \frac{3}{4}Q_1Q_3a' - P_2g' + Q_1g' - b_1P_1' - \frac{1}{2}aP_3P_1' \\ & -gP_2' - \frac{1}{2}aP_1P_3' - \frac{3a'P_3'}{2} - \frac{1}{2}aQ_3Q_1' \\ & +B_2Q_3' + \frac{1}{4}aP_2Q_3' - \frac{3}{4}aQ_1Q_3' + 2b_1'' - aP_3'' \end{aligned} \tag{A2}$$

$$\begin{aligned} 0 = & -\frac{1}{2}b_1P_1P_2 - \frac{1}{2}gP_2^2 - \frac{1}{2}B_2P_1P_3 - \frac{1}{8}aP_1P_2P_3 + \frac{1}{8}aP_1P_3Q_1 \\ & -\frac{1}{2}b_1Q_1Q_2 - \frac{1}{2}gQ_2^2 - \frac{1}{2}B_2Q_1Q_3 - \frac{1}{8}aP_2Q_1Q_3 + \frac{1}{8}aQ_1^2Q_3 \\ & -\frac{3}{4}P_2P_3a' - \frac{3}{4}Q_2Q_3a' + P_2b_1' - Q_1b_1' - \frac{1}{2}aP_3P_2' \\ & -B_2P_3' - \frac{3}{4}aP_2P_3' + \frac{1}{4}aQ_1P_3' - b_1Q_1' - gQ_2' \\ & -\frac{1}{2}aQ_3Q_2' - \frac{1}{2}aQ_2Q_3' - \frac{3a'Q_3'}{2} + 2g'' - aQ_3'' \end{aligned} \tag{A3}$$

and

$$\begin{aligned} 0 = & B_2P_1P_2 + \frac{1}{4}aP_1P_2^2 - \frac{1}{4}aP_1P_2Q_1 + B_2Q_1Q_2 + \\ & \frac{1}{4}aP_2Q_1Q_2 - \frac{1}{4}aQ_1^2Q_2 - \frac{1}{2}P_1^2a' + \frac{1}{2}P_2^2a' - \frac{1}{2}Q_1^2a' + \\ & \frac{1}{2}Q_2^2a' - \frac{1}{2}aP_1P_1' - a'P_1' + B_2P_2' + \frac{3}{4}aP_2P_2' - \frac{1}{4}aQ_1P_2' + \\ & B_2Q_1' + \frac{1}{4}aP_2Q_1' - \frac{3}{4}aQ_1Q_1' + \frac{1}{2}aQ_2Q_2' + a'Q_2' - \frac{1}{2}aP_1'' + \frac{1}{2}aQ_2'' \end{aligned} \tag{A4}$$

Appendix B. Nonautonomous Two-Dimensional Black-Scholes

In this Appendix we give the differential equations which the functions $a(t)$, $b_1(t)$, $f(t)$ and $g(t)$ of the generic symmetry vector Equation (51) of the nonautonomous two-dimensional Black-Scholes Equation satisfy.

The system is:

$$\begin{aligned} 0 = & -\frac{1}{2}b_1Q_1Q_3 - \frac{1}{2}fQ_2Q_3 - 2ka' - \frac{1}{4}P_1^2a' \\ & +\frac{1}{2}Q_2a' - \frac{1}{4}Q_3^2a' - P_1b_1' - Q_3f' + 2g' - 2ak' \\ & -\frac{1}{2}aP_1P_1' + \frac{1}{2}aQ_2' - \frac{1}{2}aQ_3Q_3' + a'' \end{aligned} \tag{B1}$$

$$
\begin{aligned}
0 &= -\frac{1}{2}b_1Q_1^2 - \frac{1}{2}fQ_1Q_2 + \frac{1}{2}B_2Q_2Q_3 + \frac{1}{8}aQ_1Q_2Q_3 - \frac{3}{4}Q_1Q_3a' \\
&\quad -Q_1f' + \frac{3a'P_1'}{2} - \frac{1}{2}aQ_3Q_1' - B_2Q_3' - \frac{3}{4}aQ_1Q_3' + 2b_1'' + aP_1''
\end{aligned} \tag{B2}
$$

$$
\begin{aligned}
0 &= -\frac{1}{2}b_1Q_1Q_2 - \frac{1}{2}fQ_2^2 - \frac{1}{2}B_2Q_1Q_3 - \frac{1}{8}aQ_1^2Q_3 - \frac{3}{4}Q_2Q_3a' \\
&\quad +Q_1b_1' + B_2P_1' + \frac{1}{4}aQ_1P_1' + b_1Q_1' + fQ_2' \\
&\quad -\frac{1}{2}aQ_3Q_2' - \frac{1}{2}aQ_2Q_3' + \frac{3a'Q_3'}{2} + 2f'' + aQ_3''
\end{aligned} \tag{B3}
$$

$$
\begin{aligned}
0 &= -\frac{1}{2}B_2Q_1^2 - \frac{1}{8}aQ_1^3 + \frac{1}{2}B_2Q_2^2 + \frac{1}{8}aQ_1Q_2^2 \\
&\quad -Q_1Q_2a' - \frac{1}{2}aQ_2Q_1' + a'Q_1' - B_2Q_2' - \frac{3}{4}aQ_1Q_2' + \frac{1}{2}aQ_1''
\end{aligned} \tag{B4}
$$

and

$$
\begin{aligned}
0 &= -B_2Q_1Q_2 - \frac{1}{4}aQ_1^2Q_2 + \frac{1}{2}Q_1^2a' - \frac{1}{2}Q_2^2a' + \\
&\quad B_2Q_1' + \frac{3}{4}aQ_1Q_1' - \frac{1}{2}aQ_2Q_2' + a'Q_2' + \frac{1}{2}aQ_2''
\end{aligned} \tag{B5}
$$

References

1. Black, F.; Scholes, M. The valuation of option contracts and a test of market efficiency. *J. Finance* **1972**, *27*, 399–417.
2. Black, F.; Scholes, M. The pricing of options and corporate liabilities. *J. Political Econ.* **1973**, *81*, 637–654.
3. Merton, R.C. On the pricing of corporate data: The risk structure of interest rates. *J. Finance* **1974**, *29*, 449–470.
4. Schwartz, E.S. The stochastic behaviour of commodity prices: Implications for valuation and hedging. *J. Finance* **1997**, *52*, 923–973.
5. Morozov, V.V. Classification of six-dimensional nilpotent Lie algebras. *Izvestia Vysshikh Uchebn Zavendeniĭ Matematika* **1958**, *5*, 161–171.
6. Mubarakzyanov, G.M. On solvable lie algebras. *Izvestia Vysshikh Uchebn Zavendeniĭ Matematika* **1963**, *32*, 114–123.
7. Mubarakzyanov, G.M. Classification of real structures of five-dimensional Lie algebras. *Izvestia Vysshikh Uchebn Zavendeniĭ Matematika* **1963**, *34*, 99–106.
8. Mubarakzyanov, G.M. Classification of solvable six-dimensional Lie algebras with one nilpotent base element. *Izvestia Vysshikh Uchebn Zavendeniĭ Matematika* **1963**, *35*, 104–116.
9. Gazizov, R.K.; Ibragimov, N.H. Lie symmetry analysis of Differential equations in Finance. *Nonlinear Dyn.* **1997**, *17*, 387–407.
10. Sophocleuous, C.; Leach, P.G.L.; Andriopoulos, K. Algebraic properties of evolution partial differential equations modelling prices of commodities. *Math. Methods Appl. Sci.* **2008**, *31*, 679–694.
11. Tamizhmani, K.M.; Krishnakumar, K.; Leach, P.G.L. Algebraic resolution of equations of the Black-Scholes type with arbitrary time-dependent parameters. *Appl. Math. Comput.* **2014**, *247*, 115–124.
12. Paliathanasis, A.; Morris, R.M.; Leach, P.G.L. The algebraic properties of the space- and time-dependent one-factor problem of commodities. *Quaest. Math.*, submitted for publication, 2016.
13. Werner, A.; Eliezer, C.J. The lengthening pendulum. *J. Aust. Math. Soc.* **1969**, *9*, 331–336.
14. Lewis, H.R., Jr. Classical and quantum systems with time-dependent harmonic oscillator-type Hamiltonians. *Phys. Rev. Lett.* **1967**, *18*, 510–512.

15. Lewis, H.R., Jr. Motion of a time-dependent harmonic oscillator and of a charged particle in a time-dependent, axially symmetric, electromagnetic field. *Phys. Rev.* **1968**, *172*, 1313–1315.
16. Ross, D.K. The behaviour of a simple pendulum with uniformly shortening string length. *Int. J. Nonlinear Mech.* **1979**, *14*, 175–182.
17. Ermakov, V. Second order differential equations: Conditions of complete integrability. *Appl. Anal. Discret. Math.* **2008**, *2*, 123–145.
18. Pinney, E. The nonlinear equation $y''(x) + p(x)y + cy^{-3} = 0$. *Proc. Am. Math. Soc.* **1950**, *1*, 681.
19. Lutzky, M. Symmetry groups and conserved quantities for the harmonic oscillator. *J. Phys. A Math. Gen.* **1978**, *11*, 249–258.
20. Leach, P.G.L. SL(2,R) and the repulsive oscillator. *J. Phys. A Math. Gen.* **1980**, *13*, 1991–2000.
21. Leach, P.G.L.; Andriopoulos, K.S. The Ermakov equation: A commentary. *Appl. Anal. Discret. Math.* **2008**, *2*, 146–157.
22. Lemmer, R.L.; Leach, P.G.L. A classical viewpoint on quantum chaos. *Arab J. Math. Soc.* **1999**, *5*, 1–17.
23. Paliathanasis, A.; Tsamparlis, M. The geometric origin of Lie point symmetries of the Schrödinger and the Klein-Gordon equations. *Int. J. Geom. Methods Mod. Phys.* **2014**, *14*, 1450037.
24. Maharaj, A.; Andriopoulos, K.S.; Leach, P.G.L.; Abdalla, M.S. The Lie algebraic solution of a time-depedent Schrödinger Equation invariant under the $\{sl(2,R) \oplus_s W_3\} \oplus_s \infty A_1$. *Il Nuovo Cimento B* **2010**, *125*, 1–13.
25. Leach, P.G.L.; Andriopoulos, K.S. Newtonian economics. In *Group Analysis of Differential Equations*; Ibragimov, N.H., Sophocleous, C., Damianou, P.A., Eds.; University of Cyprus: Nicosia, Cyprus, 2005; pp. 134–142.
26. Cimpoiasu, R.; Constantinescu, R. New Symmetries and Particular solutions for 2D Black-Scholes model. In Proceedings of the 7th Mathematical Physics Meeting: Summer School and Conference on Modern Mathematical Physics, Belgrade, Serbia, 9–12 September 2012.
27. Olver, P.J. Applications of Lie Groups to Differential Equations. *Graduate Texts in Mathematics*; Springer-Verlag: New York, NY, USA, 1993.
28. Ibragimov, N.H. Transformation groups applied to mathematical physics. In *Mathematics and Its Applications (Soviet Series)*; D Reidel Publishers: Dordrecht, The Netherlands, 1985.
29. Achdou, Y.; Pironneau, O. Computational Methods for Option Pricing. *Front. Appl. Math.* **2008**, *2*, 123–145.
30. Bozhkov, Y.; Dimas, S. Group classification of a generalized Black-Scholes-Merton equation. *Commun. Nonlinear Sci. Numer. Simul.* **2014**, *19*, 2200–2211.
31. Paliathanasis, A.; Tsamparlis, M. Lie point symmetries of a general class of PDEs: The heat equation. *J. Geom. Phys.* **2012**, *62*, 2443–2456.
32. Dimas, S.; Tsoubelis, D. SYM: A new symmetry-finding package for Mathematica. In *Group Analysis of Differential Equations*; Ibragimov, N.H., Sophocleous, C., Damianou, P.A., Eds.; University of Cyprus: Nicosia, Cyprus, 2005; pp. 64–70.
33. Dimas, S.; Tsoubelis, D. A new Mathematica-based program for solving overdetermined systems of PDEs. In Proceedings of the 8th International Mathematica Symposium, Avignon, France, 19–23 June 2006.
34. Dimas, S. Partial Differential Equations, Algebraic Computing and Nonlinear Systems. Ph.D. Thesis, University of Patras, Patras, Greece, 2008.

9

Existence Results for a New Class of Boundary Value Problems of Nonlinear Fractional Differential Equations

Meysam Alvan [1], Rahmat Darzi [2] and Amin Mahmoodi [3,*]

[1] Department of mathematics, Central Tehran Branch Islamic Azad university, Tehran 13185/768, Iran; m.alvan.r.math.t@gmail.com

[2] Department of Mathematics, Neka Branch Islamic Azad University, Neka 48411-86114, Iran; r.darzi@iauneka.ac.ir

[3] Department of mathematics, Central Tehran Branch Islamic Azad univesity, Tehran 13185/768, Iran

* Correspondence: a_mahmoodi@iauctb.ac.ir

Academic Editor: Hari M. Srivastava

Abstract: In this article, we study the following fractional boundary value problem

$$^{c}D_{0+}^{\alpha}u(t)+2r\,^{c}D_{0+}^{\alpha-1}u(t)+r^{2}\,^{c}D_{0+}^{\alpha-2}u(t)=f(t,u(t)),\quad r>0,\quad 0<t<1,$$

$$u(0)=u(1),\quad u'(0)=u'(1),\quad u'(\xi)+ru(\xi)=\eta,\quad \xi\in(0,1)$$

Where $2\leqslant\alpha<3$, $^{c}D_{0+}^{\alpha-i}\,(i=0,1,2)$ are the standard Caputo derivative and η is a positive real number. Some new existence results are obtained by means of the contraction mapping principle and Schauder fixed point theorem. Some illustrative examples are also presented.

Keywords: Fractional boundary value problem; Contraction mapping principle; Schauder fixed point theorem; Mathematics Subject Classification 2000; 26A33; 34A08; 34B18

1. Introduction

In the recent years, fractional calculus has been one of the most interesting issues that have attracted many scientists, specially in mathematics and engineering sciences. Many natural phenomena can be presented by boundary value problems of fractional differential equations. Many authors in different fields such as chemical physics, fluid flows, electrical networks, viscoelasticity, try to model these phenomena by boundary value problems of fractional differential equations [1–4]. To achieve extra information in fractional calculus, specially boundary value problems, readers can refer to valuable papers or books [5–27].

In this paper, we investigate the existence and uniqueness of solution for the following new class of fractional boundary value problem

$$^{c}D_{0+}^{\alpha}u(t)+2r\,^{c}D_{0+}^{\alpha-1}u(t)+r^{2}\,^{c}D_{0+}^{\alpha-2}u(t)=f(t,u(t)),\quad r>0,\quad 0<t<1 \tag{1}$$

with the boundary conditions

$$u(0)=u(1),\quad u'(0)=u'(1),\quad u'(\xi)+ru(\xi)=\eta,\quad \xi\in(0,1) \tag{2}$$

where $^{c}D_{0+}^{\alpha-i}\,(i=0,1,2)$ are the standard Caputo derivative and $f:[0,1]\times\mathbb{R}\to\mathbb{R}$ is a continuously differentiable function satisfying the following assumptions:

(A_0) $f \in C([0,1] \times \mathbb{R}, \mathbb{R})$ and there exists a constant $L > 0$ so that

$$|f(t,u) - f(t,v)| \leqslant L|u-v|, \quad t \in (0,1), \quad \forall u,v \in \mathbb{R},$$

in which L satisfies the condition $L < \dfrac{r^2 G(\alpha-1)}{2e^r}$.

(A_1) $f \in C([0,1] \times \mathbb{R}, \mathbb{R})$, $p \in C([0,1])$ and A is a constant, so that

$$|f(t,u)| \leqslant p(t) + A|u|, \quad t \in (0,1), \quad \forall u \in \mathbb{R},$$

it satisfies the condition $0 < A < \dfrac{r^2 G(\alpha-1)}{2}$.

Because the boundary conditions $u(0) = u(1)$ and $u'(0) = u'(1)$ in (1.2) involve periodicity, it is not possible to directly transform the boundary value into integral equation. To overcome this problem, presenting a suitable substitution is needed. It is worth saying that Lemma 2.7 (see Lemma 2.3 in [17] and Lemma 2.6 in [21]) is an important and valuable tool to achieve the new result. The contraction mapping principle and fixed point theorem play the main role in finding new existence results for the problem.

The main result of this paper can be seen in two Theorems; 3.1 and 3.2. In Theorem 3.1, the uniqueness of solution is proved by using Banach contraction principle. In Theorem3.2, we present an existence theorem by means of Schauder fixed point theorem.

We can extend the result even for the following boundary value problem

$$\sum_{k=0}^{n-1} \binom{n-1}{k} r^k \, {}^cD_{0+}^{\alpha-k} u(t) = f(t,u(t)), \quad r > 0, \quad 0 < t < 1 \tag{3}$$

where $n-1 \leqslant \alpha < n$, $n > 4$, with the boundary conditions

$$\begin{cases} u(0) = u(1), \quad u'(0) = u'(1), \; \ldots, \quad u^{(n-2)}(0) = u^{(n-2)}(1) \\ \sum_{0}^{n-2} \binom{n-2}{k} r^k u^{n-k-2}(\xi) = \eta, \quad r > 0, \quad \xi \in (0,1) \end{cases} \tag{4}$$

The plan of this paper is as follows:

In Section 2, we give some basic definitions and technical lemmas. Section 3 contains the proofs of our main results. Finally, we provide two examples to show the applicability of the results.

2. Basic Definitions and Preliminaries

In this section, we present some definitions and technical lemmas which will be used in the remainder of this paper. These and the related results and proofs can be found in the literature [6–8,17,21].

Definition 2.1. ([7,8]) The Riemann-Liouville fractional integral of order $\alpha > 0$, of a function $u : \mathbb{R}^+ \to \mathbb{R}$ is defined by

$$I_{0+}^{\alpha} u(t) = \frac{1}{G(\alpha)} \int_0^t (t-s)^{\alpha-1} u(s)\, ds, \quad n-1 < \alpha \leqslant n \tag{5}$$

whenever the right-hand side is defined on $\mathbb{R}^+$.

Definition 2.2. ([7,8]) The Riemann-Liouville fractional derivative of order $\alpha > 0$, of a function $u : \mathbb{R}^+ \to \mathbb{R}$ is given by

$$D_{0+}^{\alpha} u(t) = \frac{1}{G(n-\alpha)} \frac{d^n}{dt^n} \int_0^t (t-s)^{n-\alpha-1} u(s)\, ds, \quad n-1 < \alpha \leqslant n \tag{6}$$

where $n = [\alpha] + 1$, and $[\alpha]$ denotes the integer part of real number α.

Definition 2.3. ([7,8]) The Caputo fractional derivative of order $\alpha > 0$, of a function $u : \mathbb{R}^+ \to \mathbb{R}$ is defined by

$$ {}^cD_{0+}^{\alpha} u(t) = \frac{1}{G(n-\alpha)} \int_0^t (t-s)^{n-\alpha-1} u^{(n)}(s)\, ds, \quad n-1 < \alpha \leqslant n \tag{7}$$

whenever the right-hand side is defined on $\mathbb{R}^+$.

Definition 2.4. ([7,8]) The Caputo fractional derivative of order $\alpha > 0$, of a function $u : \mathbb{R}^+ \to \mathbb{R}$ is defined via the Riemann-Liouville fractional derivative by

$$\left({}^cD_{0+}^{\alpha} u\right)(t) = \left(D_{0+}^{\alpha} \left[u(s) - \sum_{k=0}^{n-1} \frac{u^{(k)}(0^+)}{k!} s^k \right] \right)(t) \tag{8}$$

Where $n = \alpha$ for $\alpha \in \mathbb{N}; n = [\alpha] + 1$ for $\alpha \notin \mathbb{N}$.

Lemma 2.5. ([6]) Let $\text{l} \in \mathbb{N}, \alpha > 0$. If $\left(D_{0+}^{\alpha} u\right)(t)$ and $\left(D_{0+}^{\alpha+\text{l}} u\right)(t)$ exist, then

$$\left(D^{\text{l}} D_{0+}^{\alpha} u\right)(t) = \left(D_{0+}^{\alpha+\text{l}} u\right)(t) \tag{9}$$

Lemma 2.6. ([6]) Let $n \in \mathbb{N}$, $\alpha \in (n-1, n]$. If $u \in C^n [0, b)$ $(b > 0$ is real number), then

$$\left(I_{0+}^{\alpha}\, {}^cD_{0+}^{\alpha} u\right)(t) = u(t) - \sum_{k=0}^{n-1} \frac{u^{(k)}(0^+)}{k!} t^k \tag{10}$$

holds on $(0, b)$.

Lemma 2.7. ([17,21]) Let $n \in \mathbb{N}, \alpha \in (n-1, n]$. If $u \in C^{n-1} [0, b)$ and ${}^cD_{0+}^{\alpha} u \in C(0, b)$, then

$$\left(I_{0+}^{\alpha}\, {}^cD_{0+}^{\alpha} u\right)(t) = u(t) - \sum_{k=0}^{n-1} \frac{u^{(k)}(0^+)}{k!} t^k \tag{11}$$

holds on $(0, b)$.

Lemma 2.8. Let $r > 0$, $g \in C[0, 1]$. If $u \in C^3 [0, 1]$ is a solution of BVP

$${}^cD_{0+}^{\alpha} u(t) + 2r\, {}^cD_{0+}^{\alpha-1} u(t) + r^2\, {}^cD_{0+}^{\alpha-2} u(t) = g(t), \quad r > 0, \quad 0 < t < 1 \tag{12}$$

$$u(0) = u(1), \quad u'(0) = u'(1), \quad u'(\xi) + ru(\xi) = \eta, \quad \xi \in (0, 1) \tag{13}$$

Then v satisfies

$$\begin{aligned} & v(t) \\ &= \frac{\eta}{r} e^{rt} - \frac{e^{r(t-\xi)}}{r(e^r-1)\Gamma(\alpha-2)} \int_0^1 \int_0^m e^{rt} (m-\tau)^{\alpha-3} g(\tau) d\tau dm \\ &- \frac{e^{r(t-\xi)}}{r\Gamma(\alpha-2)} \int_0^{\xi} \int_0^m e^{rt} (m-\tau)^{\alpha-3} g(\tau) d\tau dm \\ &+ \frac{1 + t(e^r-1)}{(e^r-1)^2 \Gamma(\alpha-2)} \int_0^1 \int_0^m e^{rt} (m-\tau)^{\alpha-3} g(\tau) d\tau dm \\ &+ \int_0^1 \int_0^1 \int_0^m \frac{e^{rt} (m-\tau)^{\alpha-3}}{(e^r-1)\Gamma(\alpha-2)} g(\tau) d\tau dm dn \\ &+ \int_0^t \int_0^n \int_0^m \frac{e^{rt} (m-\tau)^{\alpha-3}}{\Gamma(\alpha-2)} g(\tau) d\tau dm dn \end{aligned} \tag{14}$$

Conversely, if $v(t)$ is given by (14), then $u = ve^{-rt} \in C^2 [0, 1]$ and u is a solution of $BVP\, (12) - (13)$.

Proof. Let $u \in C^2[0,1]$ be a solution of $BVP(12)-(13)$. Since $u'' \in C[0,1]$, Def. (2.3) show that ${}^cD_{0+}^{\alpha-2}u \in C[0,1]$ and ${}^cD_{0+}^{\alpha-1}u \in C^1[0,1]$

From the relation ${}^cD_{0+}^{\alpha}u = g(t) - 2r^2\ {}^cD_{0+}^{\alpha-1}u - r\ {}^cD_{0+}^{\alpha-2}u$ and $g \in C[0,1]$, we have ${}^cD_{0+}^{\alpha}u \in C(0,1)$. Thus, by Lemma 2.7, we have the following relations

$$I_{0+}^{\alpha}\ {}^cD_{0+}^{\alpha}u(t) = u(t) - a_0 - a_1t - a_2t^2, \quad t \in (0,1) \tag{15}$$

$$I_{0+}^{\alpha-1}\ {}^cD_{0+}^{\alpha-1}u(t) = u(t) - b_1 - b_2t, \quad t \in (0,1) \tag{16}$$

and

$$I_{0+}^{\alpha-2}\ {}^cD_{0+}^{\alpha-2}u(t) = u(t) - c_2, \quad t \in (0,1) \tag{17}$$

therefore

$$\begin{aligned} I_{0+}^{\alpha}\ {}^cD_{0+}^{\alpha-1}u(t) &= I_{0+}^{1} I_{0+}^{\alpha-1}\ {}^cD_{0+}^{\alpha-1}u(t) \\ &= \textstyle\int_0^t u(s)\,ds - b_0 - b_1t - b_2\frac{t^2}{2} \end{aligned} \tag{18}$$

$$\begin{aligned} I_{0+}^{\alpha}\ {}^cD_{0+}^{\alpha-2}u(t) = &\ I_{0+}^{2} I_{0+}^{\alpha-2}\ {}^cD_{0+}^{\alpha-2}u(t) \\ &= \textstyle\int_0^t\int_0^r u(s)\,dsdr - c_0 - c_1t - c_2\frac{t^2}{2} \end{aligned} \tag{19}$$

now, from $(12), (13), (18)$ and (19), we have

$$(t) + 2r\int_0^t u(s)\,ds + r^2\int_0^t\int_0^r u(s)\,dsdr = d_0 + d_1t + d_2\frac{t^2}{2} + I_{0+}^{\alpha}g(t) \tag{20}$$

where $d_0, d_1, d_2 \in \mathbb{R}$.

It is easy to see that $\phi(t) = \int_0^t (t-s)^{\alpha-3} g(s)\,ds \in C[0,1]$. Since $u'' \in C[0,1]$, it follows from (20) that

$$\begin{aligned} u''(t) + 2ru'(t) + r^2u(t) &= d_2 + \frac{d^2}{dt^2} I_{0+}^{\alpha}g(t) \\ &= d_2 + \frac{1}{G(\alpha-2)} \textstyle\int_0^t (t-s)^{\alpha-3} g(s)\,ds \end{aligned} \tag{21}$$

assuming $v = ue^{rt}$, the formulas (21) yield

$$v''(t) = d_2e^{rt} + \frac{e^{rt}}{G(\alpha-2)}\int_0^t (t-s)^{\alpha-3} g(s)\,ds \tag{22}$$

by integrating both sides of (22) twice, we obtain

$$v(t) = v(0) + v'(0)t + \frac{d_2(e^{rt} - rt - 1)}{r^2} + \int_0^t\int_0^n\int_0^m \frac{e^{rm}(m-\tau)^{\alpha-3}}{G(\alpha-2)} g(\tau)\,d\tau dm dn \tag{23}$$

thus, it follows boundary conditions $u(0) = u(1)$ and $u'(0) = u'(1)$, that

$$v(1) = v(0)e^r \tag{24}$$

$$v'(1) = v'(0)e^r \tag{25}$$

now, the formulas (24) and (25) imply that

$$\begin{aligned} v(0) = &\frac{d_2}{r^2} + \frac{1}{(e^r-1)^2 G(\alpha-2)} \textstyle\int_0^1\int_0^m e^{rm}(m-\tau)^{\alpha-3} g(\tau)\,d\tau dm \\ &+ \frac{1}{(e^r-1)G(\alpha-2)} \textstyle\int_0^1\int_0^n\int_0^m e^{rm}(m-\tau)^{\alpha-3} g(\tau)\,d\tau dm dn \end{aligned} \tag{26}$$

and

$$v(0) = \frac{d_2}{r} + \frac{1}{(e^r - 1)\Gamma(\alpha - 2)} \int_0^1 \int_0^m e^{rm} (m - \tau)^{\alpha - 3} g(\tau)\, d\tau dm \tag{27}$$

respectively. Substituting (26), and (27) into (23), we have

$$\begin{aligned} v(t) &= \frac{d_2}{r^2} e^{rt} + \frac{1 + t(e^r - 1)}{(e^r - 1)^2 \Gamma(\alpha - 2)} \int_0^1 \int_0^m e^{rm} (m - \tau)^{\alpha - 3} g(\tau)\, d\tau dm \\ &+ \frac{1}{(e^r - 1)\Gamma(\alpha - 2)} \int_0^1 \int_0^1 \int_0^m \frac{e^{rm}(m - \tau)^{\alpha - 3}}{(e^r - 1)\Gamma(\alpha - 2)} g(\tau)\, d\tau dm dn \\ &+ \frac{1}{\Gamma(\alpha - 2)} \int_0^t \int_0^n \int_0^m e^{rm} (m - \tau)^{\alpha - 3} g(\tau)\, d\tau dm dn \end{aligned} \tag{28}$$

by differentiating both sides of (23) and using the condition $u'(\xi) + ru(\xi)$, $\xi \in (0, 1)$, we have

$$v'(\xi) = \left(ue^{rt}\right)'_{t=\xi} = u'(\xi) e^{r\xi} + ru(\xi) e^{r\xi} = \left[u'(\xi) + ru(\xi)\right] e^{r\xi} = \eta e^{r\xi} \tag{29}$$

thus,

$$\begin{aligned} &\frac{d_2}{r} e^{r\xi} + \frac{1}{(e^r - 1) G(\alpha - 2)} \int_0^1 \int_0^m e^{rm} (m - \tau)^{\alpha - 3} g(\tau)\, d\tau dm \\ &+ \frac{1}{G(\alpha - 2)} \int_0^\xi \int_0^m e^{rm} (m - \tau)^{\alpha - 3} g(\tau)\, d\tau dm \\ &= \eta e^{r\xi}. \end{aligned} \tag{30}$$

Hence, it follows from (30) that

$$\begin{aligned} d_2 &= r\eta - \frac{re^{-r\xi}}{(e^r - 1)\Gamma(\alpha - 2)} \int_0^1 \int_0^m e^{rm} (m - \tau)^{\alpha - 3} g(\tau)\, d\tau dm \\ &- \frac{re^{-r\xi}}{\Gamma(\alpha - 2)} \int_0^\xi \int_0^m e^{rm} (m - \tau)^{\alpha - 3} g(\tau)\, d\tau dm. \end{aligned} \tag{31}$$

Substituting, (31) into (28), the relation (14) is obtained.

Conversely, since $\int_0^t (t - s)^{\alpha - 3} g(s)\, ds$ is continuous on $[0, 1]$, by differentiating both sides of (14), we obtain

$$\begin{aligned} v'(t) &= \eta e^{rt} - \frac{e^{r(t - \xi)}}{(e^r - 1)\Gamma(\alpha - 2)} \int_0^1 \int_0^m e^{rm} (m - \tau)^{\alpha - 3} g(\tau)\, d\tau dm \\ &- \frac{e^{r(t - \xi)}}{\Gamma(\alpha - 2)} \int_0^\xi \int_0^m e^{rm} (m - \tau)^{\alpha - 3} g(\tau)\, d\tau dm \\ &+ \frac{1}{(e^r - 1)\Gamma(\alpha - 2)} \int_0^1 \int_0^m e^{rm} (m - \tau)^{\alpha - 3} g(\tau)\, d\tau dm \\ &+ \frac{1}{\Gamma(\alpha - 2)} \int_0^t \int_0^m e^{rm} (m - \tau)^{\alpha - 3} g(\tau)\, d\tau dm. \end{aligned} \tag{32}$$

By differentiating both sides of (32), we will get

$$\begin{aligned} v''(t) &= r\eta e^{rt} - \frac{re^{r(t - \xi)}}{(e^r - 1)\Gamma(\alpha - 2)} \int_0^1 \int_0^m e^{rm} (m - \tau)^{\alpha - 3} g(\tau)\, d\tau dm \\ &- \frac{re^{r(t - \xi)}}{\Gamma(\alpha - 2)} \int_0^1 \int_0^m e^{rm} (m - \tau)^{\alpha - 3} g(\tau)\, d\tau dm \\ &+ \frac{e^{rt}}{\Gamma(\alpha - 2)} \int_0^t (t - \tau)^{\alpha - 3} g(\tau)\, d\tau \\ &= e^{rt}\left(d_2 + I^{\alpha - 2} g(t)\right), \end{aligned} \tag{33}$$

where d_2 is described as in (31), and so $v \in C^2[0, 1]$. Furthermore, from (32) together with (23) and (31), we ensure that (24) holds on $[0, 1]$, and

$$v(1) = v(0) e^r,\ v'(1) = v'(0) e^r,\ v'(\xi) = \eta e^{r\xi}, \quad \xi \in (0, 1), \tag{34}$$

Now, assume that $u = ve^{-rt}$. Keeping in mind that $u'' \in C[0,1]$, because $v \in C^2[0,1]$, it follows (32) and (33) that

$$u'' + 2ru' + r^2 u = d_2 + I_{0+}^{\alpha-2} g(t). \tag{35}$$

Therefore,

$${}^cD_{0+}^{\alpha-2} u'' + 2r\,{}^cD_{0+}^{\alpha-2} u' + r^2\,{}^cD_{0+}^{\alpha-2} u = {}^cD_{0+}^{\alpha-2} I_{0+}^{\alpha-2} g(t) = g(t), \tag{36}$$

on $(0,1)$. From the fact that $u'' \in C[0,1]$, and Def. (2.4) we get

$$\begin{aligned} {}^cD_{0+}^{\alpha} u(t) &= \left[D_{0+}^{\alpha}\left\{u(s) - u(0) - u'(0)s - \frac{u(0)}{2}s^2\right\}\right](t) \\ &= \left[D_{0+}^{\alpha} u\right](t) - \frac{u(0)}{G(1-\alpha)}t^{-\alpha} - \frac{u'(0)}{G(2-\alpha)}t^{1-\alpha} - \frac{u''(0)}{G(3-\alpha)}t^{2-\alpha}, \end{aligned} \tag{37}$$

and

$$\begin{aligned} \left[D_{0+}^{\alpha} u\right](t) &= \frac{1}{G(3-\alpha)}\frac{d^3}{dt^3}\int_0^t (t-s)^{2-\alpha} u(s)\,ds \\ &= \frac{1}{(\alpha-3)G(3-\alpha)}\frac{d^3}{dt^3}\int_0^t u(s)\,d(t-s)^{3-\alpha} \\ &= \frac{1}{(3-\alpha)G(3-\alpha)}\frac{d^3}{dt^3}\left\{u(0)t^{3-\alpha} + \int_0^t (t-s)^{3-\alpha} u'(s)\,ds\right\} \\ &= \frac{1}{(\alpha-4)(3-\alpha)G(3-\alpha)}\frac{d^3}{dt^3}\left\{(\alpha-4)u(0)t^{3-\alpha}\right. \\ &\quad \left. + \frac{d^3}{dt^3}\int_0^t u'(s)\,d(t-s)^{4-\alpha}\right\} \\ &= \frac{1}{G(5-\alpha)}\frac{d^3}{dt^3}\left\{(4-\alpha)u(0)t^{3-\alpha} + u'(0)t^{4-\alpha}\right. \\ &\quad \left. + \int_0^t (t-s)^{4-\alpha} u''(s)\,ds\right\} \\ &= \frac{1}{G(5-\alpha)}\left\{\sum_{j=0}^{1}\left[\prod_{i=j+1}^{4}(i-\alpha)\right] u^{(j)}(0)t^{j-\alpha} + (4-\alpha)(3-\alpha)\right. \\ &\quad \left. + \frac{d}{dt}\int_0^t (t-s)^{2-\alpha} u''(s)\,ds\right\}. \end{aligned}$$

Consequently,

$$\left[D_{0+}^{\alpha} u\right](t) = \frac{u(0)}{\Gamma(1-\alpha)}t^{-\alpha} - \frac{u'(0)}{\Gamma(2-\alpha)}t^{1-\alpha} + \frac{1}{\Gamma(3-\alpha)}\frac{d}{dt}\int_0^t (t-s)^{2-\alpha} u''(s)\,ds. \tag{38}$$

It follows (37) and (38) that

$$\begin{aligned} \left[{}^cD_{0+}^{\alpha} u(s)\right](t) &= \left[D_{0+}^{\alpha-2} u''(s)\right](t) - \frac{u''(0)}{G(3-\alpha)}t^{2-\alpha} \\ &= D_{0+}^{\alpha-2}\left[u''(s) - u''(0)\right](t) \\ &= \left[{}^cD_{0+}^{\alpha} u''(s)\right](t). \end{aligned}$$

Similarly, we can show that $\left[{}^cD_{0+}^{\alpha-1} u\right](t) = \left[{}^cD_{0+}^{\alpha-2} u'\right](t)$. Moreover, it follows (36) that

$${}^cD_{0+}^{\alpha} u(t) + 2r\,{}^cD_{0+}^{\alpha-1} u(t) + r^2\,{}^cD_{0+}^{\alpha-2} u(t) = g(t), \quad t \in (0,1).$$

On the other hand, the relation (34) implies that

$$u(0) = u(1), \quad u'(0) = u'(1), \quad u'(\xi) + ru(\xi) = \eta.$$

Then, $u \in C^2[0,1]$ is a solution of BVP (12) – (13). Thus, this ends the proof.

3. Main Result

Let $U = C[0,1]$ be a Banach space with the norm $||u|| = max_{t\in[0,1]} \{u(t)\}$. Consider the space U with the norm $||u||_* = max_{t\in[0,1]} \{u(t) e^{-rt}\}$ in which r is described as in (1). It is well known that the norm $||u||_*$ is equivalent to the norm $||u||$.

For the forthcoming analysis, we need the assumptions $(A0)$ and $(A1)$.

Theorem 3.1. Let the assumption $(A0)$ hold. Then, the boundary value problem $(1)-(2)$ has a unique solution.

Proof. Define the operator $T : U \to U$ by

$$\begin{aligned} Tv(t) &= \frac{\eta}{r} e^{rt} - \frac{e^{r(t-\xi)}}{r(e^r-1)\Gamma(\alpha-2)} \int_0^1 \int_0^m e^{rm} (m-\tau)^{\alpha-3} g(\tau)\, d\tau dm \\ &- \frac{e^{r(t-\xi)}}{r\Gamma(\alpha-2)} \int_0^{\xi} \int_0^m e^{rm} (m-\tau)^{\alpha-3} g(\tau)\, d\tau dm \\ &+ \frac{1+t(e^r-1)}{(e^r-1)^2 \Gamma(\alpha-2)} \int_0^1 \int_0^m e^{rm} (m-\tau)^{\alpha-3} g(\tau)\, d\tau dm \\ &+ \int_0^1 \int_0^1 \int_0^m \frac{e^{rm}(m-\tau)^{\alpha-3}}{(e^r-1)\Gamma(\alpha-2)} g(\tau)\, d\tau dm dn \\ &+ \int_0^t \int_0^n \int_0^m \frac{(m-\tau)^{\alpha-3}}{\Gamma(\alpha-2)} g(\tau)\, d\tau dm dn, \end{aligned}$$

where the function $g(t) = f(t, v(t) e^{-rt})$ is continuous on $[0,1]$, for any $v \in U$ (from (A_0)). It is easy to see that the operator T maps U into U.

In view of Lemma (2.10), the operator T has a fixed point $v \in V$ if and only if $u = ve^{-rt}$ is a solution of $BVP\,(1.1)-(1.2)$ with $u \in C^2[0,1]$. So, it is sufficient to show that the operator T has a fixed point on U. For $v_1, v_2 \in U$ and for $s \in C[0,1]$, we obtain

$$\begin{aligned} \left|f\left(s, v_2(s) e^{-rs}\right) - f\left(s, v_1(s) e^{-rs}\right)\right| &\leqslant L\left|v_2(s) e^{-rs} - v_1(s) e^{-rs}\right| \\ &\leqslant L||v_2 - v_1||_* \end{aligned} \tag{39}$$

Hence, from (39), we have the following inequality

$$\begin{aligned} |Tv_2(t) - Tv_1(t)| &\leqslant \frac{L}{G(\alpha-2)} ||v_2 - v_1||_* \\ &\left[\frac{e^{r(t-\xi)}}{r(e^r-1)} \int_0^1 \int_0^m e^{rm} (m-\tau)^{\alpha-3} d\tau dm \right. \\ &+ \frac{e^{r(t-\xi)}}{r} \int_0^{\xi} \int_0^m e^{rm} (m-\tau)^{\alpha-3} d\tau dm \\ &+ \frac{1+t(e^r-1)}{(e^r-1)^2} \int_0^1 \int_0^m e^{rm} (m-\tau)^{\alpha-3} d\tau dm \\ &+ \frac{1}{e^r-1} \int_0^1 \int_0^n \int_0^m e^{rm} (m-\tau)^{\alpha-3} d\tau dm dn \\ &\left. + \int_0^t \int_0^n \int_0^m e^{rm} (m-\tau)^{\alpha-3} d\tau dm dn \right] \\ &\leqslant \frac{2Le^{rt}}{r^2\Gamma(\alpha-1)} ||v_2 - v_1||_*. \end{aligned}$$

Consequently,

$$||Tv_2(t) - Tv_1(t)||_* \leqslant \frac{2Le^{rt}}{r^2\Gamma(\alpha-1)} ||v_2 - v_1||_*. \tag{40}$$

By the Banach contraction principle, it follows that T has an unique fixed point $v \in U$. Therefore, $u = e^{-rt}v$ is a unique solution of $FBVP\,(1)-(2)$.

Now, we prove the existence of solutions of $(1)-(2)$ by applying Schauder fixed point theorem.

Theorem 3.2. Let the assumption $(A1)$ hold. Then, the boundary value problem $(1)-(2)$ has at least one solution $u \in C^2[0,1]$.

Proof. Let us consider $P = sup\{|p(t)|; t \in [0,1]\}$ and $B_R = \{v \in U; ||v - v_0||_* \leqslant R\}$ in which $v_0 = \frac{\eta}{r}e^{rt}$ and $R > \frac{2(Pr + A\eta)}{r(r^2\Gamma(\alpha-1)2A)}$. For $v \in U$, by $(A1)$, we find that

$$\left|f\left(s, v(s)e^{-rs}\right)\right| \leqslant P + A||v||_* \leqslant P + A(||v_0||_* + R) \leqslant P + A\left(\frac{\eta}{r} + R\right),$$

and so,

$$|Tv(t) - v_0(t)| \leqslant \frac{P + A\left(\frac{\eta}{r} + R\right)}{r^2\Gamma(\alpha-1)}\left[e^{r(t-\xi)} + e^{r(t-\xi)}\left(e^{r\xi} - 1\right) + e^{rt}\right]. \tag{41}$$

From (41), we have

$$||Tv - v_0||_* \leqslant \frac{2\left[P + A\left(\frac{\eta}{r} + R\right)\right]}{r^2G(\alpha-1)} < R.$$

Thus, T maps B_R into B_R, i.e. $T(B_R) \subseteq B_R$. Now, we prove that T is completely continuous on B_R. We will give the proof in the case that U is equipped with the usual norm, since the norm $||u||_*$ is equivalent to the usual norm. Since $T(B_R) \subseteq B_R$, we have $||Tu|| \leqslant ||Tu||_* e^r \leqslant (v_0 + R)e^r$ for any $u \in B_R$, and so $\{z; z \in T(B_R)\}$ is uniformly bounded. On the other hand, for any $v \in B_R$, it follows from (40) and $(A1)$ that

$$(Tv)'(t) \leqslant \left(\eta + \frac{P + A\left(\frac{\eta}{r} + R\right)}{rG(\alpha-1)}\right)e^r, \quad t \in [0,1]$$

and this shows that $T(B_R)$ is equicontinuous. Thus, by Arzella-Ascoli theorem, it implies that $T(B_R)$ is relatively compact. Finally, we show that T is continuous on B_R. Let (v_n) be an arbitrary sequence in B_R and $v \in B_R$so that $||v_n - v|| \to 0$ as $n \to \infty$. Therefore, $||v_n - v||_* \to 0$, as $n \to \infty$ and so there exists two constants k_1, k_2 so that $v_n(t)e^{-rt}$ $(n = 1, 2, \ldots)$ and $v(t)e^{-rt} \in [k_1, k_2]$, for each $t \in [0,1]$. Since f is uniformly continuous on $[0,1] \times [k_1, k_2]$, it follows that for any $\epsilon > 0$, there exists $\delta > 0$ whenever $|u_1 - u_2| < \delta$, $u_1, u_2 \in [k_1, k_2]$ then,

$$|f(t, u_2) - f(t, u_1)| \leqslant \vartheta\epsilon, \quad t \in [0,1], \tag{42}$$

where $\vartheta = \frac{r^2\Gamma(\alpha-1)}{2e^r}$. Since $v_n \to v$, there exists $N \geqslant 1$, such that the following relation

$$\left|v_n(t)e^{-rt} - v(t)e^{-rt}\right| \leqslant \delta, \quad t \in [0,1],$$

satisfies for $n \geqslant N$. Now, for any $n \geqslant N$ (3.5) yields

$$\begin{aligned}
|Tv_n(t) - Tv(t)| \leqslant \frac{\vartheta\epsilon}{G(\alpha-2)}&\left[\frac{e^{r(t-\xi)}}{r(\alpha-1)}\int_0^1\int_0^m e^{rm}(m-s)^{\alpha-3}\,d\tau dm\right.\\
&+\frac{e^{r(t-\xi)}}{r}\int_0^\xi\int_0^m e^{rm}(m-s)^{\alpha-3}\,d\tau dm\\
&+\frac{1+t(e^r-1)}{(e^r-1)^2}\int_0^1\int_0^m e^{rm}(m-\tau)^{\alpha-3}\,d\tau dm\\
&+\frac{1}{e^r-1}\int_0^1\int_0^n\int_0^m e^{rm}(m-\tau)^{\alpha-3}\,d\tau dm dn\\
&\left.+\int_0^t\int_0^n\int_0^m e^{rm}(m-\tau)^{\alpha-3}\,d\tau dm dn\right]\\
&\leqslant \frac{2\vartheta\epsilon}{r^2\Gamma(\alpha-1)}e^{rt}.
\end{aligned}$$

Consequently

$$||Tv_n(t) - Tv(t)|| < \frac{2\vartheta\epsilon}{r^2\Gamma(\alpha-1)}e^r = \epsilon.$$

Thus, all the assumptions of the Schauder fixed point theorem are satisfied. Then, there exists a point $v \in B_R$ with $v = Tv$ In view of Lemma (14), we conclude that $u = ve^{-rt}$ $(u \in C^2[0,1])$ is a solution of boundary value problem $(1)-(2)$. As a result, the proof is complete.

4. Illustrative Examples

Example 4.1. Consider the boundary value problem

$$\begin{cases} {}^cD_{0+}^{\frac{5}{2}}u(t) + 2r^cD_{0+}^{\frac{3}{2}}u(t) + r^2u(t) = f(t,u(t)), \quad 0<t<1, \\ u(0) = u(1), \quad u'(0) = u'(1), \quad u'(\xi) + ru(\xi) = \eta, \end{cases} \tag{43}$$

where $r > 0$, $f(t,u) = h(t)\dfrac{u}{1+u^2}$ with

$$|h(t)| \leqslant \frac{r^2G\left(\frac{5}{2}-1\right)}{2e^r} = \frac{r^2\sqrt{\pi}}{4e^r}.$$

It is easy to see that the assumption $(\boldsymbol{Ao})$ holds. So, by Theorem 3.1, BVP (43) has a unique solution.

Example 4.2. Consider the boundary value problem

$$\begin{cases} {}^cD_{0+}^{\frac{5}{2}}u(t) + 2r^cD_{0+}^{\frac{3}{2}}u(t) + r^2u(t) = f(t,u(t)), \quad 0<t<1, \\ u(0) = u(1), \quad u'(0) = u'(1), \quad u'(\xi) + ru(\xi) = \eta, \end{cases} \tag{44}$$

where $r > 0$, $f(t,u) = p_1(t) + p_2(t)u$ with $p_1, p_2 \in C[0,1]$ and $\max |p_2(t)|_{t\in[0,1]} \leqslant \dfrac{r^2\sqrt{\pi}}{4}$. Thus, the conclusion of Theorem 3.2 applies to the problem.

Acknowledgments: The authors are grateful to the referees for their comments and suggestions which improved the quality of the paper.

Author Contributions: All authors contributed to the technical analysis and development of the results.

Conflicts of Interest: The authors declare no conflict of interest.

References

1. Oldham, K.B.; Spanier, J. *The Fractional Calculus*; Academic press: New York, NY, USA; London, UK, 1974.
2. *The Fractional Calculus and Its Application*; Ross, B., Ed.; Lecture notes in mathematics; Springer-Verlag: Berlin, Germany, 1975.
3. Tatom, F.B. The relationship between fractional calculus and fractals. *Fractals* **1995**, *3*, 217–229. [CrossRef]
4. Nonnenmacher, T.F.; Metzler, R. On the Riemann-Liouville fractional calculus and some recent applications. *Fractals* **1995**, *3*, 557–566. [CrossRef]
5. Samko, S.G.; Kilbas, A.A.; Marichev, O.I. *Fractional Integral and Derivatives: Theory and Application*; CRC Press: Boca Raton, FL, USA, 1993.
6. Kilbas, A.A.; Srivastava, H.M.; Trujillo, J.J. *Theory and Application of Fractional Differential Equations*; Elsevier Science: Amsterdam, The Netherlands, 2006.
7. Miller, K.S.; Ross, B. *An Introduction to the Fractional Calculus and Fractional Differential Equation*; John Wiley and Sons: New York, NY, USA, 1993.
8. Podlubny, I. *Fractional Differential Equations*; Academic Press: San Diego, CA, USA, 1999.
9. Lakshmikantham, V.; Leela, S.; Vasundhara, J. *Theory of Fractional Dynamic Systems*; Cambridge Academic Publishers: Cambridge, UK, 2009.

10. Agarwal, R.P.; Benchohara, M.; Slimani, B.A. Existence results for differential equations with fractional order and impulses. *Mem. Diff. Equ. Math. Phys.* **2008**, *44*, 1–21.
11. Agarwal, R.P.; Benchohara, M.; Hamani, S. Boundary value problems for fractional differential equations. *J. Geor. Math.* **2009**, *16*, 401–411.
12. Ahmad, B.; Nieto, J.J. Existence of solutions for nonlocal boundary value problems of higher order nonlinear fractional differential equations. *Abstr. Appl. Anal.* 2009. [CrossRef]
13. Darzi, R.; Mohammadzadeh, B.; Neamaty, A.; Băleanu, D. Lower and upper solutions method for positive solutions of fractional boundary value problems. *Abstr. Appl. Anal.* 2013. [CrossRef]
14. Zhang, S. Positive solutions for boundary value problems of nonlinear fractional differential equations. *Elect. J. Diff. Equ.* **2006**, *36*, 1–12. [CrossRef]
15. Rehman, M.U.; Khan, R.A. Existence and uniqueness of solutions form multi-point for boundary value problems for fractional differential equations. *Appl. Math. Lett.* **2010**, *23*, 1038–1044. [CrossRef]
16. Darzi, R.; Mohammadzadeh, B.; Neamaty, A.; Băleanu, D. On the Existence and Uniqueness of Solution of a Nonlinear Fractional Differential Equations. *J. Comput. Anal. Appl.* **2013**, *15*, 152–162.
17. Băleanu, D.; Diethelm, K.; Scalas, E.; Trujillo, J.J. *Fractional Calculus: Models and Numerical Methods*; World Scientific: Singapore, Singapore, 2012.
18. Băleanu, D.; Trujillo, J.J. On exact solutions of a class of fractional Euler-Lagrange equations. *Nonlinear Dyn.* **2008**, *52*, 331–335. [CrossRef]
19. Băleanu, D.; Mustafa, O.G. On the global existence of solutions to a class fractional differential equations. *Comp. Math. Appl.* **2010**, *59*, 1835–1841. [CrossRef]
20. El-Shahed, M. Nontrivial solutions for a nonlinear multi-point boundary value problems of fractional order. *Comp. Math. Appl.* **2010**, *59*, 3438–3443. [CrossRef]
21. Chai, G. Existence results for boundary value problems of nonlinear fractional differential equations. *Comp. Math. Appl.* **2011**, *62*, 2374–2382. [CrossRef]
22. Anguraj, A.; Ranjini, M.C.; Rivero, M.; Trujillo, J.J. Existence results for fractional neutral functional differential equations with random impulses. *Mathematics* **2015**, *3*, 16–28. [CrossRef]
23. Sambandham, B.; Vatsala, A.S. Basic results for sequential Caputo fractional differential equations. *Mathematics* **2015**, *3*, 76–91. [CrossRef]
24. Morita, T.; Sato, K. Asymptotic expansions of fractional derivatives and their applications. *Mathematics* **2015**, *3*, 171–189. [CrossRef]
25. Diekema, E. The fractional orthogonal derivative. *Mathematics* **2015**, *3*, 273–298. [CrossRef]
26. Rogosin, S. The role of the Mittag-Leffler function in fractional modeling. *Mathematics* **2015**, *3*, 368–381. [CrossRef]
27. Nieto, J.J.; Ouahab, A.; Venktesh, V. Implicit fractional differential equations via the Liouville–Caputo derivative. *Mathematics* **2015**, *3*, 398–411. [CrossRef]

10

Solution of Differential Equations with Polynomial Coefficients with the Aid of an Analytic Continuation of Laplace Transform

Tohru Morita [1,*] and Ken-ichi Sato [2]

1 Graduate School of Information Sciences, Tohoku University, Sendai 980-8577, Japan
2 College of Engineering, Nihon University, Koriyama 963-8642, Japan; kensatokurume@ybb.ne.jp
* Correspondence: senmm@jcom.home.ne.jp

Academic Editor: Hari M. Srivastava

Abstract: In a series of papers, we discussed the solution of Laplace's differential equation (DE) by using fractional calculus, operational calculus in the framework of distribution theory, and Laplace transform. The solutions of Kummer's DE, which are expressed by the confluent hypergeometric functions, are obtained with the aid of the analytic continuation (AC) of Riemann–Liouville fractional derivative (fD) and the distribution theory in the space $\mathcal{D}'_R$ or the AC of Laplace transform. We now obtain the solutions of the hypergeometric DE, which are expressed by the hypergeometric functions, with the aid of the AC of Riemann–Liouville fD, and the distribution theory in the space $\mathcal{D}'_{r,R}$, which is introduced in this paper, or by the term-by-term inverse Laplace transform of AC of Laplace transform of the solution expressed by a series.

Keywords: Kummer's differential equation; hypergeometric differential equation; distribution theory; operational calculus; fractional calculus; Laplace transform

1. Introduction

Stimulated by Yosida's work [1,2], in which the solution of Laplace's differential equation (DE) is obtained with the aid of the operational calculus of Mikusiński [3], we are concerned in [4–6], with the DE or fractional DE of the form:

$$\sum_{l=0}^{m}(a_l t + b_l)\cdot {}_0D_R^{l\sigma}u(t) = f(t), \quad t > 0, \tag{1}$$

where $\sigma = \frac{1}{2}$ or $\sigma = 1$, $m = 2$, and $a_l \in \mathbb{C}$ and $b_l \in \mathbb{C}$ are constants.

In solving the DE, we assume that the solution $u(t)$ and the inhomogeneous part $f(t)$ for $t > 0$ are expressed as a linear combination of

$$g_\nu(t) := \frac{1}{\Gamma(\nu)}t^{\nu-1}, \tag{2}$$

for $\nu \in \mathbb{C}\backslash\mathbb{Z}_{<1}$, where $\Gamma(\nu)$ is the gamma function.

In [5,6], ${}_0D_R^{\beta}u(t)$ is the analytic continuation (AC) of Riemann–Liouville fractional derivative (fD), which was introduced in [7,8] and is reviewed in [9]. It is defined for $u(t)$ and $f(t)$ satisfying the following conditions.

Condition A. $u(t)H(t)$ and $f(t)H(t)$ are expressed as a linear combination of $g_\nu(t)H(t)$ for $\nu \in S$, where S is an enumerable set of $\nu \in \mathbb{C}\backslash\mathbb{Z}_{<1}$ satisfying Re $\nu > -M$ for some $M \in \mathbb{Z}_{>-1}$.

We now adopt Condition A. We then express $u(t)$ as follows:

$$u(t) = \sum_{\nu \in S} u_{\nu-1} g_\nu(t) = \sum_{\nu \in S} u_{\nu-1} \frac{1}{\Gamma(\nu)} t^{\nu-1}, \tag{3}$$

where $u_{\nu-1} \in \mathbb{C}$ are constants.

For $\nu \in \mathbb{C}\backslash\mathbb{Z}_{<1}$, ${}_0D_R^\beta g_\nu(t)$ is defined such that

$${}_0D_R^\beta g_\nu(t) = \begin{cases} g_{\nu-\beta}(t), & \nu - \beta \in \mathbb{C}\backslash\mathbb{Z}_{<1}, \\ 0, & \nu - \beta \in \mathbb{Z}_{<1}. \end{cases} \tag{4}$$

When $\beta = n \in \mathbb{Z}_{>-1}$, ${}_0D_R^n g_\nu(t) = \frac{d^n}{dt^n} g_\nu(t)$. Throughout the present paper, the equations involving β are valid for $\beta \in \mathbb{C}$, but in the applications given in Sections 4–7, we use them only for $\beta = n \in \mathbb{Z}_{>-1}$, when ${}_0D_R^n g_\nu(t) = \frac{d^n}{dt^n} g_\nu(t)$.

We use $\mathbb{R}$, $\mathbb{C}$ and $\mathbb{Z}$ to denote the sets of all real numbers, of all complex numbers and of all integers, respectively. We also use $\mathbb{R}_{>r} := \{x \in \mathbb{R} | x > r\}$ for $r \in \mathbb{R}$, ${}_+\mathbb{C} := \{z \in \mathbb{C} | \mathrm{Re}\, z > 0\}$, $\mathbb{Z}_{>a} := \{n \in \mathbb{Z} | n > a\}$, $\mathbb{Z}_{<b} := \{n \in \mathbb{Z} | n < b\}$ and $\mathbb{Z}_{[a,b]} := \{n \in \mathbb{Z} | a \leq n \leq b\}$ for $a, b \in \mathbb{Z}$ satisfying $a < b$. We use Heaviside's step function $H(t)$, which is defined such that (i) $H(t) = 1$ for $t > 0$ and $= 0$ for $t \leq 0$, and (ii) when $f(t)$ is defined on $\mathbb{R}_{>r}$, $f(t)H(t-r)$ is equal to $f(t)$ when $t > r$ and to 0 when $t \leq r$.

In [4–6], we take up a modified Kummer's DE as an example, which is

$$t \cdot \frac{d^2}{dt^2} u(t) + (c - bt) \cdot \frac{d}{dt} u(t) - ab \cdot u(t) = 0, \ \ t > 0, \tag{5}$$

where $a \in \mathbb{C}$, $b \in \mathbb{C}$ and $c \in \mathbb{C}$ are constants. Kummer's DE is this DE with $b = 1$ [10,11]. If $c \notin \mathbb{Z}$, the basic solutions of Equation (5) are given by

$$u_1(t) := {}_1F_1(a; c; bt), \tag{6}$$

$$u_2(t) := t^{1-c} \cdot {}_1F_1(a - c + 1; 2 - c; bt). \tag{7}$$

Here ${}_1F_1(a; c; z) = \sum_{k=0}^\infty \frac{(a)_k}{k!(c)_k} z^k$ is the confluent hypergeometric series, where $z \in \mathbb{C}$, $(a)_n = \prod_{k=0}^{n-1}(a + k)$ for $a \in \mathbb{C}$ and $n \in \mathbb{Z}_{>0}$, and $(a)_0 = 1$. These solutions are expressed as linear combinations of $g_\nu(t)$.

Remark 1. In [4–6], a, b and c in Equation (5) are expressed as $\gamma_1 + 1$, $-\alpha$ and $\gamma_1 + \gamma_2 + 2$, respectively.

In [4,5], we consider the theory of distributions in the space $\mathcal{D}'_R$, which is presented in [12,13] and is explained briefly in Section 3.3. The solution of Equation (1) with the aid of distributions in $\mathcal{D}'_R$ is presented in [5], assuming that the solution satisfies Condition A. Both of the solutions given by Equations (6) and (7), of Equation (5), satisfy Condition A, and we can obtain both of them, by solving Equation (5) by this method.

In [4], we adopt the following condition.

Condition B. $u(t)H(t)$ and $f(t)H(t)$ are expressed as a linear combination of $g_\nu(t)H(t)$ for $\nu \in S_1$, where S_1 is a set of $\nu \in {}_+\mathbb{C}$.

When Condition B is satisfied, ${}_0D_R^\beta u(t)$ of a function $u(t)$ denotes the Riemann–Liouville fD which is defined when $u(t)H(t)$ is locally integrable on $\mathbb{R}$, and hence ${}_0D_R^\beta g_\nu(t)$ is defined only for $\nu \in {}_+\mathbb{C}$, satisfying Equation (4). In this case, the DE given by Equation (1) in terms of the distribution theory in space $\mathcal{D}'_R$ is presented in [4]. The solutions of fractional DE with constant coefficients are presented in [12,13]. In [4], the solution given by Equation (6) of Equation (5), satisfies Condition B, and hence we can obtain it by solving Equation (5) by this method. However, the solution given by Equation (7) satisfies Condition B only when $1 - c > -1$, and hence we can obtain it only when $1 - c > -1$, by this method.

Condition C. There exists $\lambda \in \mathbb{R}_{>0}$ such that $u(t)e^{-\lambda t} \to 0$ as $t \to \infty$.

In [6], it was mentioned that, when Conditions B and C are satisfied, the Laplace transform of $u(t)$ exists and the DE is solved with the aid of Laplace transform, and that the solutions of Equation (5), satisfying Condition B, satisfy Condition C and hence are obtained by using the Laplace transform.

In [6], the AC of Laplace (AC-Laplace) transform is introduced as in Section 1.1 given below, and it is shown that, when Conditions A and C are satisfied, the AC-Laplace transform of $u(t)$, which is denoted by $\hat{u}(s) = \mathcal{L}_H[u(t)] = \mathcal{L}_H[u(t)](s)$, exists and the DE given by Equation (1) is solved with the aid of the AC-Laplace transform. In fact, the AC-Laplace transform of $g_\nu(t)$ and of $u(t)$ given by Equation (3) are expressed as

$$\hat{g}_\nu(s) = \mathcal{L}_H[g_\nu(t)] = s^{-\nu}, \quad \nu \in \mathbb{C}\backslash\mathbb{Z}_{<1}, \tag{8}$$

$$\hat{u}(s) = \mathcal{L}_H[u(t)] = \sum_{\nu \in S} u_{\nu-1}\hat{g}_\nu(s). \tag{9}$$

We review the solution in terms of the AC-Laplace transform in Section 2, and the solution with the aid of the distribution theory in Section 3. In Section 4 we confirm the following lemma.

Lemma 1. Both of the solutions given by Equations (6) and (7), of Equation (5), satisfy Conditions A and C, and hence we can obtain both of them, by solving Equation (5) by using distribution theory in the space $\mathcal{D}'_R$ and also by using the AC-Laplace transform.

In Section 5, we consider the hypergeometric DE, which is given by

$$t(1-t)\cdot\frac{d^2}{dt^2}u(t) + (c - (a+b+1)t)\cdot\frac{d}{dt}u(t) - ab\cdot u(t) = 0, \ \ t > 0, \tag{10}$$

where $a \in \mathbb{C}$, $b \in \mathbb{C}$ and $c \in \mathbb{C}$ are constants.

If $c \notin \mathbb{Z}$, the basic solutions of Equation (10) in [10,11] are given by

$$u_1(t) := {}_2F_1(a, b; c; t), \tag{11}$$

$$u_2(t) := t^{1-c}\cdot {}_2F_1(1+a-c, 1+b-c; 2-c; t), \tag{12}$$

where ${}_2F_1(a, b; c; z) = \sum_{n=0}^{\infty} \frac{(a)_n(b)_n}{n!(c)_n} z^n$ of $z \in \mathbb{C}$ is the hypergeometric series.

Remark 2. These solutions of Equation (10) converge only at t satisfying $|t| < 1$, and do not satisfy Condition A, and they are not obtained by the methods stated above.

We introduce the theory of distributions in the space $\mathcal{D}'_{r,R}$, in Section 3.4. We now use the step function $H_r(t)$, which is defined for $r \in \mathbb{R}_{>0}$ such that (i) $H_r(t) = 1$ for $0 < t < r$ and $= 0$ for $t \leq 0$ or $t \geq r$, and (ii) when $f(t)$ is defined on $\mathbb{R}_{>0} \cap \mathbb{R}_{<r}$, $f(t)H_r(t)$ is equal to $f(t)$ for $0 < t < r$ and to 0 for $t \leq 0$ or $t \geq r$.

Condition D. Condition A with $H(t)$ replaced by $H_r(t)$ is valid.

In Section 5, we show that when $u(t)$ satisfies Condition D, we can solve Equation (10) with the aid of distributions in $\mathcal{D}'_{r,R}$.

Definition 1. Let $u(t)$ be given by Equation (3) and satisfy Condition D. Then, we define its AC of Riemann–Liouville fD of order $\beta \in \mathbb{C}$, by

$$ {}_0D_R^{\beta}u(t) = \sum_{\nu \in S, \nu-\beta \notin \mathbb{Z}_{<1}} u_{\nu-1}g_{\nu-\beta}(t), \tag{13}$$

which satisfies Condition D.

Definition 2. Let $u(t)$ be given by Equation (3) and satisfy Condition D. Then, we define $\hat{u}(s) = \mathcal{L}_S[u(t)] = \mathcal{L}_S[u(t)](s)$ by

$$\hat{u}(s) = \mathcal{L}_S[u(t)] = \sum_{\nu \in S} u_{\nu-1}\hat{g}_{\nu}(s). \tag{14}$$

We call this the AC-Laplace transform series of $u(t)$.

For the solutions of Equation (10), the existence of the AC-Laplace transform is not guaranteed, but we can define $\hat{u}(s)$ by Equation (14). We can then set up a DE satisfied by the thus-defined $\hat{u}(s)$. In Section 5, we write the DE for the $\hat{u}(s)$, and its solution in the form of Equation (14) is obtained. We find that the obtained series converges for no value of s, and yet we obtain the solution $u(t)$ by the term-by-term inverse Laplace transform of the series $\hat{u}(s)$.

The solutions given by Equations (11) and (12), of Equation (10), satisfy Condition D for $r = 1$, and we show that they are obtained by using the distribution theory in the space $\mathcal{D}'_{r,R}$ and also by using the AC of Laplace transform series, in Section 5.

In Section 6, we show that the Bessel functions $J_{\pm\nu}(t)$ are the solusions of Bessel's DE with the aid of the AC-Laplace transform. In Section 7, some discussions are given on Hermite's DE. Concluding remarks are given in Section 8.

1.1. *Definition of the AC-Laplace Transform*

The AC-Laplace transform $\hat{f}(s) = \mathcal{L}_H[f(t)]$ of a function $f(t)$ is defined in [6] as follows.

Condition E. $f_\gamma(z)$ is expressed as $f_\gamma(z) = z^{\gamma-1}f_1(z)$ on a neighborhood of $\mathbb{R}_{>0}$, for $0 \le \arg z < 2\pi$, where $\gamma \in \mathbb{C}\backslash\mathbb{Z}_{<1}$, and $f_1(z)$ is analytic on the neighborhood of $\mathbb{R}_{>0}$.

Definition 3. Let $f_\gamma(z)$ and $u(t) = f_\gamma(t)$ satisfy Conditions E and C, respectively. Then, we define the AC-Laplace transform $\hat{f}_\gamma(s)$ for $\gamma \in \mathbb{C}\backslash\mathbb{Z}_{<1}$, by $\hat{f}_\gamma(s) = \mathcal{L}_H[f_\gamma(t)]$, where

$$\mathcal{L}_H[f_\gamma(t)] = e^{-i\pi\gamma}\frac{1}{2i\sin\pi\gamma}\int_{C_H} f_\gamma(\zeta)e^{-s\zeta}d\zeta, \quad \gamma \in \mathbb{C}\backslash\mathbb{Z}. \tag{15}$$

When $\gamma = n \in \mathbb{Z}_{>0}$, we put $\mathcal{L}_H[f_n(t)] = \lim_{\gamma_i \to n}\mathcal{L}_H[t^{\gamma_i-1}\cdot f_1(t)]$, where $\gamma_i \in \mathbb{C}\backslash\mathbb{Z}$. Here, C_H is the contour which appears in Hankel's formula giving the AC of the gamma function $\Gamma(z)$, so that C_H is the contour which starts from $\infty + i\epsilon$, goes to $\delta + i\epsilon$, encircles the origin counterclockwise, goes to $\delta - i\epsilon$, and then to $\infty - i\epsilon$, where $\delta \in \mathbb{R}_{>0}$ and $\epsilon \in \mathbb{R}_{>0}$ satisfy $\epsilon \leqq 1$ and $\delta < 1$, see [14] (Section 12.22).

Remark 3. $\hat{f}_\gamma(s)$ defined by Definition 3 is an analytic continuation of the Laplace transform defined by $\hat{f}_\gamma(s) = \int_0^\infty f_\gamma(t)e^{-st}dt$ for Re $\gamma > 0$, as a function of γ.

1.2. Remarks on Recent Developments

Here, we call attention to recent developments on the solutions of differential equations related with fractional calculus and perturbation method, which are based on He's variational iteration method (VIM) [15]. By using the VIM, He gave the fD and fI which involve the terms determined by the initial or boundary condition. Liu *et al.* [16] discussed the solution of heat conduction in a fractal medium with the aid of He's fD. Kumar *et al.* discussed the solution of partial differential equations involving time-fD by using Laplace transform and perturbation method based on the VIM; see [17,18] and references in them. In [19,20], discussions are given on the fractional complex transform, which reduces an equation involving fD to an equation involving only integer-order derivatives.

2. AC-Laplace Transform

In the present section and Section 2.1, we assume that $u(t)$ satisfies Conditions C and A, and we put $\hat{u}(s) = \mathcal{L}_H[u(t)]$.

Lemma 2. Let $\nu \in \mathbb{C}\backslash\mathbb{Z}_{<1}$ and $n \in \mathbb{Z}_{>-1}$. Then

$$t^n \cdot g_\nu(t) = (\nu)_n \cdot g_{\nu+n}(t), \tag{16}$$

$$(-1)^n \frac{d^n}{ds^n} \hat{g}_\nu(s) = (\nu)_n \cdot \hat{g}_{\nu+n}(s), \tag{17}$$

where $(\nu)_n$ is Pochhammer's symbol, so that $(\nu)_n = \frac{\Gamma(\nu+n)}{\Gamma(\nu)}$.

Proof We confirm Equation (16) by using Equation (2) on both sides of Equation (16). By Equation (8), $\mathcal{L}_H[g_\nu(t)] = \hat{g}_\nu(s) = s^{-\nu}$. By taking derivatives of this equation repeatedly, we obtain Equation (17). ■

Lemma 3. Let $\nu \in \mathbb{C}\backslash\mathbb{Z}_{<1}$, $n \in \mathbb{Z}_{>-1}$ and $\beta \in \mathbb{C}$. Then

$$\mathcal{L}_H[t^n \cdot g_\nu(t)] = (-1)^n \frac{d^n}{ds^n} \hat{g}_\nu(s), \tag{18}$$

Proof Equation (18) is obtained by comparing Equations (16) and (17) with Equation (8). ■

Applying Lemma 3 to Equation (4), with the aid of Equation (8), we obtain

$$\mathcal{L}_H[t^n \cdot {}_0D_R^\beta g_\nu(t)] = \begin{cases} (-1)^n \frac{d^n}{ds^n} \hat{g}_{\nu-\beta}(s) = (-1)^n \frac{d^n}{ds^n}[s^\beta \hat{g}_\nu(s)], & \nu - \beta \in \mathbb{C}\backslash\mathbb{Z}_{<1}, \\ 0, & \nu - \beta \in \mathbb{Z}_{<1}. \end{cases} \tag{19}$$

Lemma 4. Let $\nu \in \mathbb{C}\backslash\mathbb{Z}_{<1}$, $n \in \mathbb{Z}_{>-1}$ and $\beta \in \mathbb{C}$. Then

$$\mathcal{L}_H[t^n \cdot {}_0D_R^\beta g_\nu(t)] = (-1)^n \frac{d^n}{ds^n}[s^\beta \hat{g}_\nu(s)] - (-1)^n \frac{d^n}{ds^n} \langle s^\beta \hat{g}_\nu(s) \rangle_z, \tag{20}$$

where

$$\langle s^\beta \hat{g}_\nu(s) \rangle_z = \begin{cases} 0, & \nu - \beta \in \mathbb{C}\backslash\mathbb{Z}_{<1}, \\ s^\beta \hat{g}_\nu(s) = \hat{g}_{-k}(s) = s^k, & -k = \nu - \beta \in \mathbb{Z}_{<1}. \end{cases} \tag{21}$$

Proof Here, $\langle s^\beta \hat{g}_\nu(s) \rangle_z$ is so defined that Equation (20) with Equation (21) represents Equation (19). ■

Lemma 5. Let $u(t)$ be expressed by Equation (3), $n \in \mathbb{Z}_{>-1}$ and $\beta \in \mathbb{C}$. Then

$$\mathcal{L}_H[t^n \cdot u(t)] = (-1)^n \frac{d^n}{ds^n}\hat{u}(s), \tag{22}$$

$$\mathcal{L}_H[t^n \cdot {}_0D_R^{\beta} u(t)] = (-1)^n \frac{d^n}{ds^n}[s^{\beta}\hat{u}(s)] - \langle (-1)^n \frac{d^n}{ds^n}[s^{\beta}\hat{u}(s)]\rangle_z, \tag{23}$$

where

$$\langle (-1)^n \frac{d^n}{ds^n}[s^{\beta}\hat{u}(s)]\rangle_z = (-1)^n \frac{d^n}{ds^n}\langle s^{\beta}\hat{u}(s)\rangle_z, \quad \langle s^{\beta}\hat{u}(s)\rangle_z = \sum_{\substack{\nu \in S \\ -k=\nu-\beta \in \mathbb{Z}_{<1}}} u_{\nu-1} s^k. \tag{24}$$

Proof We confirm these with the aid of Equation (9) and Lemmas 3 and 4. ■

Remark 4. Let $\beta = n \in \mathbb{Z}_{>-1}$. Then, Equation (24) gives $\langle \hat{u}(s)\rangle_z = 0$, and

$$\langle s^n \hat{u}(s)\rangle_z = \sum_{\substack{k=0 \\ n-k \in S}}^{n-1} u_{n-k-1} s^k, \quad n \in \mathbb{Z}_{>0}, \tag{25}$$

so that

$$\langle \hat{u}(s)\rangle_z = 0, \quad \langle s \cdot \hat{u}(s)\rangle_z = u_0, \quad \langle s^2 \cdot \hat{u}(s)\rangle_z = u_1 + u_0 s, \quad \langle s^3 \cdot \hat{u}(s)\rangle_z = u_2 + u_1 s + u_0 s^2, \cdots. \tag{26}$$

Theorem 1. *Let $u(t)$ satisfy Conditions C and A, and be expressed by* Equation (3). *Then, the AC-Laplace transform $\hat{u}(s)$ is expressed as* Equation (9) *and there exists $M \in \mathbb{R}_{>0}$ such that the series given by* Equation (9) *converges for $|s| > M$.*

2.1. Recipe of Solving Differential Equation with Polynomial Coefficients

We now give a recipe of solving the DE with polynomial coefficients, which is given by

$$\sum_{l=0}^{2}(a_l t^2 + b_l t + c_l) \cdot \frac{d^l}{dt^l}u(t) = f(t), \quad t > 0, \tag{27}$$

where $a_l \in \mathbb{C}$, $b_l \in \mathbb{C}$ and $c_l \in \mathbb{C}$ for $l \in \mathbb{Z}_{[0,2]}$ are constants. We then introduce the function $p(t,s)$ by

$$p(t,s) = \sum_{l=0}^{2}(a_l t^2 + b_l t + c_l) \cdot s^l, \tag{28}$$

and express Equation (27) as

$$p(t, \frac{d}{dt})u(t) = f(t), \quad t > 0. \tag{29}$$

In Sections 4 and 5, we discuss modified Kummer's DE given by Equation (5) and the hypergeometric DE given by Equation (10).

We obtain the following theorems, with the aid of Lemma 5.

Theorem 2. *Let $u(t)$ in the form of* Equation (3) *be the solution of* Equation (29). *Then $\hat{u}(s)$ given by* Equation (9) *is a solution of the DE:*

$$p(-\frac{d}{ds}, s)\hat{u}(s) = \hat{f}(s) + \langle p(-\frac{d}{ds}, s)\hat{u}(s)\rangle_z. \tag{30}$$

Theorem 3. *Let* $\hat{u}(s)$ *in the form of* Equation (9) *be a solution of* Equation (30). *Then, the corresponding* $u(t)$ *given by* Equation (3) *is a solution of* Equation (29).

Corollary 4. *Let* $\hat{u}(s)$ *be a solution of*

$$p(-\frac{d}{ds}, s)\hat{u}(s) = \hat{f}(s). \tag{31}$$

If the obtained $\hat{u}(s)$ *satisfies*

$$\langle p(-\frac{d}{ds}, s)\hat{u}(s)\rangle_z = 0, \tag{32}$$

then $u(t)$ *given by* Equation (3) *is a solution of* Equation (29), *that is of* Equation (27).

2.2. Term-by-Term Operators for $u(t)$ and $\hat{u}(s)$

The AC of Riemann–Liouville fD: ${}_0D_R^{\beta}u(t)$ and the AC-Laplace transform series: $\mathcal{L}_S[u(t)]$, of $u(t)$ in the form of Equation (3), are defined by Equation (13) and Equation (14). We now define the operators which appear in Lemma 5.

Definition 4. Let $n \in \mathbb{Z}_{>0}$, $\beta \in \mathbb{C}$ and $\gamma \in \mathbb{C}$, and let $u(t)$ and $\hat{u}(s)$ be expressed as Equation (3) and Equation (14), respectively. Then, we adopt

$$t^n \cdot u(t) = \sum_{\nu \in S} (t^n \cdot u_{\nu-1} g_\nu(t)), \tag{33}$$

$$s^{\beta}\hat{u}(s) = \sum_{\nu \in S} (s^{\beta} \cdot u_{\nu-1}\hat{g}_\nu(s)), \quad s^{\gamma}(s^{\beta}\hat{u}(s)) = s^{\gamma+\beta}\hat{u}(s), \tag{34}$$

$$(-1)^n \frac{d^n}{ds^n}[s^{\beta}\hat{u}(s)] = \sum_{\nu \in S} [(-1)^n \frac{d^n}{ds^n}(s^{\beta}u_{\nu-1}\hat{g}_\nu(s))]. \tag{35}$$

Proposition 5. *Let the operators on the lhs of the equations in Definition 2, 3 and 4 be defined by the respective rhs. Then, Theorems 2 and 3 and Corollary 4 are valid.*

3. Operational Calculus in the Spaces $\mathcal{D}'_R$ and $\mathcal{D}'_{r,R}$

We now consider the theory of distributions in the spaces $\mathcal{D}'_R$ and $\mathcal{D}'_{r,R}$, which are explained briefly in Sections 3.3 and 3.4, respectively.

3.1. Operational Calculus in the Spaces $\mathcal{D}'_R$

In the theory of distributions in $\mathcal{D}'_R$, a regular distribution is such a distribution that can be regarded as a function which is locally integrable on $\mathbb{R}$. In the present paper, when the product $u(t)H(t) \in \mathcal{L}_{loc}(\mathbb{R})$, we consider the regular distribution $\tilde{u}(t)$ which is regarded to be equal to $u(t)H(t)$. We then denote this correspondence between a distribution and a function by $\tilde{u}(t) \bullet\!\!-\!\!\circ \; u(t)H(t)$. Here, $u(t)H(t) \in \mathcal{L}_{loc}(\mathbb{R})$ denotes that $u(t)H(t)$ is locally integrable on $\mathbb{R}$. From Equation (2), we have $g_1(t) = 1$. We put $\tilde{H}(t) = \tilde{g}_1(t)$, and then we obtain $\tilde{H}(t) \bullet\!\!-\!\!\circ \; g_1(t)H(t) = H(t)$.

Definition 5. In the space $\mathcal{D}'_R$, the differential operator D^{λ} of order $\lambda \in \mathbb{C}$ is defined as follows:

(i) If Re $\nu > 0$ and $u(t)H(t) \in \mathcal{L}_{loc}(\mathbb{R})$, then $D^{-\nu}\tilde{u}(t) \bullet\!\!-\!\!\circ \; [{}_0D_R^{-\nu}u(t)]H(t)$.
(ii) If $h(t) \in \mathcal{D}'_R$, then $D^{\lambda}h(t) \in \mathcal{D}'_R$.
(iii) The index law stated in the following lemma is valid.

Lemma 6. Let $h(t) \in \mathcal{D}'_R$. Then, the index law:

$$D^{\lambda} D^{\mu} h(t) = D^{\lambda+\mu} h(t), \tag{36}$$

is valid for every pair of $\lambda \in \mathbb{C}$ and $\mu \in \mathbb{C}$.

Dirac's delta function $\delta(t) \in \mathcal{D}'_R$ is the distribution defined by $\delta(t) = D\tilde{H}(t)$.

We now introduce the correspondence even between a distribution, which is not a regular one, and a function, as follows.

Definition 6. Let $\nu \in \mathbb{C}$, $u(t)$ and $v(t) = {}_0D_R^{\nu} u(t)$ satisfy $u(t)H(t) \in \mathcal{L}_{loc}(\mathbb{R})$ and ${}_0D_R^{-\nu} v(t) = u(t)$. Then, $\tilde{v}(t) = D^{\nu}\tilde{u}(t) \bullet\!\!-\!\!\circ\ {}_0D_R^{\nu} u(t)H(t) = v(t)H(t)$.

We now define $\tilde{g}_\nu(t)$ for $\nu \in \mathbb{C}$ by

$$\tilde{g}_\nu(t) = D^{-\nu+1}\tilde{H}(t) = D^{-\nu}\delta(t) = \hat{g}_\nu(D)\delta(t). \tag{37}$$

From Equations (2) and (4), we have $g_1(t) = 1$, $g_\nu(t) = {}_0D_R^{-\nu+1} g_1(t)$ and $g_1(t) = {}_0D_R^{\nu-1} g_\nu(t)$ for $\nu \in \mathbb{C}\backslash\mathbb{Z}_{<1}$. Since $g_1(t)H(t) = H(t) \in \mathcal{L}_{loc}(\mathbb{R})$, by Definition 6, we have

$$\tilde{g}_\nu(t) = \hat{g}_\nu(D)\delta(t) \bullet\!\!-\!\!\circ\ g_\nu(t)H(t), \quad \nu \in \mathbb{C}\backslash\mathbb{Z}_{<1}. \tag{38}$$

Remark 5. Note here that when $\mathrm{Re}\,\nu \le -1$, $\tilde{g}_\nu(t)$ is not a regular distribution.

Lemma 7. Let $\nu \in \mathbb{C}\backslash\mathbb{Z}_{<1}$ and $n \in \mathbb{Z}_{>-1}$. Then

$$(-1)^n \frac{\partial^n}{\partial D^n}\tilde{g}_\nu(t) = (-1)^n \frac{\partial^n}{\partial D^n}\hat{g}_\nu(D)\delta(t) \bullet\!\!-\!\!\circ\ [t^n \cdot g_\nu(t)]H(t), \tag{39}$$

where the derivative with respect to D is taken regarding D as a variable.

Proof By comparing Equation (18) with Equation (38), we confirm this. ∎

By Equations (37) and (36), we have

$$D^{\beta}\tilde{g}_\nu(t) = D^{\beta-\nu}\delta(t) = \tilde{g}_{\nu-\beta}(t). \tag{40}$$

Comparing this with Equations (4) and (38), we have

Lemma 8. Let $\nu \in \mathbb{C}\backslash\mathbb{Z}_{<1}$ and $\beta \in \mathbb{C}$. Then

$$D^{\beta}\tilde{g}_\nu(t) - \langle D^{\beta}\tilde{g}_\nu(t)\rangle_z = D^{\beta}\hat{g}_\nu(D)\delta(t) - \langle D^{\beta}\hat{g}_\nu(D)\rangle_z\delta(t) \bullet\!\!-\!\!\circ\ [{}_0D_R^{\beta} g_\nu(t)]H(t), \tag{41}$$

where

$$\langle D^{\beta}\tilde{g}_\nu(t)\rangle_z = \langle D^{\beta}\hat{g}_\nu(D)\rangle_z\delta(t) = \begin{cases} 0, & \nu-\beta \in \mathbb{C}\backslash\mathbb{Z}_{<1}, \\ D^{\beta}\tilde{g}_\nu(t) = \tilde{g}_{-k}(t) = D^k\delta(t), & -k = \nu-\beta \in \mathbb{Z}_{<1}. \end{cases} \tag{42}$$

Definition 7. Let $u(t)$ be expressed by Equation (3). Then, we define the distribution $\tilde{u}(t)$ by

$$\tilde{u}(t) := \hat{u}(D)\delta(t) = \sum_{\nu\in S} u_{\nu-1}\tilde{g}_\nu(t) = \sum_{\nu\in S} u_{\nu-1}\hat{g}_\nu(D)\delta(t), \tag{43}$$

in accordance with Equation (37).

Definition 8. We adopt Definition 4 with $\hat{u}(s)$, $\hat{g}_\nu(s)$, s and $\frac{d}{ds}$ replaced by $\tilde{u}(t) = \hat{u}(D)\delta(t)$, $\tilde{g}_\nu(t) = \hat{g}_\nu(D)\delta(t)$, D and $\frac{\partial}{\partial D}$, respectively.

Lemma 9. Let $u(t)$ be expressed by Equation (3), $n \in \mathbb{Z}_{>-1}$ and $\beta \in \mathbb{C}$. Then

$$(-1)^n \frac{\partial^n}{\partial D^n}\tilde{u}(t) = (-1)^n \frac{\partial^n}{\partial D^n}\hat{u}(D)\delta(t) \bullet\!\!-\!\!\circ\ [t^n \cdot u(t)]H(t), \tag{44}$$

$$(-1)^n \frac{\partial^n}{\partial D^n} D^\beta \tilde{u}(t) - \langle (-1)^n \frac{\partial^n}{\partial D^n} D^\beta \tilde{u}(t)\rangle_z = (-1)^n \frac{\partial^n}{\partial D^n}[D^\beta \hat{u}(D)]\delta(t) - \langle (-1)^n \frac{\partial^n}{\partial D^n}[D^\beta \hat{u}(D)]\rangle_z \delta(t)$$
$$\bullet\!\!-\!\!\circ\ [t^n \cdot {}_0D_R^\beta u(t)]H(t), \tag{45}$$

where $\langle D^\beta \hat{u}(D)\rangle_z$ and $\langle (-1)^n \frac{\partial^n}{\partial D^n}[D^\beta \hat{u}(D)]\rangle_z$ are defined by Equation (24) with $-\frac{d}{ds}$ and s replaced by $-\frac{\partial}{\partial D}$ and D, respectively.

Proof The first equalities in Equations (44) and (45) are due to Definition 7. When $u(t) = g_\nu(t)$ for $\nu \in \mathbb{C}\backslash\mathbb{Z}_{<-1}$, Equation (44) is Equation (39). Equation (45) for $n = 0$ in this case is Equation (41). Equation (45) for $n > 0$ in this case is derived by applying Equation (44) to Equation (41). When $u(t)$ is expressed by Equation (3), we confirm Equations (44) and (45) by using the rhs of Equations (43), (14) and (3) in place of $\tilde{u}(t)$, $\hat{u}(D)$ and $u(t)$, respectively, in Equations (44) and (45), and comparing the respective equations for the case of $u(t) = g_\nu(t)$ for $\nu \in \mathbb{C}\backslash\mathbb{Z}_{<1}$. ■

3.1.1. Recipe of Solving Linear Differential Equation with Polynomial Coefficients

By applying Lemma 9 to the DE given by Equation (29), we obtain the following lemma.

Lemma 10. The DE for $\tilde{u}(t) = \hat{u}(D)\delta(t)$ corresponding to Equation (29) is

$$p(-\frac{\partial}{\partial D}, D)\hat{u}(D)\delta(t) = \tilde{f}(t) + \langle p(-\frac{\partial}{\partial D}, D)\hat{u}(D)\rangle_z \delta(t). \tag{46}$$

Theorem 6. *When $\tilde{u}(t) = \hat{u}(D)\delta(t)$ is a solution of* Equation (46), *the corresponding $u(t)$ is a solution of* Equation (29).

Proposition 7. *Let $\hat{u}(s)$ be a solution of* Equation (30), *and $\tilde{u}(t) = \hat{u}(D)\delta(t)$ obtained from it be expressed in the form of* Equation (43). *Then $u(t)$ given by* Equation (3) *is a solution of* Equation (29).

Corollary 8. *Let $\hat{u}(s)$ be a solution of* Equation (31) *and $\tilde{u}(t) = \hat{u}(D)\delta(t)$, which is obtained from it, be expressed in the form of* Equation (43). *If the obtained $\tilde{u}(t)$ satisfies*

$$\langle p(-\frac{\partial}{\partial D}, D)\hat{u}(D)\rangle_z \delta(t) = 0, \tag{47}$$

then $u(t)$ given by Equation (3) *is a solution of* Equation (29).

3.2. Operational Calculus in the Space $\mathcal{D}'_{r,R}$

For the function $H_r(t)$ defined below Equation (12) in the Introduction section, we define the regular distribution $\tilde{H}_r(t)$ in the space $\mathcal{D}'_{r,R}$, so that $\tilde{H}_r(t) \bullet\!\!-\!\!\circ\ g_1(t)H_r(t) = H_r(t)$, and then define $\delta_r(t) \in \mathcal{D}'_{r,R}$ by $\delta_r(t) = D\tilde{H}_r(t) = \delta(t) - \delta(t-r)$. We can use the formulas presented in Section 3.1 for the space $\mathcal{D}'_R$, also in the space $\mathcal{D}'_{r,R}$, if we replace $H(t)$, $\tilde{H}(t)$ and $\delta(t)$, by $H_r(t)$, $\tilde{H}_r(t)$ and $\delta_r(t)$, respectively.

In Section 4, Theorem 1 applies. In Section 5, the rhs of Equation (44) for $n = 0$ is $u(t)H_r(t) = \sum_{\nu \in S} u_{\nu-1} g_\nu(t)H_r(t)$, which converges at $|t| < 1$. Although $\hat{u}(s)$ converges for no value of s, we regard that the rhs of Equation (44) converges at $|t| < 1$, and the lhs of Equation (44) represents the corresponding regular distribution. The result obtained for $u(t)$ is then the one obtained by the term-by-term inverse Laplace transform of $\hat{u}(s)$, which is expressed as Equation (14).

3.3. Distributions in the Space $\mathcal{D}'_R$

Distributions in the space $\mathcal{D}'$ are first introduced [21–24]. The distributions are either regular ones or their derivatives. A regular distribution in $\mathcal{D}'$ corresponds to a function which is locally integrable on $\mathbb{R}$. The space $\mathcal{D}$, that is dual to $\mathcal{D}'$, is the space of testing functions, which are infinitely differentiable on $\mathbb{R}$ and have a compact support. A distribution $h \in \mathcal{D}'$ is a functional, to which $\langle h, \phi \rangle \in \mathbb{C}$ is associated with every $\phi \in \mathcal{D}$. When $\tilde{f} \in \mathcal{D}'$ corresponds to a function f, then we put $\langle \tilde{f}, \phi \rangle = \int_{-\infty}^{\infty} f(t)\phi(t)dt$.

Definition 9. Let $\tilde{f}(t)$ be a regular distribution in $\mathcal{D}'$ and $f(t)$ be the function corresponding to it. If $h(t) \in \mathcal{D}'$ is expressed by $h(t) = D^n \tilde{f}(t)$ for $n \in \mathbb{Z}_{>-1}$, then

$$\langle h, \phi \rangle = \langle D^n \tilde{f}, \phi \rangle = \int_{-\infty}^{\infty} f(t)[D_W^n \phi(t)]dt, \tag{48}$$

for every $\phi(t) \in \mathcal{D}$, where $D_W^n \phi(t) = (-1)^n \frac{d^n}{dt^n} \phi(t)$.

Lemma 11. D^n for $n \in \mathbb{Z}_{>-1}$ are operators in the space $\mathcal{D}'$.

In the present study, we consider functions $g_\nu(t)$ and ${}_0D_R^{-\lambda} g_\nu(t)$ for $\nu \in \mathbb{R}_{>0}$ and $\lambda \in \mathbb{R}_{>0}$, and the distributions $\tilde{g}_\nu(t)$ and $D^{-\lambda} \tilde{g}_\nu(t)$ corresponding to them. We then desire to introduce the operator $D_W^{-\lambda}$ such that $\langle D^{-\lambda} h, \phi \rangle = \langle h, D_W^{-\lambda} \phi \rangle$ for $h \in \mathcal{D}'$ and $\phi \in \mathcal{D}$. However, we find that $D_W^{-\lambda} \phi$ does not belong to $\mathcal{D}$.

In this situation, we consider the problem in the space $\mathcal{D}'_R$ [12,13]. A regular distribution in $\mathcal{D}'_R$ is such a distribution that it corresponds to a function which is locally integrable on $\mathbb{R}$ and has a support bounded on the left. The space $\mathcal{D}_R$, that is dual to $\mathcal{D}'_R$, is the space of testing functions, which are infinitely differentiable on $\mathbb{R}$ and have a support bounded on the right.

Definition 10. Definition 9 with $\mathcal{D}'$ and $\mathcal{D}$ replaced by $\mathcal{D}'_R$ and $\mathcal{D}_R$, respectively, is valid.

Lemma 12. D^ν for $\nu \in \mathbb{C}$ are operators in the space $\mathcal{D}'_R$.

3.4. Distributions in the Space $\mathcal{D}'_{r,R}$

The space $\mathcal{D}_{r,R}$ for $r \in \mathbb{R}$ is such a subspace of $\mathcal{D}_R$, that if $\phi(t) \in \mathcal{D}_{r,R}$, $\phi(t) = 0$ for $t \geq r$. A regular distribution $\tilde{f}(t) \in \mathcal{D}'_{r,R}$ is such a distribution in $\mathcal{D}'_{r,R}$ that it corresponds to a function which is locally integrable on $(-\infty, r)$, and has a support bounded on the left.

Definition 11. Definition 9 with $\mathcal{D}'$ and $\mathcal{D}$ replaced by $\mathcal{D}'_{r,R}$ and $\mathcal{D}_{r,R}$, respectively, is valid.

Lemma 13. D^ν for $\nu \in \mathbb{C}$ are operators in the space $\mathcal{D}'_{r,R}$.

Now the rhs of Equation (48) may be expressed as $\int_{-\infty}^{r} f(t)[D_W^n \phi(t)]dt$.

4. Solution of Modified Kummer's DE

We now study the modified Kummer's DE given by Equation (5). We define $p_K(t,s)$ by

$$p_K(t,s) = t \cdot s^2 + (c - bt)s - ab = t(s^2 - bs) + cs - ab. \tag{49}$$

Then Equation (5) is expressed as $p_K(t, \frac{d}{dt})u(t) = 0,\ t > 0$, and the DE which corresponds to Equation (30) is

$$p_K(-\frac{d}{ds}, s)\hat{u}(s) = -\frac{d}{ds}[(s^2 - bs)\hat{u}(s)] + (cs - ab)\hat{u}(s) = -(1-c)u_0, \tag{50}$$

where the rhs is evaluated by using the first equality and Equations (24) and (26) as

$$\langle p_K(-\frac{d}{ds}, s)\hat{u}(s)\rangle_z = -\frac{d}{ds}(u_1 + u_0 \cdot s - b \cdot u_0) + c \cdot u_0 = -u_0 + c \cdot u_0. \tag{51}$$

4.1. Solution Satisfying $(1-c)u_0 = 0$

When $(1-c)u_0 = 0$, the rhs of Equation (50) is 0, and the solution of Equation (50) is given by

$$\begin{aligned}\hat{u}(s) &= Cs^{a-1}(s-b)^{-1-a+c} = Cs^{-2+c}\sum_{k=0}^{\infty}\frac{(1+a-c)_k}{k!}(\frac{b}{s})^k \\ &= Cs^{-2+c} \cdot {}_1F_0(1+a-c;;\frac{b}{s}), \quad |s| > |b|,\end{aligned} \tag{52}$$

where ${}_1F_0(a;;z) = \sum_{k=0}^{\infty}\frac{(a)_k}{k!}z^k$ and C is an arbitrary constant. If $1-c \notin \mathbb{Z}_{<1}$, $\hat{u}(s)$ is of the form of Equation (9), and we can take its inverse Laplace transform. Then, choosing $C = \Gamma(2-c)$, we obtain

$$u(t) = \Gamma(2-c)t^{1-c}\sum_{k=0}^{\infty}\frac{(1+a-c)_k}{k!\Gamma(2-c+k)}b^k t^k = t^{1-c} \cdot {}_1F_1(1+a-c; 2-c; bt), \quad t > 0, \tag{53}$$

by using the formula $\Gamma(2-c+k) = \Gamma(2-c)(2-c)_k$. This is the solution given by Equation (7).

By Corollary 4, we confirm the following lemma.

Lemma 14. If $1-c \notin \mathbb{Z}_{<0}$, then $u(t)$ given by Equation (53) is a solution of Equation (5).

The derivation of this lemma based on the distribution theory is as follows. When $1-c \notin \mathbb{Z}_{<0}$, $\tilde{u}(t) = \hat{u}(D)\delta(t)$ takes the form of Equation (43), and satisfies $p_K(-\frac{\partial}{\partial D}, D)\hat{u}(D)\delta(t) = (1-c)u_0 \cdot \delta(t) = 0$. We put $C = \Gamma(2-c)$, and then the corresponding solution $u(t)$ is given by Equation (53), by Corollary 8.

4.2. Solution Satisfying $u_0 = 1$

In [4–6], the other solution given by Equation (6) of Equation (5) is obtained by solving the inhomogeneous Equation (50) for $b \neq 0$, $1-c \neq 0$ and $u_0 = 1$. We note here that the method using the following lemma with the aid of the solution given by Equation (53), which is already obtained, is easier.

Lemma 15. If $u(a, c; bt)$ is a solution of Equation (5), $t^{1-c}u(1+a-c, 2-c; bt)$ is also a solution of the same equation.

Proof We put $u(t) = t^{\lambda}w(t)$ in Equation (5), and then we obtain Equation (5) with a, c and u replaced by $1+a-c$, $2-c$ and w, respectively, when $\lambda = 1-c$. ■

4.3. Solution by the Basic Method

We note that the solutions of Equations (5) and (50) are obtained by using the basic method of solution, which is explained in [14] (Section 10.3). In that method, the solution of Equation (5) is assumed to be given by

$$u(t) = t^{\alpha} \sum_{k=0}^{\infty} p_k t^k, \tag{54}$$

where $\alpha \in \mathbb{C}\backslash\mathbb{Z}_{<1}$, $p_k \in \mathbb{C}$ and $p_0 \neq 0$. Then, by Theorem 2, we have

$$\hat{u}(s) = s^{-\alpha-1} \sum_{k=0}^{\infty} \hat{p}_k s^{-k}, \tag{55}$$

where $\hat{p}_k = p_k \Gamma(\alpha + k + 1)$.

In solving Equation (5), Equation (54) is put in it. In solving Equation (50), Equation (55) is put in it. We then note that the condition of $\hat{p}_0 = p_0 \Gamma(\alpha + 1) = C \neq 0$ requires (i) $\alpha = 1 - c$ and $\hat{p}_0 = p_0 \Gamma(2 - c) \neq 0$, or (ii) $\alpha = 0$ and $\hat{p}_0 = p_0 = u_0 \neq 0$. Then $\hat{p}_k$ or p_k for $k \in \mathbb{Z}_{>0}$ are determined by

$$\begin{aligned} \hat{p}_k &= \hat{p}_{k-1} \frac{\alpha + a + k - 1}{\alpha + c + k - 1} b = p_0 \Gamma(\alpha + 1) \frac{(\alpha + a)_k}{(\alpha + c)_k} b^k, \\ p_k &= p_{k-1} \frac{\alpha + a + k - 1}{(\alpha + c + k - 1)(\alpha + k)} b = p_0 \frac{(\alpha + a)_k}{(\alpha + c)_k (\alpha + 1)_k} b^k. \end{aligned} \tag{56}$$

When $\alpha = 0$, we have

$$\hat{p}_k = p_0 \frac{(a)_k}{(c)_k} b^k, \quad p_k = p_0 \frac{(a)_k}{k!(c)_k} b^k. \tag{57}$$

When $\alpha = 1 - c$, we have

$$\hat{p}_k = p_0 \Gamma(2 - c) \frac{(1 + a - c)_k}{k!} b^k, \quad p_k = p_0 \frac{(1 + a - c)_k}{k!(2 - c)_k} b^k. \tag{58}$$

Using Equation (58) with $\alpha = 1 - c$ and $p_0 = 1$ in Equations (55) and (54), we obtain Equations (52) and (53), respectively. Using Equation (57) with $\alpha = 0$ and $p_0 = 1$ in Equation (54), we obtain Equation (6).

4.4. Solution by the Modified Nishimoto's Method

In [25], the solution of Kummer's DE is given by modified Nishimoto's method, where the solution is expressed by fD of a function. In that paper, the solution given by Equation (53) of Equation (5) is obtained by using Lemma 11 and Proof of Lemma 10 in [25] (Section 3.2).

Remark 6. In Lemmas 9 and 10 in [25], four of Kummer's eight solutions of Kummer's DE are given. We note here that these lemmas are applicable to the modified Kummer's DE, if we introduce the following replacements in Sections 3.1 and 3.2 in [25]. We replace (i): (11) in Lemma 9 and in Proof of Lemma 9 by (18), (ii): e^z in Equations (14), (16) and (17) by $e^{\delta \cdot z}$; (iii): 1 and -1 in the column for δ_l in Table 2 by δ and $-\delta$, respectively; (iv): ${}_1F_1(*;*;z)$ and ${}_1F_1(*;*;-z)$ in (16) and (17) by ${}_1F_1(*;*;\delta \cdot z)$ and ${}_1F_1(*;*;-\delta \cdot z)$, respectively, and (v): $\delta_4 = -\delta_3 = -1$ in Proof of Lemma 9 by $\delta_4 = -\delta_3 = -\delta$.

5. Solution of the Hypergeometric DE

In solving the hypergeometric DE given by Equation (10), we introduce $p_H(t,s)$, by

$$p_H(t,s) = t(1-t)\cdot s^2 + [c-(a+b+1)t]\cdot s - ab = -t^2\cdot s^2 + t[s^2-(a+b+1)s] + cs - ab. \quad (59)$$

Now , we put $\hat{u}(s) = \mathcal{L}_S[u(t)]$ and then the DE which corresponds to Equation (30) is

$$p_H(-\frac{d}{ds},s)\hat{u}(s) = -\frac{d^2}{ds^2}[s^2\hat{u}(s)] - \frac{d}{ds}[(s^2-(a+b+1)s)\hat{u}(s)] + (cs-ab)\cdot\hat{u}(s) = -(1-c)u_0, \quad (60)$$

where the rhs is evaluated by using the first equality and Equations (24) and (26) as

$$\langle p_H(-\frac{d}{ds},s)\hat{u}(s)\rangle_z = -u_0 + c\cdot u_0. \quad (61)$$

Lemma 16. Let $\hat{u}(s)$ be a solution of Equation (60), $\hat{u}(s) = s^{a-1}\hat{w}(s)$ and

$$p_W(t,s) = -t^2\cdot s + t(s-(1-a+b)) - (a-c). \quad (62)$$

Then , $\hat{w}(s)$ is a solution of the following DE:

$$p_W(-\frac{d}{ds},s)\hat{w}(s) = -\frac{d^2}{ds^2}[s\cdot\hat{w}(s)] - \frac{d}{ds}[(s-(1-a+b))\hat{w}(s)] - (a-c)\hat{w}(s) = -(1-c)u_0 s^{-a}. \quad (63)$$

Proof We put $\hat{u}(s) = s^\lambda\hat{v}(s)$ in Equation (60). By using $\frac{d}{ds}[s^\lambda\hat{v}(s)] = s^\lambda[\lambda s^{-1}\hat{v}(s) + \frac{d}{ds}\hat{v}(s)]$ and

$$\frac{d^2}{ds^2}[s^\lambda\hat{v}(s)] = s^\lambda\{\lambda(\lambda-1)s^{-2}\hat{v}(s) + 2\lambda\frac{d}{ds}[s^{-1}\hat{v}(s)] + \frac{d^2}{ds^2}\hat{v}(s)\}, \quad (64)$$

we obtain

$$-\frac{d^2}{ds^2}[s^2\hat{v}(s)] - \frac{d}{ds}[(s-(1-2\lambda+a+b))s\cdot\hat{v}(s)] - [(\lambda-c)s + (\lambda-a)(\lambda-b)]\hat{v}(s)$$
$$= -(1-c)u_0 s^{-\lambda}.$$

When $\lambda = a$ and $\hat{w}(s) = s\cdot\hat{v}(s)$, this gives Equation (63). ■

5.1. *Solution Satisfying* $(1-c)u_0 = 0$

In this section, we use the distribution theory in the space $\mathcal{D}'_{r,R}$ for $r = 1$, so that $\delta_1(t)$ appears.

Lemma 17. Let $a - c \notin \mathbb{Z}_{<0}$. Then, we have the following complementary solution (C-solution) of Equation (63):

$$\hat{w}(s) = Ks^{-1-a+c}\sum_{k=0}^{\infty}\frac{(1+b-c)_k(1+a-c)_k}{k!}s^{-k} = Ks^{-1-a+c}\cdot {}_2F_0(1+b-c, 1+a-c;;s^{-1}), \quad (65)$$

where K is any constant, and ${}_2F_0(a,b;;z) = \sum_{k=0}^{\infty}\frac{(a)_k(b)_k}{k!}z^k$.

Proof For $p_W(t,s)$ given by Equation (62), we choose the DE for $w(t)$ as follows:

$$p_W(t,\frac{d}{dt})w(t) = (t-t^2)\frac{d}{dt}w(t) - [(1-a+b)t + (a-c)]w(t) = 0, \quad t > 0. \quad (66)$$

The solution of this DE is

$$w(t) = C_1 \cdot t^{a-c}(1-t)^{-1-b+c} = C_1 \cdot t^{a-c} \sum_{k=0}^{\infty} \frac{(1+b-c)_k}{k!} t^k$$
$$= C_1 \cdot t^{a-c} {}_1F_0(1+b-c;;t), \quad 0 < t < 1, \tag{67}$$

where C_1 is an arbitrary constant. $\hat{w}(s)$ given by Equation (65) is the term-by-term Laplace transform of this $w(t)$. Theorem 2 and Proposition 5 show that Equation (63) for $(1-c)u_0 = 0$ is satisfied by this $\hat{w}(s)$, since the first equality of Equation (63) shows $\langle p_W(-\frac{d}{ds}, s)\hat{w}(s)\rangle_z = 0$,

If $a - c \notin \mathbb{Z}_{<0}$, this $w(t)$ takes the form of Equation (3), and the corresponding $\tilde{w}(t)$, expressed as Equation (43), is given by $\tilde{w}(t) = \hat{w}(D)\delta_1(t)$, when $\hat{w}(s)$ is given by Equation (65) and $K = C_1 \cdot \Gamma(1+a-c)$. ■

Remark 7. We note that $\hat{w}(s)$ given by Equation (65) converges for no value of s, but $\tilde{w}(t) = \hat{w}(D)\delta_1(t)$ obtained by using this $\hat{w}(s)$ corresponds to Equation (67). Moreover, we note that Equation (67) is obtained from Equation (65) by term-by-term inverse Laplace transform.

We choose $K = \Gamma(2-c)$. Then, by Lemmas 16 and 17,

$$\hat{u}(s) = s^{a-1}\hat{w}(s) = \Gamma(2-c)s^{-2+c} \cdot {}_2F_0(1+a-c, 1+b-c;;s^{-1}) \tag{68}$$

is a solution of Equation (63). Now, by its term-by-term inverse Laplace transform, or from $\tilde{u}(t) = \hat{u}(D)\delta_1(t) = D^{a-1}\hat{w}(D)\delta_1(t)$, we obtain $u(t)$ given by Equation (12), if $2 - c \notin \mathbb{Z}_{<1}$. We note that $(1-c)u_0 = 0$ for this solution $u(t)$ if $1 - c \notin \mathbb{Z}_{<1}$.

Remark 8. Since $\hat{u}(s) = s^{a-1}\hat{w}(s)$, $\tilde{u}(t) = \hat{u}(D)\delta_1(t) = D^{a-1}\hat{w}(D)\delta_1(t)$, and hence the solution given by Equation (12) is obtained by using Equation (67) in $u(t) = {}_0D_R^{a-1}w(t)$.

Remark 9. Remark 7 is valid even when w, Equations (65) and (67) are replaced by u, Equations (68) and (12), respectively.

5.2. Solution Satisfying $u_0 = 1$

The other solution given by Equation (11) is obtained by solving the inhomogeneous Equation (63) for $1 - c \neq 0$ and $u_0 = 1$. We note here that the method using the following lemma with the solution given by Equation (12), which is already obtained, is easier.

Lemma 18. If $u(a, b, c; t)$ is a solution of Equation (10), $t^{1-c} \cdot u(1+a-c, 1+b-c, 2-c; t)$ is also a solution of the same equation.

Proof We put $u(t) = t^{\lambda}w(t)$ in Equation (10), and then we obtain Equation (10) with a, b, c and u replaced by $1+a-c$, $1+b-c$, $2-c$ and w, respectively, when $\lambda = 1-c$. ■

5.3. Solution by the Basic Method

The statements in Section 4.3 are valid even if we replace Equations (50), (5), (52) and (53) by Equations (60), (10), (68) and (12), respectively. Here, Equations (56)~(58) also must be replaced by

$$\hat{p}_k = \hat{p}_{k-1}\frac{(\alpha+a+k-1)(\alpha+b+k-1)}{\alpha+c+k-1} = p_0\Gamma(\alpha+1)\frac{(\alpha+a)_k(\alpha+b)_k}{(\alpha+c)_k},$$
$$p_k = p_{k-1}\frac{(\alpha+a+k-1)(\alpha+b+k-1)}{(\alpha+c+k-1)(\alpha+k)} = p_0\frac{(\alpha+a)_k(\alpha+b)_k}{(\alpha+c)_k(\alpha+1)_k}. \tag{69}$$

When $\alpha = 0$, we have

$$\hat{p}_k = p_0\frac{(a)_k(b)_k}{(c)_k}, \quad p_k = p_0\frac{(a)_k(b)_k}{k!(c)_k}, \tag{70}$$

When $\alpha = 1 - c$, we have

$$\hat{p}_k = p_0\Gamma(2-c)\frac{(1+a-c)_k(1+b-c)_k}{k!}, \quad p_k = p_0\frac{(1+a-c)_k(1+b-c)_k}{k!(2-c)_k}. \tag{71}$$

Using Equation (71) with $\alpha = 1 - c$ and $p_0 = 1$ in Equations (54) and (55), we obtain Equations (12) and (68), respectively. Using Equation (70) with $\alpha = 0$ and $p_0 = 1$ in Equation (54), we obtain Equation (11).

5.4. Solution by the Modified Nishimoto's Method

In [26], the solution of the hypergeometric DE is given by modified Nishimoto's method, where the solution is expressed by fD of a function. In that paper, Kummer's 24 solutions of the DE are obtained by that method.

6. Solution of Bessel's DE

We now take up Bessel's DE:

$$t^2 \cdot \frac{d^2}{dt^2}u(t) + t \cdot \frac{d}{dt}u(t) + (t^2 - \nu^2) \cdot u(t) = 0, \quad t > 0, \tag{72}$$

where $\nu \in \mathbb{C}$ is a constant. The Bessel functions $J_{\pm\nu}(t)$ are solutions of this DE, where

$$J_\nu(t) = (\frac{t}{2})^\nu \sum_{k=0}^{\infty} \frac{(-1)^k}{k!\Gamma(\nu+k+1)}(\frac{t^2}{4})^k = (\frac{t}{2})^\nu \frac{1}{\Gamma(\nu+1)}{}_0F_1(;\nu+1;-\frac{t^2}{4}), \tag{73}$$

for $\nu \notin \mathbb{Z}_{<0}$, where ${}_0F_1(;c;z) = \sum_{k=0}^{\infty}\frac{1}{k!(c)_k}z^k$.

In order to obtain the solutions given by Equation (73) by the present method, we put $x = t^2$ and $u(t) = t^\lambda v(x)$ in Equation (72). Then, if $\lambda = \nu$, we obtain the DE for $v(x)$:

$$p_B(x, \frac{d}{dx})v(x) = 4x \cdot \frac{d^2}{dx^2}v(x) + 4(\nu+1) \cdot \frac{d}{dx}v(x) + v(x) = 0, \quad x > 0, \tag{74}$$

and if $\lambda = -\nu$, we obtain the same equation with ν replaced by $-\nu$.

When $\lambda = \nu$, by Theorem 2, the DE satisfied by the AC-Laplace transform $\hat{v}(s)$ of $v(x)$ is given by

$$p_B(-\frac{d}{ds}, s)\hat{v}(s) = -4\frac{d}{ds}[s^2\hat{v}(s)] + [4(\nu+1)s + 1]\hat{v}(s) = 4\nu v_0. \tag{75}$$

The solution satisfying $\nu v_0 = 0$ is given by

$$\hat{v}(s) = Cs^{\nu-1}e^{-1/(4s)} = Cs^{\nu-1}\sum_{k=0}^{\infty}\frac{(-1)^k}{k!}\frac{1}{(4s)^k}, \tag{76}$$

where C is an arbitrary constant. By the inverse Laplace transform, we obtain

$$v(x) = Cx^{-\nu}\sum_{k=0}^{\infty}\frac{(-1)^k}{k!\Gamma(k-\nu+1)}(\frac{x}{4})^k = Cx^{-\nu}\frac{1}{\Gamma(1-\nu)}\cdot {}_0F_1(;1-\nu;-\frac{x}{4}). \tag{77}$$

This satisfies $\nu v_0 = 0$ if $\nu \notin \mathbb{Z}_{>0}$. By Corollary 4, this $v(x)$ is a solution of Equation (74) if $\nu \notin \mathbb{Z}_{>0}$.

Using this in $u(t) = t^{\nu}v(t^2)$ and putting $C = 2^{\nu}$, we have $u(t) = J_{-\nu}(t)$. When $\lambda = -\nu$, we obtain $u(t) = J_{\nu}(t)$.

7. Solution of Hermite's DE

We now take up Hermite's DE:

$$\frac{d^2}{dt^2}u(t) - t\cdot\frac{d}{dt}u(t) + n\cdot u(t) = 0,\ t > 0, \tag{78}$$

where $n \in \mathbb{Z}_{>-1}$ is a constant. The Hermite polynomial $He_n(t)$ and the Hermite function of the second kind $he_n(t)$ are solutions of this DE. In [11] (p. 82), they are expressed as

$$He_{2l}(t) = \frac{(-1)^l}{2^l}\frac{(2l)!}{l!}\cdot {}_1F_1(-l;\frac{1}{2};\frac{t^2}{2}),\quad He_{2l+1}(t) = \frac{(-1)^l}{2^l}\frac{(2l+1)!}{l!}\cdot t\cdot {}_1F_1(-l;\frac{3}{2};\frac{t^2}{2}), \tag{79}$$

$$he_{2l}(t) = (-1)^l 2^l l!\cdot t\cdot {}_1F_1(\frac{1}{2}-l;\frac{3}{2};\frac{t^2}{2}),\quad he_{2l+1}(t) = (-1)^{l+1}2^l l!\cdot {}_1F_1(-\frac{1}{2}-l;\frac{1}{2};\frac{t^2}{2}), \tag{80}$$

for $l \in \mathbb{Z}_{>-1}$.

In order to obtain these solutions by the present method, we put $x = t^2$ and $u(t) = v(x)$ in Equation (78). Then, we obtain the DE for $v(x)$:

$$x\cdot\frac{d^2}{dx^2}v(x) + (\frac{1}{2}-\frac{1}{2}x)\cdot\frac{d}{dx}v(x) + \frac{1}{4}n\cdot v(x) = 0,\ x > 0. \tag{81}$$

This DE is Equation (5) with t, $u(t)$, c, b and a replaced by x, $v(x)$, $\frac{1}{2}$, $\frac{1}{2}$ and $-\frac{1}{2}n$, respectively. Corresponding to the solutions given by Equations (6) and (7) of Equation (5), we have the following solutions of Equation (81):

$$v_{n,1}(x) = {}_1F_1(-\frac{1}{2}n;\frac{1}{2};\frac{x}{2}),\quad v_{n,2}(x) = x^{1/2}\cdot {}_1F_1(\frac{1}{2}-\frac{1}{2}n;\frac{3}{2};\frac{x}{2}). \tag{82}$$

The solutions given by Equations (79) and (80) of Equation (78) are given by

$$He_{2l}(t) = C_1v_{2l,1}(t^2),\quad He_{2l+1}(t) = C_2v_{2l+1,2}(t^2), \tag{83}$$

$$he_{2l}(t) = C_3v_{2l,2}(t^2),\quad he_{2l+1}(t) = C_4v_{2l+1,1}(t^2), \tag{84}$$

for $l \in \mathbb{Z}_{>-1}$, where C_k for $k \in \mathbb{Z}_{[1,4]}$ are constants.

8. Conclusions

In the present paper, we are concerned with the problem of obtaining a solution $u(t)$ of a DE with polynomial coefficients.

We know that by the basic method of solution [14] (Section 10.4), we usually obtain solutions in the form of Equation (54) which are a power of t multiplied by a power series in t. In the present paper, we are interested in obtaining the solutions with the aid of AC of Riemann–Liouville fD along with distribution theory or the Laplace transform or its AC.

We then set up the DE satisfied by the AC-Laplace transform, $\hat{u}(s)$, of $u(t)$. The obtained DE for $\hat{u}(s)$ is found to be a DE with polynomial coefficients.

We now obtain the solution in the form of Equation (55) which is a power of s multiplied by a power series in s^{-1}, by solving the DE for $\hat{u}(s)$. When it converges at large $|s|$, it is the Laplace transform of a solution of the DE for $u(t)$ or its AC.

In Section 4, we obtain such a solution $\hat{u}(s)$ that the series in it converges at $|s| > |b|$ for $b \neq 0$. Then, we obtain the solution $u(t)$ by term-by-term inverse Laplace transform by writing the distribution associated with the solution $u(t)$, by using the obtained $\hat{u}(s)$. In Section 6, another example is given.

In Section 5, we obtain such a solution $\hat{u}(s)$ that the series in it converges for no value of s. Then, we obtain the solution $u(t)$ by writing the distribution $\tilde{u}(t) = \hat{u}(D)\delta_1(t)$ associated with the solution $u(t)$, by using the obtained non-convergent series of s^{-1} for $\hat{u}(s)$. The result is seen to be obtained by term-by-term inverse Laplace transform of $\hat{u}(s)$, as mentioned in Section 3.2.

We may conclude the study in this paper as follows. When we desire to obtain the solution $u(t)$ of a linear DE with polynomial coefficients, in the form of Equation (3), we can obtain it, by setting up the DE for the AC-Laplace transform $\hat{u}(s)$, obtaining its solution in the form of Equation (9), and then taking its term-by-term inverse Laplace transform.

We can obtain the solution of the DE for $\hat{u}(s)$ by the basic method, obtaining it in the form of Equation (55). Comparing the solutions of the DE for $u(t)$ and $\hat{u}(s)$, obtained by the basic method of solution, we find that the solution $\hat{u}(s)$ is the term-by-term Laplace transform of $u(t)$. Thus, we obtain $u(t)$ by the term-by-term inverse Laplace transform of $\hat{u}(s)$, when the latter is obtained.

Acknowledgments: Section 1.2 and the paragraphs related with the basic references [21–24] on distribution theory in Section 3.3 are written in response to the recommendations of the reviewers of this paper.

Author Contributions: Stimulated by Yosida's work [1,2], a number of papers [4–6] were written on the solution of Laplace's DE. But the research there are mainly focussed on Kummer's DE. In this situation, Tohru Morita came to the idea that distributions in the space $\mathcal{D}'_{r,R}$ must be useful in solving the hypergeometric DE. Based on the idea, Tohru Morita and Ken-ichi Sato worked together to write this paper.

Conflicts of Interest: The authors declare no conflict of interest.

References

1. Yosida, K. The Algebraic Derivative and Laplace's Differential Equation. *Proc. Jpn. Acad. Ser. A* **1983**, *59*, 1–4.
2. Yosida, K. *Operational Calculus*; Springer-Verlag: New York, NY, USA, 1982.
3. Mikusiński, J. *Operational Calculus*; Pergamon Press: London, UK, 1959.
4. Morita, T.; Sato, K. Remarks on the Solution of Laplace's Differential Equation and Fractional Differential Equation of That Type. *Appl. Math.* **2013**, *4*, 13–21.
5. Morita, T.; Sato, K. Solution of Laplace's Differential Equation and Fractional Differential Equation of That Type. *Appl. Math.* **2013**, *4*, 26–36.
6. Morita, T.; Sato, K. Solution of Differential Equations with the Aid of Analytic Continuation of Laplace Transform. *Appl. Math.* **2014**, *5*, 1309–1319.
7. Lavoie, J.L.; Tremblay, R.; Osler, T.J. Fundamental Properties of Fractional Derivatives Via Pochhammer Integrals. In *Fractional Calculus and Its Applications*, Ross, B., Ed.; Lecture Notes in Mathematics; Springer: Berlin, Germany; Heidelberg, Germany, 1975; Volume 457, pp. 327–356.
8. Lavoie, J.L.; Osler, T.J.; Tremblay, R. Fractional Derivatives and Special Functions. *SIAM Rev.* **1976**, *18*, 240–268.
9. Morita, T.; Sato, K. Liouville and Riemann-Liouville Fractional Derivatives via Contour Integrals. *Fract. Calc. Appl. Anal.* **2013**, *16*, 630–653.

10. Abramowitz, M.; Stegun, I.A. *Handbook of Mathematical Functions with Formulas, Graphs and Mathematical Tables*; Dover Publ., Inc.: New York, NY, USA, 1972.
11. Magnus, M.; Oberhettinger, F. *Formulas and Theorems for the Functions of Mathematical Physics*; Chelsea Publ. Co.: New York, NY, USA, 1949.
12. Morita, T.; Sato, K. Solution of Fractional Differential Equation in Terms of Distribution Theory. *Interdiscip. Inf. Sci.* **2006**, *12*, 71–83.
13. Morita, T.; Sato, K. Neumann-Series Solution of Fractional Differential Equation. *Interdiscip. Inf. Sci.* **2010**, *16*, 127–137.
14. Whittaker, E.T.; Watson, G.N. *A Course of Modern Analysis*; Cambridge U.P.: Cambridge, UK, 1935.
15. He, J.H. A Tutorial Review on Fractal Spacetime and Fractional Calculus. *Int. J. Theor. Phys.* **2014**, *53*, 3698–3718.
16. Liu, F.J.; Li, Z.B.; Zhang, S.; Liu, H.Y. He's Fractional Derivative for Heat Conduction in a Fractal Medium Arising in Silkworm Cocoon Hierarchy. *Therm. Sci.* **2015**, *19*, 1155–1159.
17. Kumar, D.; Singh, J.; Kumar, S. A Fractional Model of Navier-Stokes Equation Arising in Unsteady Flow of a Viscous Fluid. *J. Assoc. Arab Univ. Basic Appl. Sci.* **2015**, *17*, 14–17.
18. Kumar, D.; Singh, J.; Kumar, S.; Sushila; Singh, B.P. Numerical Computation of Nonlinear Shock Wave Equation of Factional Order. *Ain Shams Eng. J.* **2015**, *6*, 605–611.
19. He, J.H.; Elagan, S.K.; Li, Z.B. Geometrical Explanation of the Fractional Complex Transform and Derivative Chain Rule for Fractional Calculus. *Phys. Lett. A* **2013**, *376*, 257–259.
20. Jia, S.M.; Hu, M.S.; Chen, Q.L.; Jia, Z.J. Exact Solution of Fractional Nizhnik-Novikov-Veselov Equation. *Therm. Sci.* **2014**, *18*, 1715–1717.
21. Schwartz, L. *Théorie des Distributions*; Hermann: Paris, France, 1966.
22. Gelfand, I.M.; Silov, G.E. *Generalized Functions*; Academic Press Inc.: New York, NY, USA, 1964; Volume 1.
23. Vladimirov, V.S. *Methods of the Theory of Generalized Functions*; Taylor & Francis Inc.: New York, NY, USA, 2002.
24. Zemanian, A.H. *Distribution Theory and Transform Analysis*; Dover Publ., Inc.: New York, NY, USA, 1965.
25. Morita, T.; Sato, K. Asymptotic Expansions of Fractional Derivatives and Their Applications. *Mathematics* **2015**, *3*, 171–189.
26. Morita, T.; Sato, K. Kummer's 24 Solutions of the Hypergeometric Differential Equation with the Aid of Fractional Calculus. *Adv. Pure Math.* **2016**, *6*, 180–191.

11

Nevanlinna's Five Values Theorem on Annuli

Hong-Yan Xu * and Hua Wang

Department of Informatics and Engineering, Jingdezhen Ceramic Institute, Jingdezhen 333403, Jiangxi, China; hhhlucy2012@126.com

* Correspondence: xhyhhh@126.com

Academic Editor: Alexander Berkovich

Abstract: By using the second main theorem of the meromorphic function on annuli, we investigate the problem on two meromorphic functions partially sharing five or more values and obtain some theorems that improve and generalize the previous results given by Cao and Yi.

Keywords: meromorphic function; Nevanlinna theory; the annuli

Mathematical Subject Classification (2010): 30D30; 30D35

---

1. Introduction and Main Results

The purpose of this paper is to study the uniqueness of two meromorphic functions sharing five or more values. Thus, we always assumed that the reader is familiar with the notations of the Nevanlinna theory, such as $T(r,f), m(r,f), N(r,f)$, and so on (see [1–4]). We use $\mathbb{C}$ to denote the open complex plane, $\overline{\mathbb{C}}$ to denote the extended complex plane and $\mathbb{X}$ to denote the subset of $\mathbb{C}$.

In 1929, R. Nevanlinna (see [5]) first investigated the uniqueness of meromorphic functions in the whole complex plane and obtained the well-known theorem: the five IMtheorem:

Theorem 1.1. *(see [5]). If f and g are two non-constant meromorphic functions that share five distinct values a_1, a_2, a_3, a_4, a_5 IM in $\mathbb{C}$, then $f(z) \equiv g(z)$.*

After his theorem, there are vast references on the uniqueness of meromorphic functions sharing values and sets in the whole complex plane (see [3]). It is an interesting topic how to extend some important uniqueness results in the complex plane to an angular domain or the unit disc. In the past several decades, the uniqueness of meromorphic functions in the value distribution attracted many investigations. For example, I. Lahiri, H.X. Yi, X.M. Li and A. Banerjee (including [3,6–8]) studied the uniqueness for meromorphic functions on the whole complex plane sharing one, two, three or some sets; M.L. Fang, H.F. Liu, Z.Q. Mao and H.Y. Xu (including [9–11]) investigated the shared value of meromorphic functions in the unit disc; J.H. Zheng, Q.C. Zhang, T.B. Cao and W.C. Lin (including [12–16]) considered many uniqueness problem on meromorphic functions on the angular domain.

In 2009, Z.Q. Mao and H.F. Liu [10] gave a different method to investigate the uniqueness problem of meromorphic functions in the unit disc and obtained the following results.

Theorem 1.2. *(see [10]). Let f, g be two meromorphic functions in $\mathbb{D}$, $a_j \in \overline{\mathbb{C}} (j = 1, 2, \ldots, 5)$ be five distinct values and $\Delta(\theta_0, \delta) = \{z : |z| < 1\} \bigcap \{z : |\arg z - \theta_0| < \delta\}, 0 \le \theta_0 \le 2\pi, 0 < \delta < \pi$ be an angular domain, such that for some $a \in \overline{\mathbb{C}}$,*

$$\limsup_{r \to 1^-} \frac{\log n(r, \Delta(\theta_0, \delta/2), f(z) = a)}{\log \frac{1}{1-r}} = \tau > 1. \tag{1}$$

If f and g share $a_j (j = 1, 2, \ldots, 5)$ IM in $\Delta(\theta_0, \delta)$, then $f(z) \equiv g(z)$.

In the same year, T.B. Cao and H.X. Yi [12] investigated the uniqueness problem of two transcendental meromorphic functions sharing five distinct values in an angular domain and obtained the following theorem:

Theorem 1.3. *(see [12], Theorem 1.3). Let f and g be two transcendental meromorphic functions. Given one angular domain $X = \{z : \alpha < \arg z < \beta\}$ with $0 < \beta - \alpha \leq 2\pi$, we assume that f and g share five distinct values $a_j \in \overline{\mathbb{C}}(j = 1,2,3,4,5)$ IM in X. Then, $f(z) \equiv g(z)$, provided that:*

$$\lim_{r\to\infty} \frac{S_{\alpha,\beta}(r,f)}{\log(rT(r,f))} = \infty, \quad (r \notin E),$$

where $S_{\alpha,\beta}(r,f)$ is used to denote the angular characteristic function of meromorphic function f.

Remark 1.1. *We may denote Theorem 1.3 by the five IM theorem in an angular domain.*

In 2003, J.H. Zheng [15,16] firstly took into account the value distribution of meromorphic functions in an angular domain. In 2010, J.H. Zheng [17] investigated the uniqueness of the meromorphic function sharing five values in an angular domain, by using Tsjui's characteristic function.

Theorem 1.4. *(see [17]). Let f and g be two nonconstant meromorphic functions in an angular domain $\Omega(\alpha,\beta) = \{z : \alpha < \arg z < \beta\}(0 < \beta - \alpha < 2\pi)$, and:*

$$\limsup_{r\to\infty} \frac{\mathfrak{T}_{\alpha,\beta}(r,f)}{\log r} = \infty.$$

If f and g share five distinct values $a_j \in \overline{\mathbb{C}}(j = 1,2,3,4,5)$ IM in $\Omega(\alpha,\beta)$, then $f(z) \equiv g(z)$.

Remark 1.2. *$\mathfrak{T}_{\alpha,\beta}(r,f)$ is Tsjui's characteristic function of f in the angular domain $\Omega(\alpha,\beta)$, which is introduced in [17].*

However, the whole complex plane, the unit disc and the angular domain can all be regarded as a simply-connected region; in other words, the theorems stated in the above references are only regarded as the uniqueness results in a simply-connected region. In fact, there exists many other sub-regions in the whole complex plane, such as: the annuli, the m-punctured complex plane, *etc.*

Recently, there have been some results focusing on the Nevanlinna theory of meromorphic functions on the annulus (see [18–23]). The annulus can be regarded as the doubly-connected region. From the doubly-connected mapping theorem [24], we can get that each doubly-connected domain is conformally equivalent to the annulus $\{z : r < |z| < R\}, 0 \leq r < R \leq +\infty$. For two cases: $r = 0$, $R = +\infty$, simultaneously, and $0 < r < R < +\infty$; the latter case, the homothety $z \mapsto \frac{z}{\sqrt{rR}}$ reduces the given domain to the annulus $\{z : \frac{1}{R_0} < |z| < R_0\}$, where $R_0 = \sqrt{\frac{R}{r}}$. Thus, every annulus is invariant with respect to the inversion $z \mapsto \frac{1}{z}$ in two cases. In 2005, Khrystiyanyn and Kondratyuk [18,19] proposed the Nevanlinna theory for meromorphic functions on annuli (see also [25]). The basic notions of the Nevanlinna theory on annuli will be shown in the next section. Lund and Ye [21] in 2009 studied meromorphic functions on annuli with the form $\{z : R_1 < |z| < R_2\}$, where $R_1 \geq 0$ and $R_2 \leq \infty$. In 2009 and 2011, Cao [26–28] investigated the uniqueness of meromorphic functions on annuli sharing some values and some sets and obtained an analog of Nevanlinna's famous five-value theorem.

Theorem 1.5. *(see [26], Corollary 3.4). Let f_1 and f_2 be two transcendental or admissible meromorphic functions on the annulus $\mathbb{A} = \{z : \frac{1}{R_0} < |z| < R_0\}$, where $1 < R_0 \leq +\infty$. Let a_j $(j = 1,2,\ldots,q)$ be q distinct complex numbers in $\overline{\mathbb{C}}$ and $k_j(j = 1,2,\ldots,q)$ be positive integers or ∞, such that:*

$$k_1 \geq k_2 \geq \cdots \geq k_q. \tag{2}$$

and

$$\overline{E}_{k_j)}(a_j, f_1) = \overline{E}_{k_j)}(a_j, f_2), \quad (j = 1, 2, \ldots, q). \tag{3}$$

Then:

(i) if $q = 7$, then $f_1(z) \equiv f_2(z)$.
(ii) if $q = 6$ and $k_3 \geq 2$, then $f_1(z) \equiv f_2(z)$.
(iii) if $q = 5$, $k_3 \geq 3$ and $k_5 \geq 2$, then $f_1(z) \equiv f_2(z)$.
(iv) if $q = 5$ and $k_4 \geq 4$, then $f_1(z) \equiv f_2(z)$.
(v) if $q = 5$, $k_3 \geq 5$ and $k_4 \geq 3$, then $f_1(z) \equiv f_2(z)$.
(vi) if $q = 5$, $k_3 \geq 6$ and $k_4 \geq 2$, then $f_1(z) \equiv f_2(z)$.

From Theorem 1.5, we can get the following theorem immediately.

Theorem 1.6. *(see [26], Theorem 3.2). Let f_1 and f_2 be two transcendental or admissible meromorphic functions on the annulus $\mathbb{A} = \{z : \frac{1}{R_0} < |z| < R_0\}$, where $1 < R_0 \leq +\infty$. Let a_j $(j = 1, 2, 3, 4, 5)$ be five distinct complex numbers in $\overline{\mathbb{C}}$. If $\overline{E}(a_j, f_1) = \overline{E}(a_j, f_2)$ for $j = 1, 2, 3, 4, 5$, then $f_1(z) \equiv f_2(z)$.*

Remark 1.3. *Write $E(a, f) = \{z \in \mathbb{A} : f(z) - a = 0\}$, where each zero with multiplicity m is counted m times. If we ignore the multiplicity, then the set is denoted by $\overline{E}(a, f)$. We use $\overline{E}_{k)}(a, f)$ to denote the set of zeros of $f - a$ with multiplicities no greater than k, in which each zero is counted only once.*

In this paper, we will further investigate the problem on the five values for meromorphic functions on annuli. To state our main theorem, we first introduce the following definition.

Definition 1.1. *For $B \subset \mathbb{A}$ and $a \in \overline{\mathbb{C}}$, we denote by $\overline{N}_0^B(r, \frac{1}{f-a})$ the reduced counting function of those zeros of $f - a$ on $\mathbb{A}$, which belong to the set B.*

Theorem 1.7. *Let f and g be two transcendental or admissible meromorphic functions on the annulus $\mathbb{A} = \{z : \frac{1}{R_0} < |z| < R_0\}$, where $1 < R_0 \leq +\infty$. Let $a_1, \ldots, a_q (q \geq 5)$ be q distinct complex numbers or ∞. Suppose that $k_1 \geq k_2 \geq \cdots \geq k_q, m$ are positive integers or infinity; $1 \leq m \leq q$ and $\delta_j (\geq 0) (j = 1, 2, \ldots, q)$ are such that:*

$$(1 + \frac{1}{k_m}) \sum_{j=m}^{q} \frac{1}{1 + k_j} + 3 + \sum_{j=1}^{q} \delta_j < (q - m - 1)(1 + \frac{1}{k_m}) + m. \tag{4}$$

Let $B_j = \overline{E}_{k_j)}(a_j, f) \setminus \overline{E}_{k_j)}(a_j, g)$ for $j = 1, 2, \ldots, q$. If:

$$\overline{N}_0^{B_j}(r, \frac{1}{f - a_j}) \leq \delta_j T_0(r, f) \tag{5}$$

and:

$$\liminf_{r \to \infty} \frac{\sum_{j=1}^{q} \overline{N}_0^{k_j)}(r, \frac{1}{f-a_j})}{\sum_{j=1}^{q} \overline{N}_0^{k_j)}(r, \frac{1}{g-a_j})}$$
$$> \frac{k_m}{(1 + k_m) \sum_{j=m}^{q} \frac{k_j}{1+k_j} - 2(1 + k_m) + (m - 2 - \sum_{j=1}^{q} \delta_j) k_m}, \tag{6}$$

then $f(z) \equiv g(z)$.

From Theorem 1.7, we can get the following consequences.

Corollary 1.1. *Let $m = 1$, $k_j = \infty$ for $j = 1, 2, \ldots, q$ and:*

$$\gamma = \liminf_{r \to R_0} \frac{\sum_{j=1}^{q} \overline{N}_0(r, \frac{1}{f-a_j})}{\sum_{j=1}^{q} \overline{N}_0(r, \frac{1}{g-a_j})} > \frac{1}{q-3}.$$

If $\overline{N}_0^{B_j}(r, \frac{1}{f-a_j}) \leq \delta_j T_0(r, f)$ *where* $\delta_j(\geq 0)$ *satisfy* $0 \leq \sum_{j=1}^{q} \delta_j < k - 3 - \frac{1}{\gamma}$, *then* $f(z) \equiv g(z)$.

If we take $q = 5$ and $\overline{E}(a_j, f) \subseteq \overline{E}(a_j, g)$, then $B_j = \emptyset$ for $j = 1, 2, \ldots, 5$. Therefore, if we choose $\delta_j = 0$ for $j = 1, 2, \ldots, 5$ and take any constant γ, such that $0 \leq 2 - \frac{1}{\gamma}$ in Corollary 1.1; we can get that $f \equiv g$. Especially, if $q = 5$ and $\overline{E}(a_j, f) = \overline{E}(a_j, g)$, then $\gamma = 1$ and $\delta_j = 0$ for $j = 1, 2, \ldots, 5$. We can obtain $f \equiv g$. Therefore, Corollary 1.1 is an improvement of Theorem 1.6.

Corollary 1.2. *Let* f *and* g *be two transcendental or admissible meromorphic functions on the annulus* $\mathbb{A} = \{z : \frac{1}{R_0} < |z| < R_0\}$, *where* $1 < R_0 \leq +\infty$. *Let* $a_1, \ldots, a_q (q \geq 5)$ *be* q *distinct complex numbers or* ∞. *Suppose that* $k_1, k_2, \cdots, k_q$ *are positive integers or infinity with* $k_1 \geq k_2 \geq \cdots \geq k_q$, *if* $\overline{E}_{k_j)}(a_j, f) \subseteq \overline{E}_{k_j)}(a_j, g)$ *and:*

$$\sum_{j=2}^{q} \frac{k_j}{1+k_j} - \frac{k_1}{\gamma(1+k_1)} - 2 > 0,$$

where γ *is stated as in Corollary 1.1; then,* $f(z) \equiv g(z)$.

Corollary 1.3. *Under the assumptions of Corollary 1.2, if* $\overline{E}_{k_j)}(a_j, f) = \overline{E}_{k_j)}(a_j, g)$ *and:*

$$\sum_{j=2}^{q} \frac{k_j}{1+k_j} - \frac{k_1}{1+k_1} - 2 > 0,$$

then we have $f(z) \equiv g(z)$.

Corollary 1.4. *Let* f *and* g *be two transcendental or admissible meromorphic functions on the annulus* $\mathbb{A} = \{z : \frac{1}{R_0} < |z| < R_0\}$, *where* $1 < R_0 \leq +\infty$. *Let* $a_1, \ldots, a_q (q \geq 5)$ *be* q *distinct complex numbers or* ∞. *Suppose that* $k_1, k_2, \cdots, k_q$ *are positive integers or infinity with* $k_1 \geq k_2 \geq \cdots \geq k_q$, *if* $\overline{E}_{k_j)}(a_j, f) \subseteq \overline{E}_{k_j)}(a_j, g)$ *and:*

$$\sum_{j=m}^{q} \frac{k_j}{1+k_j} - 2 + \frac{(m-2-\frac{1}{\gamma})k_m}{1+k_m} > 0, \tag{7}$$

where γ *is stated as in Corollary 1.1; then,* $f(z) \equiv g(z)$.

Remark 1.4. *If* $\overline{E}_{k_j)}(a_j, f) = \overline{E}_{k_j)}(a_j, g)$ *and taking* $m = 3$ *in Corollary 1.4, thus Equation (5) becomes:*

$$\sum_{j=3}^{q} \frac{k_j}{1+k_j} > 2.$$

Then, we can get Theorem 1.5 easily. Hence, Theorem 1.7 is an improvement of Theorem 1.5.

Remark 1.5. *Throughout our article, we can see that our theorem and corollaries also hold for transcendental meromorphic function in the whole complex plane, which are also extensions of some results given by Nevanlinna, Yi and Cao [3,5,26].*

2. Preliminaries and Some Lemmas

Next, we will introduce the basic notations and conclusion about meromorphic functions on annuli.

For a meromorphic function f on whole plane $\mathbb{C}$, the classical notations of the Nevanlinna theory are denoted as follows:

$$N(r, f) = \int_0^r \frac{n(t, f) - n(0, f)}{t} dt + n(0, f) \log r,$$

$$m(r, f) = \frac{1}{2\pi} \int_0^{2\pi} \log^+ |f(re^{i\theta})| d\theta, \quad T(r, f) = N(r, f) + m(r, f),$$

where $\log^+ x = \max\{\log x, 0\}$, and $n(t,f)$ is the counting function of poles of the function f in $\{z : |z| \leq t\}$.

Let f be a meromorphic function on the annulus $\mathbb{A} = \{z : \frac{1}{R_0} < |z| < R_0\}$, where $1 < r < R_0 \leq +\infty$; the notations of the Nevanlinna theory on annuli will be introduced as follows. Let:

$$N_1(r,f) = \int_{\frac{1}{r}}^{1} \frac{n_1(t,f)}{t}dt, \quad N_2(r,f) = \int_{1}^{r} \frac{n_2(t,f)}{t}dt,$$

$$m_0(r,f) = m(r,f) + m(\frac{1}{r},f) - 2m(1,f), \quad N_0(r,f) = N_1(r,f) + N_2(r,f),$$

where $n_1(t,f)$ and $n_2(t,f)$ are the counting functions of poles of the function f in $\{z : t < |z| \leq 1\}$ and $\{z : 1 < |z| \leq t\}$, respectively. Similarly, for $a \in \overline{\mathbb{C}}$, we have:

$$\begin{aligned} \overline{N}_0(r,\frac{1}{f-a}) &= \overline{N}_1(r,\frac{1}{f-a}) + \overline{N}_2(r,\frac{1}{f-a}) \\ &= \int_{\frac{1}{r}}^{1} \frac{\overline{n}_1(t,\frac{1}{f-a})}{t}dt + \int_{1}^{r} \frac{\overline{n}_2(t,\frac{1}{f-a})}{t}dt \end{aligned}$$

in which each zero of the function $f-a$ is counted only once. In addition, we use $\overline{n}_1^{k)}(t,\frac{1}{f-a})$ (or $\overline{n}_1^{(k}(t,\frac{1}{f-a})$) to denote the counting function of poles of the function $\frac{1}{f-a}$ with multiplicities $\leq k$ (or $> k$) in $\{z : t < |z| \leq 1\}$, each point counted only once. Similarly, we have the notations $\overline{N}_1^{k)}(t,f)$, $\overline{N}_1^{(k}(t,f)$, $\overline{N}_2^{k)}(t,f)$, $\overline{N}_2^{(k}(t,f)$, $\overline{N}_0^{k)}(t,f)$, $\overline{N}_0^{(k}(t,f)$.

The Nevanlinna characteristic of f on the annulus $\mathbb{A}$ is defined by:

$$T_0(r,f) = m_0(r,f) + N_0(r,f). \tag{8}$$

For a nonconstant meromorphic function f on the annulus $\mathbb{A} = \{z : \frac{1}{R_0} < |z| < R_0\}$, where $1 < r < R_0 \leq +\infty$, the following properties will be used in this paper (see [18]):

$$(i)\ \ T_0(r,f) = T_0\left(r,\frac{1}{f}\right),$$

$$(ii)\ \ \max\{T_0(r,f_1\cdot f_2), T_0(r,\frac{f_1}{f_2}), T_0(r,f_1+f_2)\} \leq T_0(r,f_1) + T_0(r,f_2) + O(1),$$

$$(iii)\ \ T_0(r,\frac{1}{f-a}) = T_0(r,f) + O(1),\ for\ every\ fixed\ a \in \mathbb{C},$$

where (iii) can be called the first fundamental theorem on annuli.

In 2005, the lemma on the logarithmic derivative on the the annulus $\mathbb{A}$ was obtained by Khrystiyanyn and Kondratyuk [19].

Lemma 2.1. *(see [19], the lemma on the logarithmic derivative). Let f be a nonconstant meromorphic function on the annulus $\mathbb{A} = \{z : \frac{1}{R_0} < |z| < R_0\}$, where $R_0 \leq +\infty$, and let $\lambda > 0$. Then:*

$$m_0\left(r,\frac{f'}{f}\right) = S_1(r,f),$$

where (i) in the case $R_0 = +\infty$,

$$S_1(r,*) = O(\log(rT_0(r,*))) \tag{9}$$

for $r \subset (1, +\infty)$, except for the set $\triangle_r$, such that $\int_{\triangle_r} r^{\lambda-1}dr < +\infty$;
(ii) if $R_0 < +\infty$, then:

$$S_1(r,*) = O(\log(\frac{T_0(r,*)}{R_0 - r})) \tag{10}$$

for $r \in (1, R_0)$, *except for the set* $\triangle_r'$, *such that* $\int_{\triangle_r'} \frac{dr}{(R_0-r)^{\lambda-1}} < +\infty$.

Definition 2.1. *Let* $f(z)$ *be a non-constant meromorphic function on the annulus* $\mathbb{A} = \{z : \frac{1}{R_0} < |z| < R_0\}$, *where* $1 < R_0 \leq +\infty$. *The function* f *is called a transcendental or admissible meromorphic function on the annulus* $\mathbb{A}$ *provided that:*

$$\limsup_{r\to\infty} \frac{T_0(r,f)}{\log r} = \infty, \quad 1 < r < R_0 = +\infty \tag{11}$$

or:

$$\limsup_{r\to R_0} \frac{T_0(r,f)}{-\log(R_0-r)} = \infty, \quad 1 < r < R_0 < +\infty, \tag{12}$$

respectively.

Then, for a transcendental or admissible meromorphic function on the annulus $\mathbb{A}$, $S_1(r,f) = o(T_0(r,f))$ holds for all $1 < r < R_0$, except for the set $\triangle_r$ or the set $\triangle_r'$ mentioned in Lemma 2.1, respectively.

The following lemma plays an important role in the proof process of Theorem 1.6, which was given by Cao, Yi and Xu [26].

Lemma 2.2. ([26], Theorem 2.3) (The second fundamental theorem). Let f be a nonconstant meromorphic function on the annulus $\mathbb{A} = \{z : \frac{1}{R_0} < |z| < R_0\}$, where $1 < R_0 \leq +\infty$. Let a_1, $a_2, \ldots, a_q$ be q distinct complex numbers in the extended complex plane $\overline{\mathbb{C}}$. Then:

$$(q-2)T_0(r,f) < \sum_{j=1}^{q} \overline{N}_0(r, \frac{1}{f-a_j}) + S_1(r,f), \tag{13}$$

where $S_1(r,f)$ is stated as in Lemma 2.1.

Lemma 2.3. *(see [26]). Let* f *be a nonconstant meromorphic function on the annulus* $\mathbb{A} = \{z : \frac{1}{R_0} < |z| < R_0\}$, *where* $1 < R < R_0 \leq +\infty$. *Let* a *be an arbitrary complex number and* k *be a positive integer. Then:*

$$(i) \quad \overline{N}_0(R, \frac{1}{f-a}) \leq \frac{k}{k+1} \overline{N}_0^{k)}(R, \frac{1}{f-a}) + \frac{1}{k+1} N_0(R, \frac{1}{f-a}),$$

$$(ii) \quad \overline{N}_0(R, \frac{1}{f-a}) \leq \frac{k}{k+1} \overline{N}_0^{k)}(R, \frac{1}{f-a}) + \frac{1}{k+1} T_0(R,f) + O(1).$$

3. The Proof of Theorem 1.7

Proof of Theorem 1.7. Suppose that $f \not\equiv g$. Then, by Lemma 2.2 and Lemma 2.3, for any integer $m(1 \le m \le q)$, we have:

$$
\begin{aligned}
(q-2)T_0(r,f) &\le \sum_{j=1}^{q} \overline{N}_0(r,\frac{1}{f-a_j}) + S_1(r,f) \\
&= \sum_{j=1}^{q} \left\{ \overline{N}_0^{k_j)}(r,\frac{1}{f-a_j}) + \overline{N}_0^{(k_j+1}(r,\frac{1}{f-a_j}) \right\} + S_1(r,f) \\
&\le \sum_{j=1}^{q} \left\{ \overline{N}_0^{k_j)}(r,\frac{1}{f-a_j}) + \frac{1}{1+k_j} N_0^{(k_j+1}(r,\frac{1}{f-a_j}) \right\} + S_1(r,f) \\
&\le \sum_{j=1}^{q} \left\{ \frac{k_j}{1+k_j} \overline{N}_0^{k_j)}(r,\frac{1}{f-a_j}) + \frac{1}{1+k_j} N_0(r,\frac{1}{f-a_j}) \right\} + S_1(r,f) \\
&\le \sum_{j=1}^{q} \frac{k_j}{1+k_j} \overline{N}_0^{k_j)}(r,\frac{1}{f-a_j}) + \left(\sum_{j=1}^{q} \frac{1}{1+k_j} \right) T_0(r,f) + S_1(r,f) \\
&\le \sum_{j=1}^{m-1} \left(\frac{k_j}{1+k_j} - \frac{k_m}{1+k_m} \right) \overline{N}_0^{k_j)}(r,\frac{1}{f-a_j}) + \left(\sum_{j=1}^{q} \frac{1}{1+k_j} \right) T_0(r,f) \\
&\quad + \sum_{j=1}^{q} \frac{k_m}{1+k_m} \overline{N}_0^{k_j)}(r,\frac{1}{f-a_j}) + S_1(r,f) \\
&\le \sum_{j=1}^{q} \frac{k_m}{1+k_m} \overline{N}_0^{k_j)}(r,\frac{1}{f-a_j}) \\
&\quad + \left(m-1 - \frac{(m-1)k_m}{1+k_m} + \sum_{j=m}^{q} \frac{1}{1+k_j} \right) T_0(r,f) + S_1(r,f).
\end{aligned}
$$

that is,

$$
\left(\sum_{j=m}^{q} \frac{k_j}{1+k_j} - 2 + \frac{(m-1)k_m}{1+k_m} \right) T_0(r,f) \le \sum_{j=1}^{q} \frac{k_m}{1+k_m} \overline{N}_0^{k_j)}(r,\frac{1}{f-a_j}) + S_1(r,f). \tag{14}
$$

$$
\left(\sum_{j=m}^{q} \frac{k_j}{1+k_j} - 2 + \frac{(m-1)k_m}{1+k_m} \right) T_0(r,g) \le \sum_{j=1}^{q} \frac{k_m}{1+k_m} \overline{N}_0^{k_j)}(r,\frac{1}{g-a_j}) + S_1(r,g). \tag{15}
$$

Since $B_j = \overline{E}_{k_j)}(a_j,f) \setminus \overline{E}_{k_j)}(a_j,g)$, let $D_j = \overline{E}_{k_j)}(a_j,f) \setminus B_j$ for $j = 1,2,\ldots,q$. Thus, it follows from Equation (3) that:

$$
\begin{aligned}
\sum_{j=1}^{q} \overline{N}_0^{k_j)}(r,\frac{1}{f-a_j}) &= \sum_{j=1}^{q} \overline{N}_0^{B_j}(r,\frac{1}{f-a_j}) + \sum_{j=1}^{q} \overline{N}_0^{D_j}(r,\frac{1}{f-a_j}) \\
&\le \sum_{j=1}^{q} \delta_j T_0(r,f) + N_0(r,\frac{1}{f-g}) \\
&\le \left(1 + \sum_{j=1}^{q} \delta_j \right) T_0(r,f) + T_0(r,g) + O(1),
\end{aligned}
$$

and since f, g are transcendental or admissible, it follows from Equations (5) and (6) that:

$$\left(\sum_{j=m}^{q}\frac{k_j}{1+k_j}-2+\frac{(m-1)k_m}{1+k_m}+o(1)\right)\sum_{j=1}^{q}\overline{N}_0^{k_j)}(r,\frac{1}{f-a_j})$$
$$\leq\left(1+\sum_{j=1}^{q}\delta_j\right)\sum_{j=1}^{q}\frac{k_m}{1+k_m}\overline{N}_0^{k_j)}(r,\frac{1}{f-a_j})+(1+o(1))\sum_{j=1}^{q}\frac{k_m}{1+k_m}\overline{N}_0^{k_j)}(r,\frac{1}{g-a_j}), \tag{16}$$

as $r\to R_0$.

Since:

$$1\geq\frac{k_1}{k_1+1}\geq\frac{k_2}{k_2+1}\geq\cdots\geq\frac{k_q}{k_q+1}\geq\frac{1}{2}, \tag{17}$$

it follows from Equation (7) that:

$$\left\{\sum_{j=m}^{q}\frac{k_j}{1+k_j}-2+\frac{(m-1)k_m}{1+k_m}-\frac{k_m}{1+k_m}\left(1+\sum_{j=1}^{q}\delta_j\right)+o(1)\right\}\sum_{j=1}^{q}\overline{N}_0^{k_j)}(r,\frac{1}{f-a_j})$$
$$\leq(1+o(1))\frac{k_m}{1+k_m}\sum_{j=1}^{q}\overline{N}_0^{k_j)}(r,\frac{1}{g-a_j}),$$

which implies:

$$\liminf_{r\to R_0}\frac{\sum_{j=1}^{q}\overline{N}_0^{k_j)}(r,\frac{1}{f-a_j})}{\sum_{j=1}^{q}\overline{N}_0^{k_j)}(r,\frac{1}{g-a_j})}\leq\frac{\frac{k_m}{1+k_m}}{\sum_{j=m}^{q}\frac{k_j}{1+k_j}-2+(m-2-\sum_{j=1}^{q}\delta_j)\frac{k_m}{1+k_m}}.$$

This is a contradiction to Equation (4). Thus, we have $f(z)\equiv g(z)$.

Therefore, we complete the proof of Theorem 1.7. □

Acknowledgments: The first author was supported by the NSF of China (11561033, 11301233), the Natural Science Foundation of Jiangxi Province in China (20151BAB201008) and the Foundation of the Education Department of Jiangxi (GJJ14644) of China.

Author Contributions: Hong-Yan Xu and Hua Wang completed the main part of this article, Hong-Yan Xu corrected the main theorems. All authors read and approved the final manuscript.

Conflicts of Interest: The authors declare no conflict of interest.

References

1. Hayman, W. *Meromorphic Functions*; Clarendon Press: Oxford, UK, 1964.
2. Yang, L. *Value Distribution Theory*; Springer-Verlag: Berlin, Germany, 1993.
3. Yi, H.X.; Yang, C.C. *Uniqueness Theory of Meromorphic Functions*; Science Press: Beijing, China, 1995.
4. Laine, I. *Nevanlinna Theory and Complex Differential Equations*; Walter de Gruyter: Berlin, Germany, 1993.
5. Nevanlinna, R. Eindentig keitssätze in der theorie der meromorphen funktionen. *Acta Math.* **1926**, *48*, 367–391.
6. Banerjee, A. Weighted sharing of a small function by a meromorphic function and its derivativer. *Comput. Math. Appl.* **2007**, *53*, 1750–1761.
7. Lahiri, I. Weighted sharing and uniqueness of meromorphic functions. *Nagoya Math. J.* **2001**, *161*, 193–206.
8. Li, X.M.; Yi, H.X. On a uniqueness theorem of meromorphic functions concerning weighted sharing of three values. *Bull. Malays. Math. Sci. Soc.* **2010**, *33*, 1–16.
9. Fang, M.L. On the uniqueness of admissible meromorphic functions in the unit disc. *Sci. China A* **1999**, *42*, 367–381.
10. Mao, Z.Q.; Liu, H.F. Meromorphic functions in the unit disc that share values in an angular domain. *J. Math. Anal. Appl.* **2009**, *359*, 444–450.

11. Xu, H.Y.; Yi, C.F.; Cao, T.B. The uniqueness problem for meromorphic functions in the unit disc sharing values and a set in an angular domain. *Math. Scand.* **2011**, *109*, 240–252.
12. Cao, T.B.; Yi, H.X. On the uniqueness of meromorphic functions that share four values in one angular domain. *J. Math. Anal. Appl.* **2009**, *358*, 81–97.
13. Lin, W.C.; Mori, S.; Tohge, K. Uniqueness theorems in an angular domain. *Tohoku Math. J.* **2006**, *58*, 509–527.
14. Zhang, Q.C. Meromorphic Functions Sharing Values in an Angular Domain. *J. Math. Anal. Appl.* **2009**, *349*, 100–112.
15. Zheng, J.H. On uniqueness of meromorphic functions with shared values in some angular domains. *Can. J. Math.* **2004**, *47*, 152–160.
16. Zheng, J.H. On uniqueness of meromorphic functions with shared values in one angular domains. *Complex Var. Elliptic Equ.* **2003**, *48*, 777–785.
17. Zheng, J.H. *Value Distribution of Meromorphic Functions*; Tsinghua University Press: Beijing, China, 2010.
18. Khrystiyanyn, A.Y.; Kondratyuk, A.A. On the Nevanlinna theory for meromorphic functions on annuli I. *Mat. Stud.* **2005**, *23*, 19–30.
19. Khrystiyanyn, A.Y.; Kondratyuk, A.A. On the Nevanlinna theory for meromorphic functions on annuli II. *Mat. Stud.* **2005**, *24*, 57–68.
20. Korhonen, R. Nevanlinna theory in an annulus, value distribution theory and related topics. *Adv. Complex Anal. Appl.* **2004**, *3*, 167–179.
21. Lund, M.; Ye, Z. Logarithmic derivatives in annuli. *J. Math. Anal. Appl.* **2009**, *356*, 441–452.
22. Lund, M.; Ye, Z. Nevanlinna theory of meromorphic functions on annuli. *Sci. China Math.* **2010**, *53*, 547–554.
23. Xu, H.Y.; Xuan, Z.X. The uniqueness of analytic functions on annuli sharing some values. *Abstr. Appl. Anal.* **2012**, *2012*, 1–13.
24. Axler, S. Harmomic functions from a complex analysis viewpoit. *Am. Math. Mon.* **1986**, *93*, 246–258.
25. Kondratyuk, A.A.; Laine, I. *Meromorphic Functions in Multiply Connected Domains*, Proceedings of the Conference on Fourier Series Methods in Complex Analysis, Mekrijärvi, Finland, 24–29 July 2005; Laine, I., Ed.; University of Joensuu: Joensuu, Finland, 2006; Volume 10, pp. 9–111.
26. Cao, T.B.; Yi, H.X.; Xu, H.Y. On the multiple values and uniqueness of meromorphic functions on annuli. *Comput. Math. Appl.* **2009**, *58*, 1457–1465.
27. Cao, T.B.; Yi, H.X. Uniqueness theorems of meromorphic functions sharing sets *IM* on annuli. *Acta Math. Sin. (Chin. Ser.)* **2011**, *54*, 623–632. (In Chinese)
28. Cao, T.B.; Deng, Z.S. On the uniqueness of meromorphic functions that share three or two finite sets on annuli. *Proc. Indian Acad. Sci. Math. Sci.* **2012**, *122*, 203–220.

12

On the Dimension of Algebraic-Geometric Trace Codes

Phong Le [1,*] and Sunil Chetty [2]

[1] Department of Mathematics and Computer Science, Goucher College, Baltimore, MD 21204, USA
[2] Department of Mathematics, College of Saint Benedict and Saint John's University, Collegeville, MN 56321, USA; schetty@csbsju.edu
[*] Correspondence: phong.le@goucher.edu

Academic Editor: Hari M. Srivastava

Abstract: We study trace codes induced from codes defined by an algebraic curve X. We determine conditions on X which admit a formula for the dimension of such a trace code. Central to our work are several dimension reducing methods for the underlying functions spaces associated to X.

Keywords: error correcting codes; trace codes; exponential sums; number theory

MSC: 11T71

---

1. Introduction

Many good error correcting codes defined over a finite field can be constructed from other codes using the trace map. More generally, given a code C over a finite field $\mathbb{F}$, one can construct a subfield subcode by restriction (e.g., in the coordinates) to a subfield of $\mathbb{F}$. In [1], Katsman and Tsfasman prove that one can often obtain better parameters than those guaranteed by trivial bounds. Precise lower bounds on the dimension of subfield subcodes have been given in, e.g., [1–3]. Delsarte's Theorem [4] is used to describe subfield subcodes as trace codes. BCH-codes, classical and generalized Goppa codes, and alternant codes can all be realized as the dual of trace codes.

Algebraic-geometric (AG) codes arise from the evaluation of the elements of an $\mathbb{F}_{q^m}$-vector space of functions in a set of $\mathbb{F}_{q^m}$-rational points on a curve X. We shall consider trace codes associated to algebraic-geometric codes. In some cases the exact dimension can be determined. As in [5,6] a key ingredient in the present work is understanding the kernel of the trace map. Our use of Bombieri's estimate, following [7], and consideration of a more general class of codes differ from the methods and setting of [5,6].

The main result, Theorem 1, is an extension of results that appear in [7]. The bound in [7] applies for trace maps from the original field to the prime field. We modify this to include trace maps to intermediate fields. Significant modifications of the original proof are needed to accommodate the more general trace in the execution of Bombieri's estimate for exponential sums [8]. The primary modification is summarised in Proposition 17.

For a general introduction on AG codes and trace codes, see [9].

2. Definition of Code and Main Result

2.1. Background

Let p be a prime number and $q = p^r$. Given a linear code C of length n over $\mathbb{F}_{q^m}$, a trace code over $\mathbb{F}_q$ is constructed from C by applying the trace map from $\mathbb{F}_{q^m}$ to $\mathbb{F}_q$ coordinate-wise to the letters of the words of C. This q-ary code is denoted $\mathrm{Tr}_{q^m/q}C$ or simply $\mathrm{Tr}(C)$ if the base fields in question are clear.

Let X be a geometrically irreducible, non-singular projective curve of genus $\mathfrak{g}$ defined over $\mathbb{F}_{q^m}$. Consider $\mathbb{F}_{q^m}(X)$ the $\mathbb{F}_{q^m}$-rational function field of X. Throughout, by a point Q on X we mean $Q \in X(\overline{\mathbb{F}}_q)$ is a geometric point defined over $\overline{\mathbb{F}}_q$. A divisor $G = \sum n_Q Q$ defined over $\mathbb{F}_{q^m}$ is a formal sum over points Q defined over $\mathbb{F}_{q^m}$, and is invariant under the action of $\mathrm{Gal}(\overline{\mathbb{F}}_q/\mathbb{F}_{q^m})$. Any divisor $G = \sum n_Q Q$ may be split into two divisors G^+ and G^-, where $G^+ = \sum_{n_Q>0} n_Q Q$ and $G^- = \sum_{n_Q<0} n_Q Q$. Hence $G = G^+ + G^-$. The sum $\sum_i n_i$ of the coefficients of G is called the degree of G, denoted $\deg(G)$. We denote the support of G to be $\mathrm{Supp}(G) := \{Q \mid n_Q \neq 0\}$.

Define $L(G)$ to be the vector space of functions

$$L(G) = \{f \in \mathbb{F}_{q^m}(X) \mid (f) + G \geq 0\} \cup \{0\}$$

To generate a code from $L(G)$ we take a subset of n distinct $\mathbb{F}_{q^m}$-rational points away from the support of the divisor G:

$$D := \{P_1, \ldots, P_n\} \subseteq X(\mathbb{F}_{q^m}) \setminus \mathrm{Supp}(G)$$

For our purposes we will take $D = X(\mathbb{F}_{q^m}) \setminus \mathrm{Supp}(G)$ the largest possible set.

We define our AG code to be

$$C := C(D, G) = \{(f(P_1), \ldots, f(P_n)), f \in L(G)\}$$

When $2\mathfrak{g} - 2 < \deg(G)$, by Riemann-Roch we have

$$\dim_{\mathbb{F}_{q^m}} L(G) = \deg(G) + 1 - \mathfrak{g}$$

Since $\deg(G) < n$ the evaluation map

$$\begin{cases} L(G) \to \mathbb{F}_{q^n} \\ f \mapsto (f(P_1), \ldots, f(P_n)) \end{cases}$$

is injective. Hence the dimension of C as an $\mathbb{F}_{q^m}$-vector space is also k. In this way we identify $f \in L(G)$ with its image in C.

An AG trace code is defined as the coordinate-wise application of the trace map

$$\mathrm{Tr}(C) := \{(\mathrm{Tr}_{q^m/q}(f(P_1)), \ldots, \mathrm{Tr}_{q^m/q}(f(P_n))), f \in L(G)\}$$

2.2. *Main Result*

For $r \in \mathbb{R}$, let $[r]$ denote the greatest integer function. Consider the divisor

$$[G/q] := \sum_{n_Q>0} [n_Q/q]Q + \sum_{n_Q<0} n_Q Q$$

That is, we are dividing the positive coefficients by q and rounding down to the nearest integer. We are in a sense dividing the pole part of G by q. This construction will be useful in determining the kernel of the trace map.

Section 3 is devoted to the proof of the following dimension formula for $\mathrm{Tr}(C)$.

Theorem 1. *Let* $2\mathfrak{g} - 2 \leq \deg([G/q])$ *and* $\deg(G) < n$. *Consider the following two conditions:*

$$|\mathrm{Supp}(G^-)| \leq 1 \tag{1}$$

$$|X(\mathbb{F}_{q^m})| > (2\mathfrak{g} - 2 + \deg(G^+))q^{m/2} + |\mathrm{Supp}(G^+)|(q^{m/2} + 1) \tag{2}$$

Under these conditions we have an exact formula for the dimension:

$$\dim_{\mathbb{F}_q} \mathrm{Tr}C = m(\deg(G) - \deg([G/q])) + \delta$$

where

$$\delta = \begin{cases} 1 & \text{if } deg(G^-) \leq 1 \\ 0 & \text{otherwise} \end{cases}$$

If q is a prime number, this theorem reduces to the main result of [7]. Theorem 1 is applicable for a more general trace when $q = p^r$ for some prime p and $r > 1$. In this setting, complications arise in the dimension reducing argument using Bombieri's estimate used in [7]. Bombieri's estimate alone does not collapse the dimension of the kernel of the trace map enough. We have addressed these complications with the addition of a degree argument that shows that the kernel can be reduced in a way that aligns with the result in [7].

2.3. Examples

Example 1. Let $q = p^r$. Consider an elliptic curve E defined over $\mathbb{F}_{q^m}$. A formula for counting points on E is given by

$$E(\mathbb{F}_{q^m}) = q^m + 1 - \pi^m - \overline{\pi}^m$$

where $\pi\overline{\pi} = q$ and $\pi + \overline{\pi} = a_p$, the linear coefficient of the numerator of an associated zeta function as described in ([10], p. 301).

For $G = kP_\infty$ we see that condition (2) is

$$q^{m/2} - \frac{\pi^m + \overline{\pi}^m}{q^{m/2}} > k$$

Assuming this, Theorem 1 states that for any D such that $|D| > \deg(G)$ we have

$$\dim_{\mathbb{F}_q} \mathrm{Tr}(C(D,G)) = m(k - [k/q]) + 1$$

Example 2. For a smooth projective curve X defined over $\mathbb{F}_{q^m}$, let $G = kP_\infty$ for some positive integer k. Using the the Hasse-Weil bound we have

$$||X(\mathbb{F}_{q^m})| - (q^m + 1)| \leq 2\mathfrak{g}q^{m/2}$$

By condition (2) of Theorem 1 we must also require the inequality

$$|X(\mathbb{F}_{q^m})| > (2\mathfrak{g} - 2 + k)q^{m/2} + (q^{m/2} + 1)$$

Combining these two inequalities, we see that condition (2) is satisfied when

$$q^{m/2} - 4\mathfrak{g} + 1 > k$$

Using Theorem 1 we obtain the following:

Corollary 2. *For X a smooth projective curve over $\mathbb{F}_{q^m}$ and $G = kP_\infty$, if $2\mathfrak{g} - 2 \leq [k/q]$ and $k < \min(n, q^{m/2} - 4\mathfrak{g} + 1)$ then*

$$\dim_{\mathbb{F}_q} \mathrm{Tr}(C) = m(k - [k/q]) + 1$$

Example 3. This is a generalization from an example in [7]. Let $q = p^r$, $X = \mathbb{P}^1$ and $G = (g)_0 - P_\infty$ where $(g)_0$ is the zero divisor of a polynomial $g(z) \in \mathbb{F}_{q^m}[z]$ which has no zeros in $\mathbb{F}_{q^m}$. Denote the number of different zeros of $g(z)$ by s. Furthermore, we take $D = \sum_{x \in \mathbb{F}_{q^m}} P_x$. From condition (2) we obtain the inequality

$$\deg(g(z)) + s < \frac{q^m + 1}{\sqrt{q^m}} + 2$$

Write $g(z) = g_1^q g_2$, where $g_1(z), g_2(z) \in \mathbb{F}_{q^m}[z]$ of degrees r_1, r_2 respectively, and $g_2(z)$ q-th power free. With sufficiently many points as above, applying Theorem 1 we have

$$\dim_{\mathbb{F}_q} \mathrm{Tr}(C(D,G)) = m((q-1)r_1 + r_2)$$

3. Proof of Main Result

Observe C is a vector space over $\mathbb{F}_{q^m}$ and $\mathrm{Tr}(C)$ is a vector space over $\mathbb{F}_q$. From this we have the bound

$$\dim_{\mathbb{F}_{q^m}} C \leq \dim_{\mathbb{F}_q} \mathrm{Tr}(C) \leq m(dim_{\mathbb{F}_{q^m}} C)$$

Using the $\mathbb{F}_q$-linearity of the trace map we have an exact sequence

$$0 \to K' \to C \to \mathrm{Tr}(C) \to 0$$

where K' is the kernel of the trace map. Note the trace map defines an $\mathbb{F}_q$-linear subspace of C. Let K be the subspace of $L(G)$ that is the inverse image of evaluation at $D = \{P_1, \ldots, P_n\}$. That is

$$K := \{f \in L(G) | (f(P_1), \ldots, f(P_n)) \in K\}$$

Notice that K' is isomorphic to K as $\mathbb{F}_q$ vector spaces. Therefore

$$m \dim_{\mathbb{F}_{q^m}} C - \dim_{\mathbb{F}_q} K = \dim_{\mathbb{F}_q} \mathrm{Tr}(C) \tag{3}$$

We can obtain the dimension of $\mathrm{Tr}(C)$ by determining $\dim_{\mathbb{F}_q} K$. In practice, this is difficult. Consider the space of functions

$$E := \{f = h^q - h \mid f \in L(G), h \in \mathbb{F}_{q^m}(X)\}$$

Using Bombieri's estimate and a degree argument, we will determine a sufficient condition (condition 2 of Theorem 1) when $K = E$ and E is isomorphic to K'. But to make this useful we will first show Theorem 1 is sufficient to determine the dimension of E.

3.1. Dimension of E

For $f = h^q - h \in L(G)$, by definition $(h^q - h) + G \geq 0$. Counting with multiplicity, each pole in h corresponds to q poles in f. For $h \in L([G/q])$, we have $h^q - h \in L(G)$.

Consider the map $\phi : L([G/q]) \to E$ where $\phi(h) = h^q - h$. By definition, the kernel is $\mathbb{F}_q \cap L([G/q])$. Note that for a general G the map ϕ is not surjective.

Lemma 3. *When* $\deg(G^-) \leq 1$ *the map* ϕ *is surjective.*

Proof. Recall that the divisor $[G/q]$ only changes the positive coefficients of G and does not change G^-. When $G^- = \emptyset$, there is no restriction on zeros in $L(G)$. Therefore, in this case ϕ is onto.

If $|\mathrm{Supp}(G^-)| = 1$, then $G^- = n_p P$ for some point $P \in X(\mathbb{F}_{q^m})$ and negative integer n_p. Every function in $L([G/q])$ must have a zero at P. In the factorization $h^q - h = \prod_{b \in \mathbb{F}_q}(h - b)$, this zero must occur in at least one factor $h - b$. Though h may not be in $L([G/q])$, there will always exist some $b \in \mathbb{F}_q$ such that $h - b \in L([G/q])$. Observe $h^q - h = (h-b)^q - (h-b) = \phi(h-b)$. In this case, ϕ is onto. □

If $G^- = \emptyset$, then the kernel of ϕ is $\mathbb{F}_q$. If $G^- \neq \emptyset$, then ϕ is injective. Therefore the δ defined in Theorem 1 is merely $\delta = \dim_{\mathbb{F}_q} \ker \phi$. Using Lemma 3, we have the following proposition.

Proposition 4. *If $deg(G^-) \leq 1$, then the sequence*

$$0 \longrightarrow \mathbb{F}_q \cap L([G/q]) \longrightarrow L([G/q]) \xrightarrow{\phi} E \longrightarrow 0$$

is exact. Therefore we have a dimension formula for E:

$$\dim_{\mathbb{F}_q} E = \dim_{\mathbb{F}_q} L([G/q]) - \dim_{\mathbb{F}_q}(\mathbb{F}_q \cap L([G/q]))$$

In general, ϕ may not be surjective. There is still a dimension bound:

$$\dim_{\mathbb{F}_q} E \geq \dim_{\mathbb{F}_q} L([G/q]) - \dim_{\mathbb{F}_q}(\mathbb{F}_q \cap L([G/q]))$$

Note in [7], a similar result is obtained with the use of group cohomology and other auxillary constructions.

3.2. Bombieri's Estimate

A key step in determining when $K = E$ is a bound developed by Bombieri [8].

Theorem 5 (Bombieri's estimate). *Let X be a complete, geometrically irreducible, nonsingular curve of genus $\mathfrak{g}$, defined over $\mathbb{F}_{q^m}$. Let $f \in \mathbb{F}_{q^m}(X), f \neq h^p - h$ for $h \in \overline{\mathbb{F}}_q(X)$, with pole divisor $(f)_\infty$ on X. Then*

$$\left| \sum_{P \in X(\mathbb{F}_{q^m}) \setminus \mathrm{Supp}(f)_\infty} \zeta_p^{\mathrm{Tr}_{q^m/p}(f(P))} \right| \leq (2\mathfrak{g} - 2 + t + \deg(f)_\infty) q^{m/2}$$

where $\zeta_p = \exp(2\pi i/p)$ is any primitive p-th root of unity and t is the number of distinct poles of f on X.

Let $\overline{E} = \{f \in K \mid f = h^p - h \text{ for some } h \in \overline{\mathbb{F}}_q(X)\}$. On this subspace of K the conditions of Bombieri's Estimate are not met.

Lemma 6. $E \subseteq \overline{E}$.

Proof. Recall that $q = p^r$. Therefore, for $g^q - g \in E$ we have

$$\begin{aligned} g^q - g &= g^{p^r} - g \\ &= (g^{p^{r-1}} + \ldots + g)^p - (g^{p^{r-1}} + \ldots + g) \end{aligned}$$

Let $h = g^{p^{r-1}} + \ldots + g$. From this we see clearly that $g^q - g = h^p - h$. □

Lemma 7. *For each $g \in \overline{\mathbb{F}}_q(X)$, there exists an $h \in \mathbb{F}_{q^m}(X)$ and $c \in \mathbb{F}_{q^m}$ such that $g^p - g = h^p - h + c$. Therefore,*

$$\overline{E} \subseteq \{f \in \mathbb{F}_{q^m}(X) \mid f = h^p - h + c \text{ for some } h \in \mathbb{F}_{q^m}(X), c \in \mathbb{F}_{q^m}\}$$

Proof. Suppose there is an $f \in \mathbb{F}_{q^m}(X)$ and an $h \in \overline{\mathbb{F}}_q(X)$ such that $f = h^p - h$. Consider $\sigma = \mathrm{Frob}_{q^m}$, the coefficient-wise q^m-Frobenius endomorphism on $\overline{\mathbb{F}}_q^m$. Observe

$$\begin{aligned} \sigma(f) &= \sigma(h^p - h) \\ &= \sigma(h^p) - \sigma(h) \\ &= \sigma(h)^p - \sigma(h) \end{aligned}$$

Furthermore, $\sigma(f) = f$. Rearranging by exponents of p we see

$$\begin{aligned} \sigma(h)^p - \sigma(h) &= h^p - h \\ \sigma(h)^p - h^p &= \sigma(h) - h \\ (\sigma(h) - h)^p &= \sigma(h) - h \end{aligned}$$

By considering the order of poles of $\sigma(h) - h$, we determine that $\sigma(h) - h$ must be a constant $a \in \mathbb{F}_p$. There is a b in $\overline{\mathbb{F}}_q$ such that $a = b^{q^m} - b$. Then $\sigma(b) = b + a$ and $\sigma(h - b) = h + a - (b + a) = h - b$. Therefore, $h - b \in \mathbb{F}_{q^m}(X)$. Let $h_1 = h - b$. Observe $f - b^p + b = h_1^p - h_1$. Also, $\sigma(b^p - b) = b^p - b$, so $b^p - b \in \mathbb{F}_{q^m}$. Therefore, $f = h_1^p - h_1 + b^p - b$. □

Consider $f \in K$, and $P \in D = X(\mathbb{F}_{q^m}) \setminus \mathrm{Supp}(G)$ as defined in our definition of the trace code. By the definition of K, we have $\mathrm{Tr}(f(P)) = 0$. Observe that if $f \in K \setminus \overline{E}$ then f satisfies the conditions of Bombieri's Estimate. Hence, $\zeta_p^{\mathrm{Tr}(f(P))} = 1$ for each P. For such f, each term of the sum in the left-hand-side in Theorem 5 contributes 1. This is a total contribution of $|X(\mathbb{F}_q) \setminus (f)_\infty|$. Hence for $f \in K \setminus \overline{E}$, we have

$$\left| \sum_{P \in X(\mathbb{F}_{q^m}) \setminus \mathrm{Supp}(f)_\infty} \zeta_p^{\mathrm{Tr}_{q^m/p}(f(P))} \right| = |(X(\mathbb{F}_{q^m}) \setminus (f)_\infty)| \leq (2\mathfrak{g} - 2 + t + \deg(f)_\infty) q^{m/2}$$

Observe $t \leq |\mathrm{Supp}(G^+)|$ and $\deg(f)_\infty \leq \deg(G^+)$. Using these two inequalities, we obtain a more general bound:

$$|X(\mathbb{F}_{q^m})| \leq (2\mathfrak{g} - 2 + \deg(G^+)) q^{m/2} + |\mathrm{Supp}(G^+)|(q^{m/2} + 1)$$

Proposition 8. *If*

$$|X(\mathbb{F}_{q^m})| > (2\mathfrak{g} - 2 + \deg(G^+)) q^{m/2} + |\mathrm{Supp}(G^+)|(q^{m/2} + 1)$$

then $K = \overline{E}$.

The condition presented in Proposition 8 is exactly condition (2) from Theorem 1.

3.3. *E and* $\overline{E}$

Recall the definitions of E and $\overline{E}$:

$$E := \{f = h^q - h \mid f \in L(G), h \in \mathbb{F}_{q^m}(X)\}$$

$$\overline{E} := \{f \in K \mid f = h^p - h \text{ for some } h \in \overline{\mathbb{F}}_q(X)\}$$

In the case presented in [7], Van der Vlugt had $\overline{E} = E$. In the current more general case, Proposition 8 provides conditions forcing all elements of K to be of the form $h^p - h$, for $h \in \overline{\mathbb{F}}_q(X)$. However, it may be that elements of this form that are not of the form $g^q - g$, with $g \in \mathbb{F}_{q^m}(X)$. We will show that this is not the case and that condition (2) of Theorem 1 is sufficient to force $K = E$. It will be useful to develop our understanding of the interplay of K, $\overline{E}$ and E, and the nature of the degree of functions therein.

As is the case in Lemma 6, elements of the form $g^q - g$, for $g \in \mathbb{F}_{q^m}(X)$, can also be written in the form $h^p - h$, for $h \in \overline{\mathbb{F}}_q(X)$. Also notice that for any $f \in K$ and $y \in \mathbb{F}_q$, the function yf is an element of K. Furthermore, for any $f \subset E$ and $y \subset \mathbb{F}_q$, yf is in E. Consider the following:

Definition 9. For $f \in \overline{\mathbb{F}}_q(X)$, let $D(f)$ be the elements $y \in \mathbb{F}_q$ such that $yf = h^p - h$, for some $h \in \overline{\mathbb{F}}_q(X)$.

We see that for $f \in K$, when $|D(f)| < q$, there is a y such that $yf \in K \setminus \overline{E}$.

Proposition 10. *For $f \in K$, $D(f)$ is an $\mathbb{F}_p$-subspace of $\mathbb{F}_q$.*

Proposition 11. *If $D(f) = \mathbb{F}_q$ and $D(g) = \mathbb{F}_q$ then $D(af + bg) = \mathbb{F}_q$ for each $a, b \in \mathbb{F}_q$.*

Lemma 12. *Let $f = h^p - h \neq 0$ for some $h \in \overline{\mathbb{F}}_q(X)$ and $D(f) \neq \{0\}$. Then*

$$|D(f)| \leq p|D(h)|$$

Proof. Let $y \in D(f)$, $y \neq 0$. Then $yf = g^p - g$ for some $g \in \overline{\mathbb{F}}_q(X)$. Hence

$$yf = g^p - g = yh^p - yh = (y^{1/p}h)^p - (y^{1/p}h) + (y^{1/p}h) - yh$$

Rearranging terms we see that

$$(y^{1/p} - y)h = (g^p - g) - ((y^{1/p}h)^p - (y^{1/p}h))$$

Therefore $(y^{1/p} - y)$ is in $D(h)$. Hence, for every $x, y \in D(f)$, $x^{1/p} - x$ and $y^{1/p} - y$ are in $D(h)$. Suppose $x \neq y$ but $x^{1/p} - x = y^{1/p} - y$. Then

$$\begin{aligned} x^{1/p} - x &= y^{1/p} - y \\ (x-y)^{1/p} &= (x-y) \\ (x-y) &= (x-y)^p \end{aligned}$$

The only elements of $\mathbb{F}_q$ equal to their own p^{th} power are elements of $\mathbb{F}_p$. Hence $x = y + t$ for some $t \in \mathbb{F}_p$. From this we see that $D(f)/\mathbb{F}_p$ can be identified with a subgroup of $D(h)$. Hence $|D(f)| \leq p|D(h)|$. □

Definition 13. For $f \in \mathbb{F}_{q^m}(X)$, define the p-linear degree of f, denoted $e(f)$, to be the largest possible integer such that $f = a_c + a_0 g + a_1 g^p + \ldots + a_{e(f)} g^{p^{e(f)}}$, where $a_c, a_0, \ldots, a_{e(f)} \in \mathbb{F}_{q^m}$, $g \in \mathbb{F}_{q^m}(X)$.

The following properties of $e(f)$ are straightforward:

Proposition 14.

1. *For $f \in \overline{E}$, we have $e(f) \geq 1$. This is a restatement of Lemma 7.*
2. *For $g \in E$, we have $e(g) \geq r$.*
3. *For $a \in \mathbb{F}_{q^m}^*$, $b \in \mathbb{F}_{q^m}$, we have $e(f) = e(af + b)$.*

Proposition 15. *Suppose $f \in \mathbb{F}_{q^m}(X)$ such that either $e(f) = 0$ or $f = h^p - h$, for some $h \in \overline{\mathbb{F}_q}(X)$. Then we have the inequality:*

$$|D(f)| \leq p^{e(f)}$$

Proof. We proceed by induction on $e(f)$. Suppose $e(f) = 0$. By Lemma 7, we have $yf \neq h^p - h + c$, for any $y \in \mathbb{F}_{q^m}^*$, any $h \in \mathbb{F}_{q^m}(X)$, and $c \in \mathbb{F}_{q^m}$. Therefore $D(f) = \{0\}$ and $|D(f)| = 1 = p^0 = p^{e(f)} = 1$.

Now consider a positive integer k and $f \in \mathbb{F}_{q^m}(X)$ such that $e(f) = k$. Without loss of generality, we may assume that $|D(f)| > 1$. There is a nonzero $y \in D(f)$ such that $yf = h^p - h$. Therefore,

$f = y^{-1}h^p - y^{-1}h$. From this we obtain $e(f) \geq 1 + e(h)$. Since $e(f) = k$ we have $e(h) < k$. By the inductive hypothesis, $|D(h)| \leq p^{e(h)}$. Combining this with Lemma 12 we have

$$|D(f)| \leq p|D(h)| = p^{e(h)+1} \leq p^{e(f)}$$

□

Corollary 16. *If* $|D(f)| = q = p^r$*, then* $e(f) \geq r$.

Proposition 17. *Suppose condition (2) from Theorem 1 holds. That is, suppose*

$$|X(\mathbb{F}_{q^m})| > (2\mathfrak{g} - 2 + \deg(G^+))q^{m/2} + |\mathrm{Supp}(G^+)|(q^{m/2} + 1)$$

If $K \neq E$*, then* $K \setminus \overline{E}$ *is nonempty.*

Proof. Suppose $K \neq E$ and $D(f) = \mathbb{F}_q$ for each $f \in K \setminus E$. Such an $f \in \mathbb{F}_{q^m}(X)$ cannot be constant. Choose $f \in K \setminus E$ with the least number of poles. In other words, $\deg(f)_\infty$ is minimal and positive. Applying Corollary 16, there is some $l \in \mathbb{Z}_{\geq 0}$, $h \in \mathbb{F}_{q^m}(X)$ and $a_c, a_1, \ldots, a_{r+l} \in \mathbb{F}_{q^m}$ such that

$$f = a_{r+l}h^{p^{r+l}} + a_{r+l-1}h^{p^{r+l-1}} + \ldots + a_1 h + a_c$$

This may be rewritten as

$$f = f_E + f_1$$

where

$$f_E = (a_{r+l}h^{p^l})^q - (a_{r+l}h^{p^l}),$$

$$f_1 = (a_{r+l}h^{p^l}) + a_{r+l-1}h^{p^{r+l-1}} + \ldots + a_1 h + a_c \in \mathbb{F}_{q^m}(X)$$

Observe $f_E \in K$ and $D(f_E) = \mathbb{F}_q$. Hence $f_1 = f - f_E \in K$. By Proposition 11, $D(f_1) = \mathbb{F}_q$. But

$$\deg(f_1)_\infty \leq p^{r+l-1} \cdot \deg(h)_\infty$$

and

$$\deg(f)_\infty = p^{r+l} \deg(h)_\infty$$

This contradicts the choice of an f with minimal poles. Hence, when $K \neq E$, we can choose an $f \in K$ not of the form $h^p - h$. □

4. Conclusions

Proof of Theorem 1. Let $2\mathfrak{g} - 2 \leq \deg([G/q])$, $\deg(G) < n$ and also assume condition (1) and condition (2).

By condition (2), Proposition 8 and Proposition 17, we see that $K = E$. Then, using condition (1) and Proposition 4, we compute the dimension of E. We apply this to Equation (3) to obtain Theorem 1, a dimension formula for algebraic-geometric trace codes, as desired. □

Author Contributions: Both authors contributed equally to this work.

Conflicts of Interest: The authors declare no conflict of interest.

References

1. Katsman, G., Tsafsman, M. A remark on algebriac geometric codes. *Contemp. Math.* **1989**, *93*, 197–200.
2. Wirtz, M. On the parameters of Goppa codes. *IEEE Trans. Inform. Theory* **1988**, *34*, 1341–1343.
3. Stichtenoth, H. On the dimension of subfield subcodes. *IEEE Trans. Inform. Theory* **1990**, *36*, 90–93.

4. Delsarte, P. On the subfield subcodes of modified Reed-Solomon codes. *IEEE Trans. Inform. Theory* **1975**, *21*, 575–576.
5. Hernando, F.; Marshall, K.; O'Sullivan, M. The dimension of subcode-subfields of shortened generalized Reed-Solomon codes. *Des. Codes Cryptogr.* **2013**, *69*, 131–142.
6. Véron, P. True dimension of some binary quadratic trace Goppa codes. *Des. Codes Cryptogr.* **2001**, *24*, 81–97.
7. Van der Vlugt, M. A new upper bound for the dimension of trace codes. *Bull. Lond. Math. Soc.* **1991**, *23*, 395–400.
8. Bombieri, E. Exponential sums in finite fields. *Am. J. Math.* **1966**, *88*, 71–105.
9. Stichtenoth, H. *Algebraic Function Fields and Codes*; Springer-Verlag: Berlin, Germany, 1993.
10. Ireland, K.; Rosen, M. *A Classical Introduction to Modern Number Theory*; Springer-Verlag: Berlin, Germany, 1998.

13

Higher Order Methods for Nonlinear Equations and Their Basins of Attraction

Kalyanasundaram Madhu and Jayakumar Jayaraman *

Department of Mathematics, Pondicherry Engineering College, Pondicherry 605014, India; kalyan742@pec.edu
* Correspondence: jjayakumar@pec.edu

Academic Editor: Michel Chipot

Abstract: In this paper, we have presented a family of fourth order iterative methods, which uses weight functions. This new family requires three function evaluations to get fourth order accuracy. By the Kung–Traub hypothesis this family of methods is optimal and has an efficiency index of 1.587. Furthermore, we have extended one of the methods to sixth and twelfth order methods whose efficiency indices are 1.565 and 1.644, respectively. Some numerical examples are tested to demonstrate the performance of the proposed methods, which verifies the theoretical results. Further, we discuss the extraneous fixed points and basins of attraction for a few existing methods, such as Newton's method and the proposed family of fourth order methods. An application problem arising from Planck's radiation law has been verified using our methods.

Keywords: nonlinear equation; iterative methods; basins of attraction; extraneous fixed points; efficiency index

MSC: 65H05, 65D05, 41A25

1. Introduction

One of the best root-finding methods for solving nonlinear scalar equation $f(x) = 0$ is Newton's method. In recent years, numerous higher order iterative methods have been developed and analyzed for solving nonlinear equations that improve classical methods, such as Newton's method (NM), Halley's iteration method, *etc.*, which are respectively given below:

$$x_{n+1} = x_n - \frac{f(x_n)}{f'(x_n)} \tag{1}$$

and:

$$x_{n+1} = x_n - \frac{2f(x_n)f'(x_n)}{2f'(x_n)^2 - f(x_n)f''(x_n)}. \tag{2}$$

The convergence order of Newton's method is two, and it is optimal with two function evaluations. Halley's iteration method has third order convergence with three function evaluations. Frequently, f'' is difficult to calculate and computationally more costly, and therefore, f'' in Equation (2) is approximated using the finite difference; still, the convergence order and total number function evaluation are maintained [1]. Such a third order method similar to Equation (2) after approximating f'' in Halley's iteration method is given below:

$$y_n = x_n - \beta\frac{f(x_n)}{f'(x_n)}, \quad x_{n+1} = x_n - \frac{2\beta f(x_n)}{(2\beta - 1)f'(x_n) + f'(y_n)}, \quad \beta \neq 0. \tag{3}$$

In the past decade, a few authors have proposed third order methods with three function evaluations free from f''; for example, [2,3] and the references therein. The efficiency index (EI) of an iterative method is measured using the formula $p^{\frac{1}{d}}$, where p is the local order of convergence and d is the number of function evaluations per full iteration cycle. Kung–Traub [4] conjectured that the order of convergence of any multi-point without the memory method with d function evaluations cannot exceed the bound 2^{d-1}, the "optimal order". Thus, the optimal order for three evaluations per iteration would be four. Jarratt's method [5] is an example of an optimal fourth order method. Recently, some optimal and non-optimal multi-point methods have been developed in [6–15] and the references therein. A non-optimal method [16] has been recently rediscovered based on a quadrature formula, which can also be obtained by giving $\beta = \frac{2}{3}$ in Equation (3). In fact, each iterative fixed-point method produces a unique basins of attraction and fractal behavior, which can be used in the evaluation of algorithms [17]. Polynomiography is defined to be the art and science of visualization in the approximation of zeros of complex polynomials, where the created polynomiography images satisfy the mathematical convergence properties of iteration functions.

This paper considers a new family of optimal fourth order methods, which is an improvement of the method given in [16]. We study extraneous fixed points and basins of attraction for two particular cases of the new family of methods and a few equivalent available methods. The rest of the paper is organized as follows. Section 2 presents the development of the methods, their convergence analysis and the extension of new fourth order methods to sixth and twelfth order. Section 3 includes some numerical examples and results for the new family of methods along with some equivalent methods, including Newton's method. In Section 4, we obtain all possible extraneous fixed points for these methods as a special study. In Section 5, we study basins of attraction for the proposed fourth order methods, Newton's method and some existing methods. Section 6 discusses an application on Planck's radiation law problem. Finally, Section 7 gives the conclusions of our work.

2. Development of the Methods and Convergence Analysis

Noor *et al.* [16] consider the following third order method for the value of $\beta = \frac{2}{3}$ in Equation (3):

$$y_n = x_n - \frac{2}{3}\frac{f(x_n)}{f'(x_n)}, \quad x_{n+1} = x_n - \frac{4f(x_n)}{f'(x_n) + 3f'(y_n)}. \tag{4}$$

This Method (4) is of order three with three evaluations per full iteration having EI = 1.442. To improve the order of the above method with the same number of function evaluations leading to an optimal method, we propose the following without memory method, which includes weight functions:

$$\begin{aligned} y_n &= x_n - \frac{2}{3}\frac{f(x_n)}{f'(x_n)} \\ x_{n+1} &= x_n - \frac{4f(x_n)}{f'(x_n) + 3f'(y_n)} \times \Big(H(\tau) \times G(\eta)\Big), \end{aligned} \tag{5}$$

where $H(\tau)$ and $G(\eta)$ are two weight functions with $\tau = \frac{f'(y_n)}{f'(x_n)}$ and $\eta = \frac{f'(x_n)}{f'(y_n)}$.

2.1. Convergence Analysis

The proofs for Theorems 1 and 2 are worked out with the help of Mathematica.

Theorem 1. *Let $f : D \subset \mathbb{R} \to \mathbb{R}$ be a sufficiently smooth function having continuous derivatives up to fourth order. If $f(x)$ has a simple root x^* in the open interval D and x_0 is chosen in a sufficiently small neighborhood of x^*, then the family of Method (5) is of local fourth-order convergence, when:*

$$H(1) = G(1) = 1,\ H'(1) = G'(1) = 0,\ H''(1) = \frac{5}{8},\ G''(1) = \frac{1}{2},\ |H'''(1)| = |G'''(1)| < \infty \tag{6}$$

and it satisfies the error equation:

$$e_{n+1} = \frac{1}{81}\left(-81c_2c_3 + 9c_4 + c_2^3\left(147 + 32H'''(1) - 32G'''(1)\right)\right)e_n^4 + O(e_n^5),$$

where $c_j = \dfrac{f^{(j)}(x^*)}{j!f'(x^*)}$, $j = 2, 3, 4, \ldots$ *and* $e_n = x_n - x^*$.

Proof. Taylor expansion of $f(x_n)$ and $f'(x_n)$ about x^* gives:

$$f(x_n) = f'(x^*)\left[e_n + c_2e_n^2 + c_3e_n^3 + c_4e_n^4 + \ldots\right] \tag{7}$$

and:

$$f'(x_n) = f'(x^*)\left[1 + 2c_2e_n + 3c_3e_n^2 + 4c_4e_n^3 + \ldots\right] \tag{8}$$

so that:

$$y_n = x - \frac{2}{3}\frac{f(x_n)}{f'(x_n)} = x^* + \frac{e_n}{3} + \frac{2}{3}c_2e_n^2 - \frac{4}{3}\left(c_2^2 - c_3\right)e_n^3 + \frac{2}{3}\left(4c_2^3 - 7c_2c_3 + 3c_4\right)e_n^4 + \ldots \quad . \tag{9}$$

Again, using Taylor expansion of $f'(y_n)$ about x^* gives:

$$f'(y_n) = f'(x^*)\left[1 + \frac{2}{3}c_2e_n + \frac{1}{3}\left(4c_2^2 + c_3\right)e_n^2 + \frac{4}{27}\left(-18c_2^3 + 27c_2c_3 + c_4\right)e_n^3 + \ldots\right]. \tag{10}$$

Using Equations (8) and (10), we have:

$$\tau = 1 - \frac{4}{3}c_2e_n + \left(4c_2^2 - \frac{8}{3}c_3\right)e_n^2 - \frac{8}{27}\left(36c_2^3 - 45c_2c_3 + 13c_4\right)e_n^3 + \ldots \tag{11}$$

and:

$$\eta = 1 + \frac{4}{3}c_2e_n + \frac{4}{9}\left(-5c_2^2 + 6c_3\right)e_n^2 + \frac{8}{27}\left(8c_2^3 - 21c_2c_3 + 13c_4\right)e_n^3 + \ldots \tag{12}$$

Using Equations (7), (8) and (10), then we have:

$$\frac{4f(x_n)}{f'(x_n) + 3f'(y_n)} = e_n - c_2^2e_n^3 + \left(3c_2^3 - 3c_2c_3 - \frac{1}{9}c_4\right)e_n^4 + \ldots \quad . \tag{13}$$

Expanding the weight function $H(\tau)$ and $G(\eta)$ about 1 using Taylor series, we get:

$$\begin{aligned} H(\tau) &= H(1) + (\tau - 1)H'(1) + \frac{1}{2}(\tau - 1)^2H''(1) + \frac{1}{6}(\tau - 1)^3H'''(1) + O(H^{(4)}(1)), \\ G(\eta) &= G(1) + (\eta - 1)G'(1) + \frac{1}{2}(\eta - 1)^2G''(1) + \frac{1}{6}(\eta - 1)^3G'''(1) + O(G^{(4)}(1)). \end{aligned} \tag{14}$$

Using Equations (13) and (14) in Equation (5), such that the conditions in Equation (6) are satisfied, we obtain:

$$e_{n+1} = \frac{1}{81}\left(-81c_2c_3 + 9c_4 + c_2^3\left(147 + 32H'''(1) - 32G'''(1)\right)\right)e_n^4 + O(e_n^5). \tag{15}$$

Equation (15) shows that Method (5) has fourth order convergence. □

Note that for each choice of $|H'''(1)| < \infty$ and $|G'''(1)| < \infty$ in Equation (15) will give rise to a new optimal fourth order method. Method (5) has efficiency index EI = 1.587, better than Method (4). Two members in the family of Method (5) satisfying Condition (6), with corresponding weight functions, are given in the following:

By choosing $H'''(1) = G'''(1) = 0$, we get a new Proposed method called as PM1 :

$$\begin{aligned} y_n &= x_n - \frac{2}{3}\frac{f(x_n)}{f'(x_n)}, \\ x_{n+1} &= x_n - \frac{4f(x_n)}{f'(x_n)+3f'(y_n)}\left(1+\frac{5}{16}\left(\tau-1\right)^2\right)\left(1+\frac{1}{4}\left(\eta-1\right)^2\right), \end{aligned} \tag{16}$$

where its error equation is:

$$e_{n+1} = \left(\frac{49}{27}c_2^3 - c_2c_3 + \frac{1}{9}c_4\right)e_n^4 + O(e_n^5).$$

By choosing $H'''(1) = 0, G'''(1) = 1$, we get another new Proposed method called as PM2 :

$$\begin{aligned} y_n &= x_n - \frac{2}{3}\frac{f(x_n)}{f'(x_n)}, \\ x_{n+1} &= x_n - \frac{4f(x_n)}{f'(x_n)+3f'(y_n)}\left(1+\frac{5}{16}\left(\tau-1\right)^2\right)\left(1+\frac{1}{4}\left(\eta-1\right)^2+\frac{1}{6}\left(\eta-1\right)^3\right), \end{aligned} \tag{17}$$

where its error equation is:

$$e_{n+1} = \left(\frac{115}{81}c_2^3 - c_2c_3 + \frac{1}{9}c_4\right)e_n^4 + O(e_n^5).$$

Remark 1. By this way, we can propose many such fourth order methods similar to PM1 and PM2. Further, the methods PM1 and PM2 are equally good, since they have the same order of convergence and efficiency. Based on the analysis done using basins of attraction, we find that PM1 is marginally better than PM2, and hence, we have considered PM1 to propose a higher order method, namely PM3.

2.2. *Higher Order Methods*

We extend the method PM1 to a new sixth order method called as PM3:

$$\begin{aligned} y_n &= x_n - \frac{2}{3}\frac{f(x_n)}{f'(x_n)}, \\ z_n &= x_n - \frac{4f(x_n)}{f'(x_n)+3f'(y_n)}\left(1+\frac{5}{16}\left(\tau-1\right)^2\right)\left(1+\frac{1}{4}\left(\eta-1\right)^2\right), \\ x_{n+1} &= z_n - \frac{1}{2}\frac{f(z_n)}{f'(x_n)}\left(3\eta-1\right). \end{aligned} \tag{18}$$

The following theorem gives the proof of convergence for Method (18).

Theorem 2. *Let $f : D \subset \mathbb{R} \to \mathbb{R}$ be a sufficiently smooth function having continuous derivatives up to fourth order. If $f(x)$ has a simple root x^* in the open interval D and x_0 is chosen in a sufficiently small neighborhood of x^*, then Method (18) is of local sixth order convergence, and it satisfies the error equation:*

$$e_{n+1} = \frac{1}{81}\left(10c_2^2 - 3c_3\right)\left(49c_2^3 - 27c_2c_3 + 3c_4\right)e_n^6 + O(e_n^7).$$

Proof. Taylor expansion of $f(z_n)$ about x^* gives:

$$\begin{aligned} f(z_n) = f'(x^*)\Bigg[&\left(\tfrac{49}{27}c_2^3 - c_2c_3 + \tfrac{1}{9}c_4\right)e_n^4 - \tfrac{2}{81}\left(403c_2^4 - 522c_2^2c3 + 81c_3^2 + 90c_2c_4 - 12c_5\right)e_n^5 \\ &+ \tfrac{2}{243}\left(4529c_2^5 - 8835c_2^3c_3 + 2343c_2^2c_4 - 891c_3c_4 + 135c_2(25c_3^2 - 3c_5) + 63c_6\right)e_n^6\Bigg]. \end{aligned} \tag{19}$$

By using Equations (8), (12) and (19) in Equation (18), we obtain:

$$e_{n+1} = \frac{1}{81}\left(10c_2^2 - 3c_3\right)\left(49c_2^3 - 27c_2c_3 + 3c_4\right)e_n^6 + O(e_n^7). \tag{20}$$

Equation (20) shows that Method (18) has sixth order convergence. □

Babajee *et al.* [7] improved a sixth order Jarratt method to a twelfth order method. Using their technique, we obtain a new twelfth order method called as PM4:

$$\begin{aligned} y_n &= x_n - \frac{2}{3}\frac{f(x_n)}{f'(x_n)}, \\ z_n &= x_n - \frac{4f(x_n)}{f'(x_n) + 3f'(y_n)}\left(1 + \frac{5}{16}\left(\tau - 1\right)^2\right)\left(1 + \frac{1}{4}\left(\eta - 1\right)^2\right), \\ w_n &= z_n - \frac{1}{2}\frac{f(z_n)}{f'(x_n)}\left(3\eta - 1\right), \\ x_{n+1} &= w_n - \frac{f(w_n)}{f'(w_n)}, \end{aligned} \tag{21}$$

where $f'(w_n)$ is approximated as follows: in order to reduce one function evaluation, we replace:

$$\begin{aligned} f'(w_n) \approx \frac{1}{z_n - w_n}\Big(& f'(x_n)(z_n - w_n) + 2f[w_n, x_n, x_n](z_n - x_n)(w_n - x_n) \\ & + (f[z_n, x_n, x_n] - 3f[w_n, x_n, x_n])(w_n - x_n)^2\Big), \end{aligned}$$

$$\begin{aligned} f[z_n, x_n, x_n] &= \frac{f[z_n, x_n] - f'(x_n)}{z_n - x_n}, \quad f[z_n, x_n] = \frac{f(z_n) - f(x_n)}{z_n - x_n}, \\ f[w_n, x_n, x_n] &= \frac{f[w_n, x_n] - f'(x_n)}{w_n - x_n}, \quad f[w_n, x_n] = \frac{f(w_n) - f(x_n)}{w_n - x_n}. \end{aligned}$$

The following theorem is given without proof, which can be worked out with the help of Mathematica.

Theorem 3. *Let $f : D \subset \mathbb{R} \to \mathbb{R}$ be a sufficiently smooth function having continuous derivatives up to fourth order. If $f(x)$ has a simple root x^* in the open interval D and x_0 is chosen in a sufficiently small neighborhood of x^*, then Method* (21) *is of local twelfth order convergence, and it satisfies the error equation:*

$$e_{n+1} = \frac{1}{6561}\left(10c_2^2 - 3c_3\right)\left(49c_2^3 - 27c_2c_3 + 3c_4\right)^2\left(10c_2^3 - 3c_2c_3 + 3c_4\right)e_n^{12} + O(e_n^{13}).$$

Remark 2. The efficiency indices for the methods PM3 and PM4 are EI = 1.565 and EI = 1.644, respectively.

2.3. Some Existing Fourth Order Methods

Consider the following fourth order optimal methods for the purpose of comparing results:

Jarratt method (JM) [5]:

$$y_n = x_n - \frac{2}{3}\frac{f(x_n)}{f'(x_n)}, \quad x_{n+1} = x_n - \frac{3f'(y_n) + f'(x_n)}{6f'(y_n) - 2f'(x_n)}\frac{f(x_n)}{f'(x_n)}. \tag{22}$$

Method of Sharifi-Babajee-Soleymani (SBS1) [12]:

$$\begin{aligned} y_n &= x_n - \frac{2}{3}\frac{f(x_n)}{f'(x_n)}, \\ x_{n+1} &= x_n - \frac{f(x_n)}{4}\left(\frac{1}{f'(x_n)} + \frac{3}{f'(y_n)}\right)\left(1 + \frac{3}{8}\left(\frac{f'(y_n)}{f'(x_n)} - 1\right)^2 - \frac{69}{64}\left(\frac{f'(y_n)}{f'(x_n)} - 1\right)^3 + \left(\frac{f(x_n)}{f'(y_n)}\right)^4\right). \end{aligned} \tag{23}$$

Method of Sharifi-Babajee-Soleymani (SBS2) [12]:

$$y_n = x_n - \frac{2}{3}\frac{f(x_n)}{f'(x_n)}, \quad x_{n+1} = x_n - \frac{f(x_n)}{4}\left(\frac{1}{f'(x_n)} + \frac{3}{f'(y_n)}\right)\left(1 + \frac{3}{8}\left(\frac{f'(y_n)}{f'(x_n)} - 1\right)^2 + \frac{1}{81}\left(\frac{f(x_n)}{f'(y_n)}\right)^3\right). \tag{24}$$

Method of Soleymani-Khratti-Karimi (SKK) [15]:

$$\begin{aligned} y_n &= x_n - \frac{2}{3}\frac{f(x_n)}{f'(x_n)}, \\ x_{n+1} &= x_n - \frac{2f(x_n)}{f'(x_n) + f'(y_n)}\left(1 + \left(\frac{f(x_n)}{f'(x_n)}\right)^4\right)\left(2 - \frac{7}{4}\frac{f'(y_n)}{f'(x_n)} + \frac{3}{4}\left(\frac{f'(y_n)}{f'(x_n)}\right)^2\right). \end{aligned} \tag{25}$$

Method of Singh-Jaiswal (SJ) [14]:

$$\begin{aligned} y_n &= x_n - \frac{2}{3}\frac{f(x_n)}{f'(x_n)}, \\ x_{n+1} &= x_n - \left(\frac{17}{8} - \frac{9}{4}\frac{f'(y_n)}{f'(x_n)} + \frac{9}{8}\left(\frac{f'(y_n)}{f'(x_n)}\right)^2\right)\left(\frac{7}{4} - \frac{3}{4}\frac{f'(y_n)}{f'(x_n)}\right)\frac{f(x_n)}{f'(x_n)}. \end{aligned} \tag{26}$$

Method of Sharma-Kumar-Sharma (SKS) [13]:

$$\begin{aligned} y_n &= x_n - \frac{2}{3}\frac{f(x_n)}{f'(x_n)}, \\ x_{n+1} &= x_n - \left(-\frac{1}{2} + \frac{9}{8}\frac{f'(x_n)}{f'(y_n)} + \frac{3}{8}\frac{f'(y_n)}{f'(x_n)}\right)\frac{f(x_n)}{f'(x_n)}. \end{aligned} \tag{27}$$

Furthermore, consider the following non-optimal method found in Divya Jain (DJ) [10]:

$$\begin{aligned} y_n &= x_n - \frac{f(x_n)}{f'(x_n)}, \\ z_n &= x_n - \frac{2f(x_n)}{f'(x_n) + f'(y_n)}, \\ x_{n+1} &= z_n - \frac{z_n - x_n}{f(z_n) - f(x_n)} f(z_n). \end{aligned} \tag{28}$$

3. Numerical Examples

In this section, we give numerical results on some test functions to compare the efficiency of the proposed family of methods with some known methods. Numerical computations have been carried out in the MATLAB software, rounding to 500 significant digits. Depending on the precision of the computer, we use the stopping criteria for the iterative process: $error = |x_N - x_{N-1}| < \epsilon$, where $\epsilon = 10^{-50}$ and N is the number of iterations required for convergence. d_1 represents the total number of function evaluations. The computational order of convergence (COC) denoted as ρ is given by (see [18]):

$$\rho = \frac{\ln |(x_N - x_{N-1})/(x_{N-1} - x_{N-2})|}{\ln |(x_{N-1} - x_{N-2})/(x_{N-2} - x_{N-3})|}.$$

Functions taken for our study are mostly used in the literature [7,11], and their simple zeros are given below:

$$f_1(x) = \sin(2\cos x) - 1 - x^2 + e^{\sin(x^3)}, \qquad x^* = -0.7848959876612125352...$$
$$f_2(x) = xe^{x^2} - \sin^2 x + 3\cos x + 5, \qquad x^* = -1.2076478271309189270...$$
$$f_3(x) = x^3 + 4x^2 - 10, \qquad x^* = 1.3652300134140968457...$$
$$f_4(x) = sin(x) + cos(x) + x, \qquad x^* = -0.4566247045676308244...$$
$$f_5(x) = \frac{x}{2} - \sin x, \qquad x^* = 1.8954942670339809471...$$
$$f_6(x) = \sqrt{x^2 + 2x + 5} - 2\sin x - x^2 + 3, \qquad x^* = 2.3319676558839640103...$$
$$f_7(x) = \sqrt{x} - \cos x, \qquad x^* = 0.6417143708728826583...$$
$$f_8(x) = x^2 + \sin\left(\frac{x}{5}\right) - \frac{1}{4}, \qquad x^* = 0.4099920179891371316...$$
$$f_9(x) = e^{-x}\sin x + \log(1 + x^2) - 2, \qquad x^* = 2.4477482864524245021...$$
$$f_{10}(x) = \sqrt{x^3} + \sin x - 30, \qquad x^* = 9.7165019933652005655...$$

From Tables 1 and 2, we observe that PM1 and PM2 converge in a lesser number of iterations and with low error when compared to Methods (1) and (4). For equivalent fourth order methods, PM1 and PM2 converge in a lesser number of iterations for certain functions, for example PM2 performs better compared to Method (25) for the functions f_2, f_7, f_8, f_9 and f_{10}. In terms of the number of iterations for convergence, PM1 and PM2 are equivalent to JM. Tables 3 and 4 displays the total number of function evaluations (d_1) and the computational order of convergence (COC) ρ for the methods taken for our study.

Table 1. Comparison of the results for some known methods and proposed methods.

f	x_0	**NM** (1)		**Noor *et al.*** (4)		**JM** (22)		**SBS1** (23)		**SBS2** (24)		**PM1** (16)	
		N	*Error*	N	*Error*	N	*Error*	N	*Error*	N	*Error*	N	*Error*
f_1	−0.9	7	7.7e − 074	5	2.5e − 093	4	1.6e − 067	4	2.1e − 074	4	4.0e − 063	4	4.4e − 065
	−0.7	7	1.0e − 074	5	3.3e − 094	4	1.4e − 070	4	1.2e − 083	4	4.2e − 063	4	7.2e − 066
f_2	−1.7	9	4.3e − 054	6	7.2e − 051	5	1.4e − 085	5	2.4e − 072	6	4.3e − 179	5	4.2e − 058
	−1.0	8	1.1e − 064	6	6.0e − 123	5	2.0e − 199	5	5.0e − 081	5	1.4e − 097	5	3.8e − 116
f_3	1.6	7	7.7e − 063	5	2.0e − 079	4	2.4e − 065	4	5.3e − 059	4	5.6e − 057	4	1.2e − 059
	1.0	8	2.8e − 088	5	5.5e − 056	5	1.4e − 187	5	1.3e − 161	5	2.7e − 135	5	2.5e − 149
f_4	−0.2	7	6.8e − 096	5	6.5e − 121	4	2.1e − 077	4	1.0e − 058	4	2.9e − 070	4	5.4e − 076
	−0.6	6	1.5e − 061	4	6.9e − 052	4	4.3e − 100	4	8.8e − 079	4	2.4e − 090	4	1.2e − 099
f_5	1.6	8	6.8e − 087	5	5.4e − 055	5	5.7e − 169	5	4.9e − 148	5	5.5e − 124	5	7.9e − 137
	2.0	7	1.8e − 080	5	1.0e − 101	4	7.4e − 079	4	1.3e − 096	4	2.9e − 072	4	1.2e − 074
f_6	2.1	6	1.5e − 055	5	1.2e − 142	4	6.5e − 096	4	2.7e − 062	4	5.3e − 080	4	6.3e − 097
	2.5	6	9.6e − 055	5	1.3e − 138	4	4.5e − 094	4	7.1e − 073	4	6.1e − 087	4	7.8e − 096
f_7	0.2	7	2.0e − 074	5	2.2e − 090	4	8.7e − 063	5	9.5e − 142	4	2.8e − 054	4	2.5e − 060
	0.9	7	3.0e − 094	5	7.3e − 121	4	3.5e − 079	4	5.9e − 055	4	1.8e − 073	4	6.9e − 081
f_8	0.2	8	8.2e − 076	6	2.8e − 143	5	7.4e − 151	5	1.0e − 118	5	1.0e − 100	5	3.8e − 114
	1.5	9	2.7e − 074	6	1.5e − 070	5	3.1e − 074	5	1.3e − 104	5	1.8e − 061	5	5.6e − 065
f_9	1.9	7	2.9e − 088	5	4.5e − 110	4	1.0e − 084	5	4.7e − 119	4	9.7e − 057	4	3.1e − 108
	2.7	6	5.9e − 058	5	6.1e − 149	4	5.8e − 102	4	7.8e − 066	4	2.9e − 078	4	1.3e − 100
f_{10}	9.9	6	9.5e − 059	5	1.9e − 149	4	3.3e − 100	4	7.7e − 072	4	3.1e − 086	4	1.7e − 101
	9.2	6	3.1e − 052	5	3.3e − 131	4	1.9e − 078	5	5.3e − 128	4	3.0e − 058	4	9.4e − 079

Table 2. Comparison of the results for some known methods and proposed methods.

f	x_0	**SKK** (25)		**SJ** (26)		**SKS** (27)		**PM2** (17)	
		N	*Error*	N	*Error*	N	*Error*	N	*Error*
f_1	−0.9	4	4.6e − 062	4	3.0e − 062	4	9.7e − 064	4	2.7e − 066
	−0.7	4	9.9e − 060	4	2.9e − 062	4	5.4e − 064	4	1.5e − 067
f_2	−1.7	6	6.4e − 147	6	3.9e − 153	6	2.9e − 186	5	6.9e − 075
	−1.0	5	1.3e − 075	5	1.9e − 149	5	4.1e − 106	5	9.8e − 127
f_3	1.6	4	9.6e − 061	4	1.6e − 054	4	4.6e − 057	4	2.0e − 062
	1.0	5	9.5e − 102	5	4.2e − 142	5	1.7e − 140	5	1.7e − 157
f_4	−0.2	4	3.5e − 056	4	2.1e − 074	4	3.0e − 075	4	1.3e − 076
	−0.6	4	1.7e − 078	4	5.0e − 099	4	2.2e − 099	4	7.4e − 100
f_5	1.6	5	3.6e − 101	5	1.0e − 134	5	2.4e − 129	5	1.5e − 143
	2.0	4	2.0e − 070	4	1.6e − 070	4	1.2e − 072	4	1.2e − 076
f_6	2.1	4	3.8e − 063	4	1.3e − 098	4	1.2e − 097	4	1.9e − 096
	2.5	4	1.5e − 075	4	2.0e − 099	4	3.2e − 097	4	6.7e − 095
f_7	0.2	5	4.6e − 159	4	1.2e − 057	4	5.2e − 059	4	1.2e − 061
	0.9	4	4.8e − 057	4	1.4e − 084	4	2.8e − 082	4	5.9e − 080
f_8	0.2	5	1.6e − 082	5	3.7e − 132	5	3.5e − 107	5	1.2e − 120
	1.5	6	2.1e − 150	5	3.0e − 054	5	1.2e − 059	5	8.6e − 071
f_9	1.9	5	3.2e − 126	4	3.1e − 080	4	3.5e − 086	4	5.9e − 089
	2.7	4	2.5e − 064	4	5.7e − 099	4	7.6e − 100	4	3.2e − 101
f_{10}	9.9	4	1.3e − 073	4	7.2e − 104	4	1.9e − 102	4	8.0e − 101
	9.2	5	8.0e − 127	4	2.9e − 079	4	5.7e − 079	4	1.3e − 078

Table 3. Total number of function evaluations (d_1) and COC (ρ).

f	x_0	**NM** (1)		**Noor *et al.*** (4)		**JM** (22)		**SBS1** (23)		**SBS2** (24)		**PM1** (16)	
		d_1	ρ	d_1	ρ	d_1	ρ	d_1	ρ	d_1	ρ	d_1	ρ
f_1	−0.9	14	1.99	15	2.99	12	3.99	12	4.00	12	3.99	12	3.99
	−0.7	14	1.99	15	2.98	12	3.99	12	3.99	12	3.99	12	3.99
f_2	−1.7	18	2.00	18	2.99	15	4.00	15	4.00	18	3.99	15	3.99
	−1.0	16	2.00	18	3.00	15	3.99	15	4.00	15	4.00	15	3.99
f_3	1.6	14	2.00	15	2.98	12	3.99	12	3.99	12	3.99	12	3.99
	1.0	16	1.99	15	2.99	15	4.00	15	3.98	15	3.99	15	3.99
f_4	−0.2	14	2.00	15	3.00	12	3.99	12	4.00	12	3.99	12	3.99
	−0.6	12	2.01	12	3.01	12	3.98	12	4.00	12	3.99	12	3.99
f_5	1.6	16	1.98	15	3.00	15	4.00	15	3.99	15	3.99	15	3.99
	2.0	14	1.99	15	2.99	12	3.99	12	3.98	12	4.00	12	3.99
f_6	2.1	12	1.99	15	3.00	12	3.99	12	3.99	12	3.99	12	4.00
	2.5	12	1.98	15	3.00	12	4.00	12	4.00	12	3.99	12	3.99
f_7	0.2	14	2.00	15	2.99	12	4.00	15	4.00	12	3.98	12	3.99
	0.9	14	2.00	15	2.98	12	3.98	12	3.99	12	4.00	12	4.01
f_8	0.2	16	2.00	18	3.00	15	3.99	15	3.99	15	4.01	15	3.98
	1.5	18	1.99	18	2.99	15	3.99	15	4.00	15	3.98	15	3.99
f_9	1.9	14	1.98	15	2.99	12	3.98	15	4.00	12	3.99	12	3.99
	2.7	12	2.00	15	2.98	12	4.00	12	3.99	12	3.99	12	3.99
f_{10}	9.9	12	1.99	15	3.00	12	4.00	12	3.98	12	3.99	12	3.99
	9.2	12	2.00	15	2.99	12	3.99	15	3.99	12	4.00	12	3.99

Table 4. Total number of function evaluations (d_1) and COC (ρ).

f	x_0	SKK (25)		SJ (26)		SKS (27)		PM2 (17)	
		d_1	ρ	d_1	ρ	d_1	ρ	d_1	ρ
f_1	−0.9	12	3.99	12	3.99	12	3.99	12	3.99
	−0.7	12	3.99	12	3.99	12	3.99	12	3.99
f_2	−1.7	18	3.99	18	3.99	18	3.99	15	3.99
	−1.0	15	3.99	15	4.00	15	3.99	15	3.99
f_3	1.6	12	3.99	12	4.00	12	3.99	12	4.00
	1.0	15	4.00	15	3.99	15	4.00	15	3.99
f_4	−0.2	12	4.00	12	4.00	12	3.98	12	3.99
	−0.6	12	3.98	12	3.99	12	3.99	12	4.00
f_5	1.6	15	3.99	15	3.99	15	3.99	15	3.99
	2.0	12	3.99	12	3.99	12	3.99	12	3.99
f_6	2.1	12	3.98	12	4.00	12	3.99	12	3.99
	2.5	12	4.00	12	4.00	12	3.99	12	3.99
f_7	0.2	15	3.99	12	3.99	12	3.98	12	3.99
	0.9	12	3.98	12	3.98	12	3.99	12	4.00
f_8	0.2	15	4.01	15	3.98	15	4.00	15	3.98
	1.5	18	4.00	15	3.99	15	3.99	15	3.99
f_9	1.9	15	3.99	12	3.99	12	3.98	12	3.99
	2.7	12	3.99	12	4.00	12	3.99	12	3.99
f_{10}	9.9	12	3.99	12	3.99	12	3.99	12	3.99
	9.2	15	3.99	12	3.99	12	3.98	12	3.99

Table 5 displays the results for the "$fzero$" command in MATLAB, where N_1 is the number of iterations to find the interval containing the root and $f(x_n)$ is the error after N number of iterations. For the $fzero$ command, zeros are considered to be points where the function actually crosses, not just touches the x-axis. It is observed that the present methods (PM1 and PM2) converge with a lesser number of total function evaluations than the $fzero$ solver.

Table 5. Results for the $fzero$ command in MATLAB.

f	x_0	N_1	N	d_1	$f(x_n)$	x^*
f_1	−0.7	6	6	19	−1.1102e − 016	−0.7849
f_2	−1.7	8	7	23	−2.6645e − 015	−1.2076
f_3	1.0	9	6	25	0	1.3652
f_4	−0.2	13	6	33	−5.5511e − 017	−0.4566
f_5	1.6	7	6	21	0	1.8955
f_6	2.1	5	5	16	8.8818e − 016	2.3320
f_7	0.9	8	4	20	0	0.6417
f_8	0.2	12	7	32	−2.7756e − 017	0.4100
f_9	1.9	8	4	21	0	2.4477
f_{10}	9.2	3	3	10	0	9.7165

4. A Study on Extraneous Fixed Points

Definition 4. A point z_0 is a fixed point of R if $R(z_0) = z_0$.

Definition 5. A point z_0 is called attracting if $|R'(z_0)| < 1$, repelling if $|R'(z_0)| > 1$ and neutral if $|R'(z_0)| = 1$. If the derivative is also zero, then the point is super attracting.

It is interesting to note that all of the above discussed methods can be written as:

$$x_{n+1} = x_n - G_f(x_n)u(x_n),\ u = \frac{f}{f'}. \tag{29}$$

As per the definition, x^* is a fixed point of this method, since $u(x^*) = 0$. However, the points $\xi \neq x^*$ at which $G_f(\xi) = 0$ are also fixed points of the method, since $G_f(\xi) = 0$; the second term on the right side of Equation (29) vanishes. Hence, these points ξ are called extraneous fixed points.

Moreover, for a general iteration function given by:

$$R_p(z) = z - G_f(z)u(z),\ z \in \mathbb{C}, \tag{30}$$

the nature of extraneous fixed points can be discussed. Based on the nature of the extraneous fixed points, the convergence of the iteration process will be determined. For more details on this aspect, the paper by Vrcsay *et al.* [19] will be useful. In fact, they investigated that if the extraneous fixed points are attractive, then the method will give erroneous results. If the extraneous fixed points are repelling or neutral, then the method may not converge to a root near the initial guess.

In this section, we will discuss the extraneous fixed points of each method for the polynomial $z^3 - 1$. As G_f does not vanish in Theorem 6, there are no extraneous fixed points.

Theorem 6. *There are no extraneous fixed points for Newton's Method (1) and Method (4).*

Theorem 7. *There are six extraneous fixed points for Jarratt Method (22).*

Proof. The extraneous fixed point of Jarratt method for which

$$G_f = \frac{3f'(y(z)) + f'(z)}{6f'(y(z)) - 2f'(z)}$$

are found. Upon substituting $y(z) = z - \frac{2f(z)}{3f'(z)}$, we get the equation $\frac{1+7z^3+19z^6}{2+14z^3+11z^6} = 0$. The extraneous fixed points are found to be $0.411175 \pm 0.453532i,\ -0.598358 \pm 0.129321i,\ 0.187183 \pm 0.582854i$. All of these fixed points are repelling (since $|R'(z_0)| > 1$). □

Theorem 8. *There are fifty two extraneous fixed points for Method (23).*

Proof. We found for Method (23),

$$G_f = \left(1 + 3\frac{f'(z)}{f'(y(z))}\right)\left(f'(z) + \frac{69}{64}\frac{(f'(z) - f'(y(z)))^3}{f'(z)^2} + \frac{f(z)^4 f'(z)}{f'(y(z))^4} + \frac{3}{8}\frac{(f'(y(z)) - f'(z))^2}{f'(z)}\right).$$

The extraneous fixed points are at found to be

$0.385139 \pm 0.301563i,\ -0.453731 \pm 0.182759i,\ 0.0685914 \pm 0.484322i,$
$-0.461227,\ 0.690937,\ -1.38146 \pm 1.63298i,\ -0.888193 \pm 0.382434i,$
$-0.626419 \pm 0.447214i,\ -0.616918 \pm 0.228042i,\ -0.546519 \pm 0.138633i,$
$-0.504031 \pm 0.0757213i,\ -0.483094 \pm 0.0349619i,\ -0.345468 \pm 0.598527i,$
$-0.0785635 \pm 0.774853i,\ 0.0935008 \pm 0.707441i,\ 0.140045 \pm 0.571188i,$
$0.177037 \pm 0.495057i,\ 0.200558 \pm 0.445297i,\ 0.205838 \pm 1.20668i,$
$0.229093 \pm 0.403647i,\ 0.253758 \pm 0.407757i,\ 0.299419 \pm 0.396038i,$

$0.365488 \pm 0.402274i$, $0.461153 \pm 0.411868i$, $0.659879 \pm 0.470766i$,
$0.704721 \pm 0.329989i$, $1.56532 \pm 0.938337i$.

All of these fixed points are repelling (since $|R'(z_0)| > 1$). □

Theorem 9. *There are thirty nine extraneous fixed points for Method (24).*

Proof. For Method (24),

$$G_f = \left(1 + 3\frac{f'(z)}{f'(y(z))}\right)\left(f'(z) + +\frac{f(z)^3 f'(z)}{81 f'(y(z))^3} + \frac{3}{8}\frac{(f'(y(z)) - f'(z))^2}{f'(z)}\right).$$

The extraneous fixed points are at

$0.385139 \pm 0.301563i$, $-0.453731 \pm 0.182759i$, $0.0685914 \pm 0.484322i$
$3.98917 \pm 6.90945i$, -7.97834, $0.41942 \pm 0.726456i$, -0.838839,
$0.277253 \pm 0.480215i$, -0.554505, $0.46341 \pm 0.53288i$, $-0.693192 \pm 0.134885i$,
$0.229782 \pm 0.667764i$, $0.367096 \pm 0.467142i$, $-0.588105 \pm 0.0843435i$,
$0.221009 \pm 0.551486i$, $0.280074 \pm 0.381388i$, $-0.470329 \pm 0.0518574i$,
$0.190255 \pm 0.433246i$, $0.615945 \pm 0.214444i$, $-0.493687 \pm 0.426202i$,
$-0.122258 \pm 0.640646i$.

All of these fixed points are repelling (since $|R'(z_0)| > 1$). □

Theorem 10. *There are twenty four extraneous fixed points for Method (25).*

Proof. We found for Method (25),

$$G_f = \left(\frac{1}{1 + \frac{f'(y(z))}{f'(z)}}\right)\left(f'(z) + +\frac{f(z)^4}{f'(z)^3}\right)\left(2f'(z) - \frac{7}{4}f'(y(z)) + \frac{3}{4}\frac{f'(y(z))^2}{f'(z)}\right).$$

The extraneous fixed points are found to be

$0.272187 \pm 0.394392i$, $0.20546 \pm 0.432916i$, $-0.477646 \pm 0.0385246i$,
$0.676726 \pm 0.202542i$, $-0.513769 \pm 0.484791i$, $-0.162957 \pm 0.687333i$,
$-2.12619 \pm 2.22671i$, $-0.51922 \pm 0.277607i$, $-0.217805 \pm 0.487789i$,
$0.210804 \pm 0.604566i$, $0.524089 \pm 0.172222i$, $2.12832 \pm 2.00454i$.

All of these fixed points are repelling (since $|R'(z_0)| > 1$). □

Theorem 11. *There are eighteen extraneous fixed points for Method (26).*

Proof. For Method (26),

$$G_f = \left(\frac{17}{8} - \frac{9}{4}\frac{f'(y(z))}{f'(z)} + \frac{9}{8}\left(\frac{f'(y(z))}{f'(z)}\right)^2\right)\left(\frac{7}{4} - \frac{3}{4}\frac{f'(y(z))}{f'(z)}\right).$$

The extraneous fixed points are at

$-0.333371 \pm 0.577415i$, 0.666742, $0.229257 \pm 0.397085i$, -0.458515,
$0.710065 \pm 0.231721i$, $-0.555709 \pm 0.499074i$, $-0.154356 \pm 0.730795i$,
$0.275117 \pm 0.402579i$, $-0.486202 \pm 0.0369693i$, $0.211085 \pm 0.439548i$.

All of these fixed points are repelling (since $|R'(z_0)| > 1$). □

Theorem 12. *There are twelve extraneous fixed points for Method (27).*

Proof. For Method (27),

$$G_f = \Big(-\tfrac{1}{2} + \tfrac{9}{8}\tfrac{f'(z)}{f'(y(z))} + \tfrac{3}{8}\tfrac{f'(y(z))}{f'(z)} \Big).$$

The extraneous fixed points are at

$0.289483 \pm 0.382811i,\ 0.186782 \pm 0.442105i,\ -0.476265 \pm 0.0592945i,$
$0.605298 \pm 0.2466i,\ -0.516211 \pm 0.400903i,\ -0.0890867 \pm 0.647503i.$ All of these fixed points are repelling (since $|R'(z_0)| > 1$). □

Theorem 13. *There are twenty four extraneous fixed points for Method (16).*

Proof. For Method (16),

$$G_f = \Big(\frac{1}{1+3\frac{f'(y(z))}{f'(z)}}\Big)\Big(f'(z) + \tfrac{5}{16}\tfrac{(f'(y(z))-f'(z))^2}{f'(z)}\Big)\Big(f'(z) + \tfrac{1}{4}f'(z)\tfrac{(f'(z)-f'(y(z)))^2}{f'(y(z))^2}\Big).$$

The extraneous fixed points are at

$0.622907 \pm 0.52714i,\ -0.767969 \pm 0.275883i,\ 0.145063 \pm 0.803023i,$
$0.310217 \pm 0.445061i,\ -0.540543 \pm 0.0461255i,\ 0.230326 \pm 0.491187i,$
$0.280277 \pm 0.377418i,\ 0.186715 \pm 0.431436i,\ -0.466992 \pm 0.0540183i,$
$0.602147 \pm 0.210285i,\ -0.483186 \pm 0.416332i,\ -0.118961 \pm 0.626617i.$

All of these fixed points are repelling (since $|R'(z_0)| > 1$). □

Theorem 14. *There are thirty extraneous fixed points for Method (17).*

Proof. For Method (17),

$$\begin{aligned} G_f &= \Big(\frac{1}{1+3\frac{f'(y(z))}{f'(z)}}\Big)\Big(f'(z) + \frac{5}{16}\frac{(f'(y(z))-f'(z))^2}{f'(z)}\Big) \\ &\quad \Big(f'(z) + \frac{1}{4}f'(z)\frac{(f'(z)-f'(y(z)))^2}{f'(y(z))^2} + \frac{1}{6}f'(z)\frac{(f'(z)-f'(y(z)))^3}{f'(y(z))^3}\Big). \end{aligned}$$

The extraneous fixed points are at

$0.280277 \pm 0.377418i,\ 0.186715 \pm 0.431436i,\ -0.466992 \pm 0.0540183i,$
$0.602147 \pm 0.210285i,\ -0.483186 \pm 0.416332i,\ -0.118961 \pm 0.626617i,$
$0.701957 \pm 0.574647i,\ -0.848638 \pm 0.320589i,\ 0.146681 \pm 0.895237i,$
$0.296076 \pm 0.447202i,\ -0.535326 \pm 0.0328086i,\ 0.23925 \pm 0.48001i,$
$0.414766 \pm 0.407081i,\ 0.145159 \pm 0.562739i,\ -0.559926 \pm 0.155658i.$

All of these fixed points are repelling (since $|R'(z_0)| > 1$). □

5. Basins of Attraction

Sections 2 and 3 discussed methods whose roots are in the real domain, that is $f : D \subset \mathbb{R} \to \mathbb{R}$. The study can be extended to functions defined in the complex plane $f : D \subset \mathbb{C} \to \mathbb{C}$ having complex zeros. From the fundamental theorem of algebra, a polynomial of degree n with real or complex coefficients has n roots, which may or may not be distinct. In such a case, a complex initial guess is needed for the convergence of complex zeros. Note that we need some basic definitions in order to study functions for the complex domain with complex zeros. We give below some definitions required for our study, which are found in [20–22]. Let $R : \mathbb{C} \to \mathbb{C}$ be a rational map on the Riemann sphere.

Definition 15. For $z \in \mathbb{C}$, we define its orbit as the set $orb(z) = \{z, R(z), R^2(z), ..., R^n(z), ...\}$.

Definition 16. A periodic point z_0 of the period m is such that $R^m(z_0) = z_0$, where m is the smallest integer.

Definition 17. The Julia set of a nonlinear map $R(z)$ denoted by $J(R)$ is the closure of the set of its repelling periodic points. The complementary of $J(R)$ is the Fatou set $F(R)$.

Definition 18. If O is an attracting periodic orbit of period m, we define the basins of attraction to be the open set $A \in \mathbb{C}$ consisting of all points $z \in \mathbb{C}$ for which the successive iterates $R^m(z), R^{2m}(z), ...$ converge towards some point of O.

Lemma 19. *Every attracting periodic orbit is contained in the Fatou set of R. In fact, the entire basins of attraction A for an attracting periodic orbit is contained in the Fatou set. However, every repelling periodic orbit is contained in the Julia set.*

In the following subsections, we produce some beautiful graphs obtained for the proposed methods and for some existing methods using MATLAB [23,24]. In fact, an iteration function is a mapping of the plane into itself. The common boundaries of these basins of attraction constitute the Julia set of the iteration function, and its complement is the Fatou set. This section is necessary in this paper to show how the proposed methods could be considered in polynomiography. In the following section, we describe the basins of attraction for Newton's method and some higher order Newton type methods for finding complex roots of polynomials $p_1(z) = z^3 - 1$ and $p_2(z) = z^4 - 1$.

5.1. Polynomiographs of $p_1(z) = z^3 - 1$

We consider the square region $[-2, 2] \times [-2, 2]$, and in this region, we have 160,000 equally-spaced grid points with mesh $h = 0.01$. It is composed of 400 columns and 400 rows, which can be related to the pixels of a computer display, which would represent a region of the complex plane [25]. Each grid point is used as an initial point z_0, and the number of iterations until convergence is counted for each point. Now, we draw the polynomiographs of $p_1(z) = z^3 - 1$ with roots $\alpha_1 = 1$, $\alpha_2 = -0.5000 - 0.8660i$ and $\alpha_3 = -0.5000 + 0.8660i$. We assign "red color" if each grid point converges to the root α_1, "green color" if they converge to the root α_2 and "blue color" if they converge to the root α_3 in at most 200 iterations and if $|z_n - \alpha_j| < 10^{-4}, j = 1, 2, 3$. In this way, the basins of attraction for each root would be assigned a characteristic color. If the iterations do not converge as per the above condition for some specific initial points, we assign "black color".

Figure 1a–j shows the polynomiographs of the methods for the cubic polynomial $p_1(z)$. There are diverging points for the method of Noor *et al.*, SBS1, SBS2 and SKK. All starting points are converging for the methods NM, JM, SJ, SKS, PM1 and PM2. In Table 6, we classify the number of converging and diverging grid points for each iterative method. Note that a point z_0 belongs to the Julia set if and only if the dynamics in a neighborhood of z_0 displays sensitive dependence on the initial conditions, so that nearby initial conditions lead to wildly different behavior after a number of iterations. For this reason, some of the methods are getting many divergent points. The common boundaries of these basins of attraction constitute the Julia set of the iteration function.

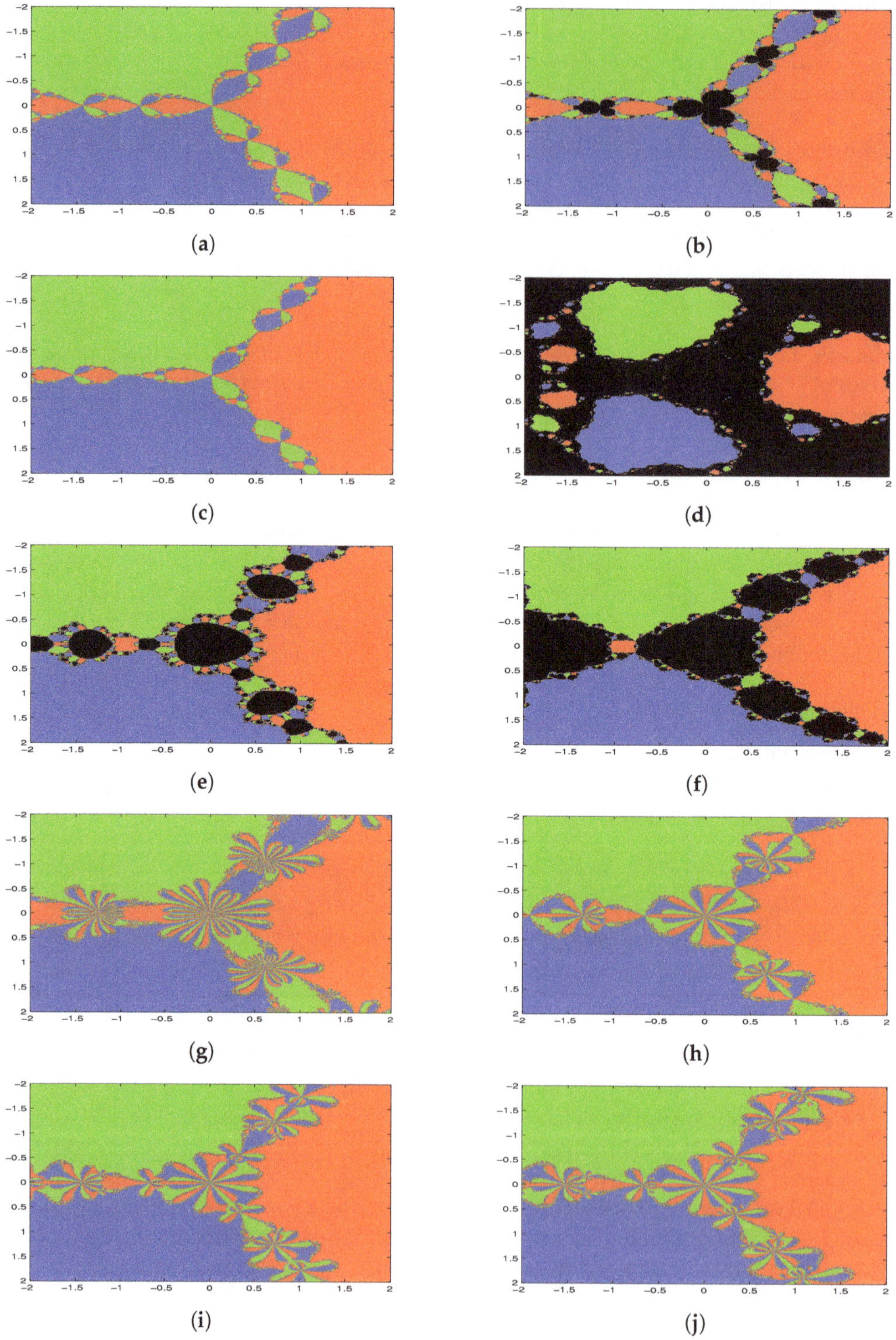

Figure 1. Polynomiographs of $p_1(z)$. (**a**) Newton's method (NM) (1); (**b**) method of Noor *et al.* (4); (**c**) Jarratt method (JM) (22); (**d**) method of Sharifi *et al.* (SBS1) (23); (**e**) method of Sharifi *et al.* (SBS2) (24); (**f**) method of Soleymani *et al.* (SKK) (25); (**g**) method of Singh *et al.* (SJ) (26); (**h**) method of Sharma *et al.* (SKS) (27); (**i**) proposed method (PM1) (16); (**j**) proposed method (PM2) (17).

Table 6. Comparison of convergent and divergent grids for polynomiographs of $p_1(z)$.

Methods	Convergent Grid Points		Divergent Grid Points
	Real Root (α_1)	Complex Roots (α_2 and α_3)	
NM (1)	56,452	103,548	0
Noor *et al.* (4)	52,372	98,670	8958
JM (22)	56,474	103,526	0
SBS1 (23)	23,174	44,160	92,666
SBS2 (24)	48,018	91,308	20,674
SKK (25)	34,722	82,590	42,688
SJ (26)	52,587	107,143	0
SKS (27)	55,178	104,822	0
PM1 (16)	55,892	104,108	0
PM2 (17)	54,622	105,378	0

5.2. *Polynomiographs of* $p_2(z) = z^4 - 1$

Next, we draw the polynomiographs of $p_2(z) = z^4 - 1$ with roots $\alpha_1 = 1$, $\alpha_2 = -1$, $\alpha_3 = i$ and $\alpha_4 = -i$. We assign yellow color if each grid point converges to the root α_1, red color if they converge to the root α_2, green color if they converge to the root α_3 and blue color if they converge to the root α_4 in at most 200 iterations and if $|z_n - \alpha_j| < 10^{-4}, j = 1,2,3,4$. Therefore, the basins of attraction for each root would be assigned a corresponding color. If the iterations do not converge as per the above condition for some specific initial points, we assign black color.

Figure 2a–j shows the polynomiographs of the methods for the quartic polynomial $p_2(z)$. There are diverging points for the method of Noor *et al.*, SBS1, SBS2, SKK, SJ, SKS, PM1 and PM2. All starting points are convergent for NM and JM. In Table 7, we classify the number of converging and diverging grid points for each iterative methods. Furthermore, we observe that the SKS, PM1 and PM2 methods are divergent at a lesser number of grid points than the method of Noor *et al.*, SBS1, SBS2, SKK and SJ. Table 8 shows that the proposed methods are better than or equal to other comparable methods with respect to the number of iterations, computational order convergence and error. All of the methods applied on the cubic and quartic polynomials $p_1(z)$ and $p_2(z)$ are convergent with real roots as the starting point.

Table 7. Comparison of convergent and divergent grids for polynomiographs of $p_2(z)$.

Methods	Convergent Grid Points		Divergent Grid Points
	Real Roots (α_1 and α_2)	Complex Roots (α_3 and α_4)	
NM (1)	80,010	79,990	0
Noor *et al.* (4)	68,133	68,120	23,747
JM (22)	80,001	79,999	0
SBS1 (23)	53,792	53,792	52,416
SBS2 (24)	60,098	60,466	39,436
SKK (25)	54,584	54,584	50,832
SJ (26)	79,427	79,427	1146
SKS (27)	79,961	79,959	80
PM1 (16)	79,962	79,979	59
PM2 (17)	79,968	79,954	78

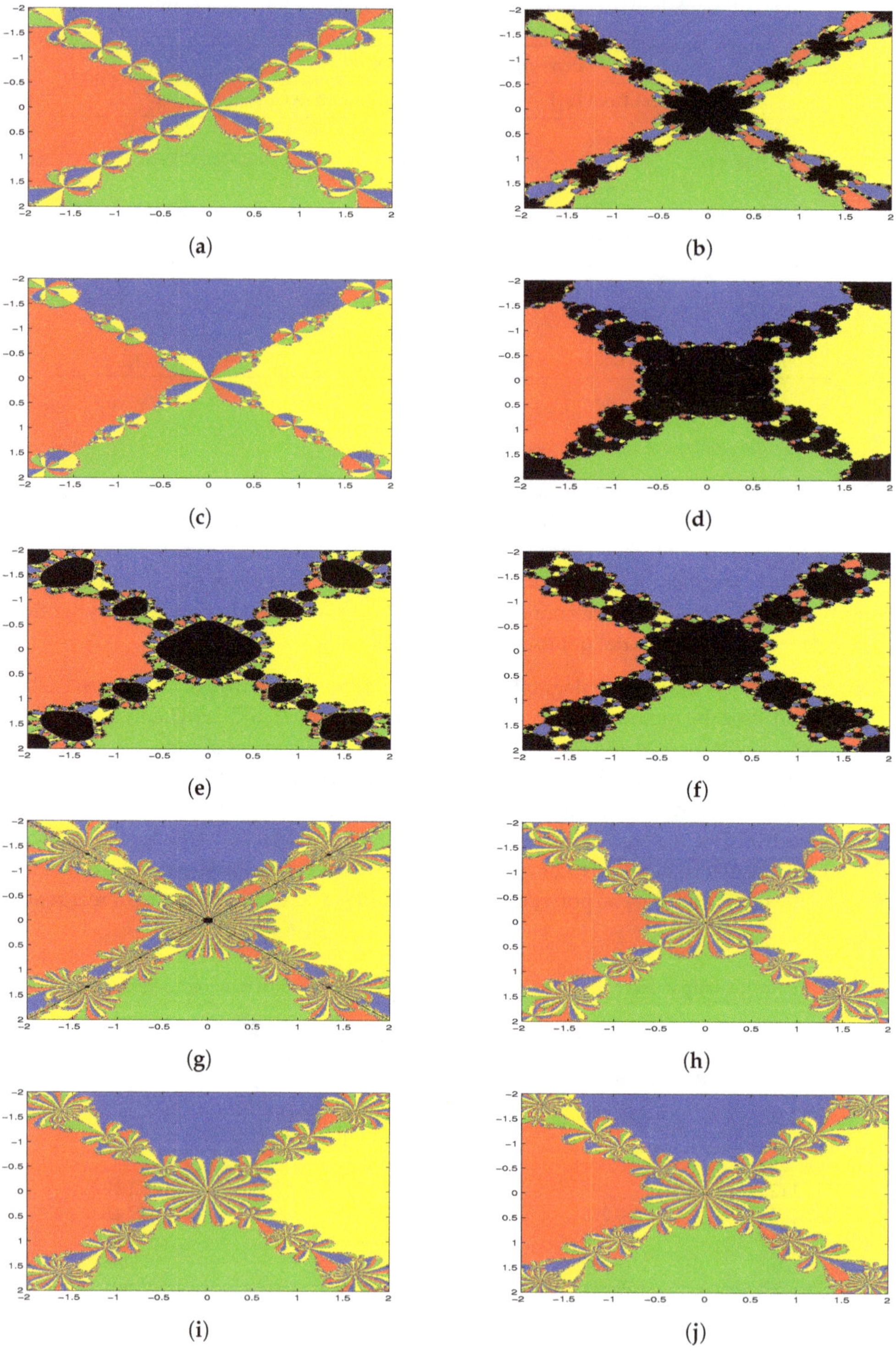

Figure 2. Polynomiographs of $p_2(z)$. (**a**) Newton's method (NM) (1); (**b**) method of Noor *et al.* (4); (**c**) Jarratt method (JM) (22); (**d**) method of Sharifi *et al.* (SBS1) (23); (**e**) method of Sharifi *et al.* (SBS2) (24); (**f**) method of Soleymani *et al.* (SKK) (25); (**g**) method of Singh *et al.* (SJ) (26); (**h**) method of Sharma *et al.* (SKS) (27); (**i**) proposed method (PM1) (16); (**j**) proposed method (PM2) (17).

Table 8. Results for polynomials $p_1(z)$, $p_2(z)$ with real roots.

Methods	$p_1(z) = z^3 - 1$				$p_2(z) = z^4 - 1$			
	z_0	M	ρ	*Error*	z_0	M	ρ	*Error*
NM (1)	1.6	9	1.99	3.3751e − 098	0.7	9	1.99	3.1014e − 066
Noor *et al.* (4)	1.6	6	3.00	3.5077e − 093	0.7	6	3.00	6.2225e − 061
JM (22)	1.6	5	3.99	2.1482e − 108	0.7	6	4.00	2.0563e − 089
SBS1 (23)	1.6	5	4.00	9.3378e − 099	0.7	7	4.00	3.3321e − 110
SBS2 (24)	1.6	5	3.99	9.4664e − 083	0.7	6	3.99	6.2610e − 055
SKK (25)	1.6	5	3.99	1.6780e − 081	0.7	7	3.99	1.7649e − 092
SJ (26)	1.6	5	3.99	8.5156e − 076	0.7	8	3.99	3.9861e − 076
SKS (27)	1.6	5	3.99	9.3481e − 084	0.7	6	3.99	4.7052e − 093
PM1 (16)	1.6	5	3.99	3.1169e − 092	0.7	6	3.99	2.4833e − 099
PM2 (17)	1.6	5	3.99	2.5579e − 102	0.7	6	3.99	3.4840e − 120

From this comparison based on the basins of attractions for cubic and quartic polynomials, we could generally say that NM, JM, PM1 and PM2 are more reliable in solving nonlinear equations. Furthermore, by observing the polynomiographs of $p_1(z)$ and $p_2(z)$, we find certain symmetrical patterns for the x-axis and y-axis, where the starting point z_0 leads to convergent real or complex pair of roots of the respective polynomials.

6. An Application Problem

To test our methods, we consider the following Planck's radiation law problem found in [10,26]:

$$\varphi(\lambda) = \frac{8\pi ch\lambda^{-5}}{e^{ch/\lambda kT} - 1}, \tag{31}$$

which calculates the energy density within an isothermal blackbody. Here, λ is the wavelength of the radiation; T is the absolute temperature of the blackbody; k is Boltzmann's constant; h is the Planck's constant; and c is the speed of light. Suppose we would like to determine wavelength λ, which corresponds to maximum energy density $\varphi(\lambda)$. From Equation (31), we get:

$$\varphi'(\lambda) = \left(\frac{8\pi ch\lambda^{-6}}{e^{ch/\lambda kT} - 1}\right)\left(\frac{(ch/\lambda kT)e^{ch/\lambda kT}}{e^{ch/\lambda kT} - 1} - 5\right) = A \cdot B.$$

It can be checked that a maxima for φ occurs when $B = 0$, that is when:

$$\left(\frac{(ch/\lambda kT)e^{ch/\lambda kT}}{e^{ch/\lambda kT} - 1}\right) = 5.$$

Here, putting $x = ch/\lambda kT$, the above equation becomes:

$$1 - \frac{x}{5} = e^{-x}. \tag{32}$$

Define:

$$f(x) = e^{-x} - 1 + \frac{x}{5}. \tag{33}$$

The aim is to find a root of the equation $f(x) = 0$. Obviously, one of the roots $x = 0$ is not taken for discussion. As argued in [26], the left-hand side of Equation (32) is zero for $x = 5$ and $e^{-5} \approx 6.74 \times 10^{-3}$. Hence, it is expected that another root of the equation $f(x) = 0$ might occur near $x = 5$. The approximate root of the Equation (33) is given by $x^* \approx 4.965114231744276303699$.

Consequently, the wavelength of radiation (λ) corresponding to which the energy density is maximum is approximated as:

$$\lambda \approx \frac{ch}{(kT)4.96511423174427630369}.$$

We apply the methods NM, DJ, PM1, PM2, PM3 and PM4 to solve Equation (33) and compared the results in Tables 9 and 10. From these tables, we note that the root x^* is reached faster by the method PM4 than by other methods. This is due to the fact that PM4 has the highest efficiency index EI = 1.644.

Table 9. Comparison of the results.

x_0	**NM** (1)				**DJ** (28)				**PM1** (16)			
	N	d_1	ρ	*Error*	N	d_1	ρ	*Error*	N	d_1	ρ	*Error*
4.0	7	14	2.00	1.4e − 101	4	16	4.00	3.3e − 086	4	12	3.99	2.2e − 069
4.5	6	12	1.99	4.5e − 063	4	16	4.00	9.1e − 110	4	12	3.99	2.1e − 093
5.0	6	12	2.00	1.4e − 101	4	16	3.99	8.5e − 185	4	12	3.99	1.3e − 168
5.5	6	12	1.99	5.4e − 066	4	16	4.00	9.9e − 112	4	12	3.99	1.6e − 095

Table 10. Comparison of the results.

x_0	**PM2** (17)				**PM3** (18)				**PM4** (21)			
	N	d_1	ρ	*Error*	N	d_1	ρ	*Error*	N	d_1	ρ	*Error*
4.0	4	12	3.99	1.4e − 069	4	16	5.99	3.3e − 224	3	15	12.11	7.7e − 144
4.5	4	12	3.99	1.6e − 093	4	16	5.99	3.3e − 306	3	15	12.03	6.4e − 198
5.0	4	12	3.99	1.1e − 168	3	12	5.99	3.7e − 093	3	15	12.00	0
5.5	4	12	3.99	1.4e − 095	3	12	5.95	3.2e − 052	3	15	11.96	3.2e − 203

Results for the *fzero* command in MATLAB for this application problem are given in Table 11.

Table 11. Results for Planck's radiation law problem in *fzero*.

x_0	N_1	N	d_1	$f(x_n)$	x^*
4.0	8	6	23	1.1102e − 016	4.9651
4.5	5	5	16	−1.1102e − 016	4.9651
5.0	1	4	6	1.1102e − 016	4.9651
5.5	5	5	15	−1.1102e − 016	4.9651

7. Conclusions

In this work, we have proposed a family of fourth order methods using weight functions. The fourth order methods are found to be optimal as per the Kung–Traub conjuncture. Further, we have extended one of the methods to sixth and twelfth order methods with four and five function evaluations, respectively. The extraneous fixed points for the fourth order methods and for some existing methods are discussed in detail. By analysis using basins of attraction, our methods PM1 and PM2 are found to be superior to the methods of Noor *et al.* [16], SBS1, SBS2, SKK and SJ; specifically, the methods of SBS1, SBS2 and SKK are very badly scaled in both cubic and quartic polynomials. Moreover, PM1 and PM2 are better than other compared methods, except Newton's method and Jaratt's method, which perform equally well. We have also verified our methods (PM1, PM2, PM3, PM4), NM and DJ on Planck's radiation law problem, and the results show that PM4 is more efficient than other compared methods.

Acknowledgments: The authors would like to thank the editors and referees for the valuable comments and for the suggestions to improve the readability of the paper.

Author Contributions: The contributions of both of the authors have been similar. Both of them have worked together to develop the present manuscript.

Conflicts of Interest: The authors declare no conflict of interest.

References

1. Ezquerro, J.A.; Hernandez, M.A. A uniparametric halley-type iteration with free second derivative. *Int. J. Pure Appl. Math.* **2003**, *6*, 99–110.
2. Babajee, D.K.R.; Dauhoo, M.Z. An Analysis of the Properties of the Variants of Newton's Method with Third Order Convergence. *Appl. Math. Comput.* **2006**, *183*, 659–684.
3. Chun, C.; Yong-Il, K. Several new third-order iterative methods for solving nonlinear equations. *Acta Appl. Math.* **2010**, *109*, 1053–1063.
4. Kung, H.T.; Traub, J.F. Optimal order of one-point and multipoint iteration. *J. Assoc. Comput. Mach.* **1974**, *21*, 643–651.
5. Jarratt, P. Some efficient fourth order multipoint methods for solving equations. *BIT Numer. Math.* **1969**, *9*, 119–124.
6. Ardelean, G. A new third-order newton-type iterative method for solving nonlinear equations. *Appl. Math. Comput.* **2013**, *219*, 9856–9864.
7. Babajee, D.K.R.; Madhu, K.; Jayaraman, J. A family of higher order multi-point iterative methods based on power mean for solving nonlinear equations. *Afr. Mat.* **2015**, doi:10.1007/s13370-015-0380-1.
8. Cordero, A.; Hueso, J.L.; Martinez, E.; Torregrosa, J.R. Efficient three-step iterative methods with sixth order convergence for nonlinear equations. *Numer. Algor.* **2010**, *53*, 485–495.
9. Cordero, A.; Hueso, J.L.; Martinez, E.; Torregrosa, J.R. A family of iterative methods with sixth and seventh order convergence for nonlinear equations. *Math. Comput. Model.* **2010**, *52*, 1490–1496.
10. Jain, D. Families of newton-like methods with fourth-order convergence. *Int. J. Comput. Math.* **2013**, *90*, 1072–1082.
11. Madhu, K.; Jayaraman, J. Class of modified Newton's method for solving nonlinear equations. *Tamsui Oxf. J. Inf. Math. Sci.* **2014**, *30*, 91–100.
12. Sharifi, M.; Babajee, D.K.R.; Soleymani, F. Finding the solution of nonlinear equations by a class of optimal methods. *Comput. Math. Appl.* **2012**, *63*, 764–774.
13. Sharma, J.R.; Kumar, G.R; Sharma, R. An efficient fourth order weighted-newton method for systems of nonlinear equations. *Numer. Algor.* **2013**, *62*, 307–323.
14. Singh, A.; Jaiswal, J.P. Several new third-order and fourth-order iterative methods for solving nonlinear equations. *Int. J. Eng. Math.* **2014**, *2014*, 828409.
15. Soleymani, F.; Khratti, S.K.; Karimi, V.S. Two new classes of optimal Jarratt-type fourth-order methods. *Appl. Math. Lett.* **2011**, *25*, 847–853.
16. Noor, M.A.; Waseem, M. Some iterative methods for solving a system of nonlinear equations. *Comput. Math. Appl.* **2009**, *57*, 101–106.
17. Kalantari, B. *Polynomial Root-Finding and Polynomiography*; World Scientific Publishing Co. Pte. Ltd.: Singapore, 2009.
18. Cordero, A.; Torregrosa, J.R. Variants of Newton's method using fifth-order quadrature formulas. *Appl. Math. Comput.* **2007**, *190*, 686–698.
19. Vrscay, E.R.; Gilbert, W.J. Extraneous fixed points, basin boundaries and chaotic dynamics for Schroder and Konig rational iteration functions. *Numer. Math.* **1988**, *52*, 1–16.
20. Amat, S.; Busquier, S.; Plaza, S. Review of some iterative root-finding methods from a dynamical point of view. *Sci. Ser. A Math. Sci.* **2004**, *10*, 3–35.
21. Blanchard, P. Complex Analytic Dynamics on the Riemann sphere. *Bull. Am. Math. Soc.* **1984**, *11*, 85–141.
22. Scott, M.; Neta, B.; Chun, C. Basin attractors for various methods. *Appl. Math. Comput.* **2011**, *218*, 2584–2599.
23. Chicharro, F.I.; Cordero, A.; Torregrosa, J.R. Drawing Dynamical and Parameters Planes of Iterative Families and Methods. *Sci. World J.* **2013**, doi:10.1155/2013/780153.

24. Introduction to Computational Engineering. Available online: http://www.caam.rice.edu (accessed on 10 May 2015).
25. Soleymani, F.; Babajee, D.K.R.; Sharifi, M. Modified jarratt method without memory with twelfth-order convergence. *Ann. Univ. Craiova Math. Comput. Sci. Ser.* **2012**, *39*, 21–34.
26. Bradie, B. *A Friendly Introduction to Numerical Analysis*; Pearson Education Inc.: New Delhi, India, 2006.

14

Inverse Eigenvalue Problems for Two Special Acyclic Matrices

Debashish Sharma [1,*,†] and Mausumi Sen [2,†]

1 Department of Mathematics, Gurucharan College, College Road, Silchar 788004, India
2 Department of Mathematics, National Institute of Technology Silchar, Silchar 788010, India; senmausumi@gmail.com
* Correspondence: debashish0612@gmail.com
† These authors contributed equally to this work.

Academic Editors: Lokenath Debnath, Indranil SenGupta and Carsten Schneider

Abstract: In this paper, we study two inverse eigenvalue problems (IEPs) of constructing two special acyclic matrices. The first problem involves the reconstruction of matrices whose graph is a path, from given information on one eigenvector of the required matrix and one eigenvalue of each of its leading principal submatrices. The second problem involves reconstruction of matrices whose graph is a broom, the eigen data being the maximum and minimum eigenvalues of each of the leading principal submatrices of the required matrix. In order to solve the problems, we use the recurrence relations among leading principal minors and the property of simplicity of the extremal eigenvalues of acyclic matrices.

Keywords: inverse eigenvalue problem; leading principal minors; graph of a matrix

MSC: 65F18, 05C50

---

1. Introduction

The problems of reconstruction of specially structured matrices from a prescribed set of eigen data are collectively known as inverse eigenvalue problems (IEPs). The level of difficulty of an IEP depends on the structure of the matrices which are to be reconstructed and on the type of eigen information available. M.T. Chu in [1] gave a detailed characterization of inverse eigenvalue problems. A few special types of inverse eigenvalue problems have been studied in [2–8]. Inverse problems for matrices with prescribed graphs have been studied in [9–14]. Inverse eigenvalue problems arise in a number of applications such as control theory, pole assignment problems, system identification, structural analysis, mass spring vibrations, circuit theory, mechanical system simulation and graph theory [1,12,15,16].

In this paper, we study two IEPs, namely IEPP (inverse eigenvalue problem for matrices whose graph is a path) and IEPB (inverse eigenvalue problem for matrices whose graph is a broom). Similar problems were studied in [5], for arrow matrices. The usual process of solving such problems involves the use of recurrence relations among the leading principal minors of $\lambda I - A$ where A is the required matrix. However, we have included graphs in our analysis by bringing in the requirement of constructing matrices which are described by graphs. In particular, we have considered paths and brooms. Thus, in addition to recurrence relations among leading principal minors, we have used spectral properties of acyclic matrices to solve the problems. Particularly, the strict interlacing of the eigenvalues in IEPB could be proved because of the fact that the minimal and maximal eigenvalues of an acyclic matrix are simple.

The paper is organized as follows : In Section 2, we discuss some preliminary concepts and clarify the notations used in the paper. In Section 3, we define the inverse problems to be studied, namely IEPP and IEPB. Section 4 deals with the analysis of IEPP, the main result being presented as Theorem 4. Section 5 deals with the analysis of IEPB, the main result being presented as Theorem 9. In Section 6, we present some numerical examples to illustrate the solutions of IEPP and IEPB.

2. Preliminary Concepts

Let V be a finite set and let P be the set of all subsets of V which have two distinct elements. Let $E \subset P$. Then $G = (V, E)$ is said to be a *graph* with vertex set V and edge set E. To avoid confusion, the vertex set of a graph G is denoted by $V(G)$ and the edge set is denoted by $E(G)$. Our choice of P implies that the graphs under consideration are free of multiple edges or loops and are undirected. If $u, v \in V$ and $\{u, v\} \in E$, then we say that uv is an *edge* and u and v are then called *adjacent* vertices. The degree of a vertex u is the number of edges which are incident on u. A vertex of degree one is called a *pendant vertex*. A *path* P of G is a sequence of distinct vertices $v_1, v_2, \ldots, v_n$ such that consecutive vertices are adjacent. The path on n vertices is denoted by P_n. A graph is said to be *connected* if there exists a path between every pair of its vertices. A *cycle* is a connected graph in which each vertex is adjacent to exactly two other vertices. A connected graph without any cycles is called a *tree*.

Given an $n \times n$ symmetric matrix A, the graph of A, denoted by $G(A)$, has vertex set $V(G) = \{1, 2, 3, \ldots, n\}$ and edge set $\{ij : i \neq j, a_{ij} \neq 0\}$. For a graph G with n vertices, $S(G)$ denotes the set of all $n \times n$ symmetric matrices which have G as their graph. A matrix whose graph is a tree is called an *acyclic* matrix. Some simple examples of acyclic matrices are the matrices whose graphs are paths or brooms (Figure 1).

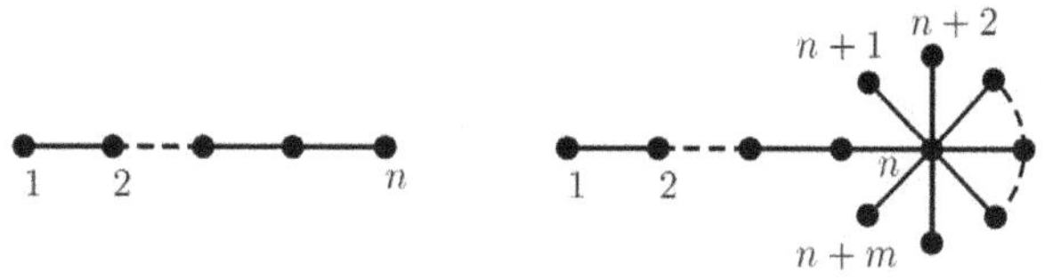

Figure 1. Path P_n and Broom B_{n+m}.

Throughout this paper, we shall use the following notation :

1. Matrix of a path P_n will be a tridiagonal matrix with non zero off-diagonal entries :

$$A_n = \begin{pmatrix} a_1 & b_1 & 0 & \ldots & 0 & 0 \\ b_1 & a_2 & b_2 & \ldots & 0 & 0 \\ 0 & b_2 & a_3 & \ddots & 0 & 0 \\ \vdots & \vdots & \ddots & \ddots & \ddots & \vdots \\ 0 & 0 & 0 & \ddots & a_{n-1} & b_{n-1} \\ 0 & 0 & 0 & \ldots & b_{n-1} & a_n \end{pmatrix}_{n \times n},$$

where the b_is are non-zero.

2. Matrix of a broom B_{n+m} will be of the following form :

$$A_{n+m} = \begin{pmatrix} a_1 & b_1 & 0 & \dots & 0 & 0 & 0 & 0 & \dots & 0 \\ b_1 & a_2 & b_2 & \dots & 0 & 0 & 0 & 0 & \dots & 0 \\ 0 & b_2 & a_3 & \ddots & 0 & 0 & 0 & 0 & \dots & 0 \\ \vdots & \vdots & \ddots & \ddots & \vdots & \vdots & \vdots & \vdots & \vdots & \vdots \\ 0 & 0 & 0 & \dots & a_{n-1} & b_{n-1} & 0 & 0 & \dots & 0 \\ 0 & 0 & 0 & \dots & b_{n-1} & a_n & b_n & b_{n+1} & \dots & b_{n+m-1} \\ 0 & 0 & 0 & \dots & 0 & b_n & a_{n+1} & 0 & \dots & 0 \\ 0 & 0 & 0 & \dots & 0 & b_{n+1} & 0 & a_{n+2} & \ddots & 0 \\ \vdots & \vdots & \vdots & \vdots & \vdots & \vdots & \vdots & \ddots & \ddots & \vdots \\ 0 & 0 & 0 & \dots & 0 & b_{n+m-1} & 0 & 0 & \dots & a_{n+m} \end{pmatrix}_{(n+m)\times(n+m)},$$

where the b_is are non zero.

3. A_i will denote the ith leading principal submatrix of the required matrix (A_n or A_{n+m}).
4. $P_i(\lambda) = det(\lambda I_i - A_i), i = 1, 2, \dots, n$ (respectively $i = 1, 2, \dots n + m$) *i.e.*, the ith leading principal minor of $\lambda I_n - A_n$ (respectively $\lambda I_{n+m} - A_{n+m}$), I_i being the identity matrix of order i. For the sake of writing the recurrence relations with ease, we define $P_0(\lambda) = 1, b_0 = 0$.

3. IEPs to be Studied

In this paper we shall study the following two inverse eigenvalue problems :

IEPP Given n real numbers $\lambda_j, 1 \leq j \leq n$ and a real vector $X_n = (x_1, x_2, \dots, x_n)^T$ find an $n \times n$ matrix $A_n \in S(P_n)$ such that λ_j is an eigenvalue of $A_j, j = 1, 2, \dots, n$ and (λ_n, X_n) is an eigenpair of A_n.

IEPB Given $2n + 2m - 1$ real numbers $\lambda_1^{(j)}, 1 \leq j \leq n + m$ and $\lambda_j^{(j)}, 2 \leq j \leq n + m$, find an $(n + m) \times (n + m)$ matrix $A_{n+m} \in S(B_{n+m})$ such that $\lambda_1^{(j)}$ and $\lambda_j^{(j)}$ are respectively the minimal and maximal eigenvalues of $A_j, j = 1, 2, \dots, n + m$.

4. Solution of IEPP

The following Lemma gives the relation between successive leading principal minors of $\lambda I_n - A_n$:

Lemma 1. *The sequence* $\{P_j(\lambda) = det(\lambda I_j - A_j)\}_{j=1}^n$ *of characteristic polynomials of* A_j *satisfies the following recurrence relations :*

1. $P_1(\lambda) = (\lambda - a_1)$
2. $P_j(\lambda) = (\lambda - a_j)P_{j-1}(\lambda) - b_{j-1}^2 P_{j-2}(\lambda), 2 \leq j \leq n.$

Here A_j *denotes the jth leading principal submatrix of* A_n, *the matrix corresponding to the path on n vertices.*

Lemma 2. *For any* $\lambda \in \mathbb{R}$ *and* $1 \leq j \leq n$, $P_j(\lambda)$ *and* $P_{j+1}(\lambda)$ *cannot be simultaneously zero.*

Proof. If $P_1(\lambda) = 0 = P_2(\lambda)$, then $(\lambda - a_2)P_1(\lambda) - b_1^2 = 0$, which implies $b_1 = 0$, but this contradicts the restriction on A_n that $b_1 \neq 0$. Once again, for $2 < j \leq n$, if $P_{j-1}(\lambda) = 0 = P_j(\lambda)$, then the recurrence relation (ii) from Lemma 1, $(\lambda - a_{j+1})P_j(\lambda) - b_j^2 P_{j-1}(\lambda) = 0$, which gives $P_{j-1}(\lambda) = 0$. This will in turn imply that $P_{j-2}(\lambda) = 0$. Thus, we will end up with $P_2(\lambda) = 0$, implying that $b_1 = 0$ which is a contradiction. □

Lemma 3. *If* $X = (x_1, x_2, \dots, x_n)^T$ *is an eigenvector of* A_n *corresponding to an eigenvalue* λ, *then* $x_1 \neq 0$ *and the components of this eigenvector are given by*

$$x_j = \frac{P_{j-1}(\lambda)}{\prod_{i=1}^{j-1} b_i} x_1, j = 2, 3, \dots, n.$$

Proof. Since (λ, X) is an eigenpair of A_n, we have $A_nX = \lambda X$. Comparing the first $n-1$ rows of this matrix equation on both sides, we have

$$(a_1 - \lambda)x_1 + b_1x_2 = 0, \tag{1}$$

$$b_{j-1}x_{j-1} + (a_j - \lambda)x_j + b_jx_{j+1} = 0, j = 2, \ldots, n-1. \tag{2}$$

By the second recurrence relation from Lemma 1,

$$P_j(\lambda) = (\lambda - a_j)P_{j-1}(\lambda) - b_{j-1}^2P_{j-2}(\lambda), j = 2, 3, \ldots, n. \tag{3}$$

We define the quantities $v_1, v_2, \ldots, v_n$ as

$$v_1 = x_1, v_j = x_j\prod_{i=1}^{j-1} b_i, 2 \leq j \leq n.$$

Multiplying Equation (2) by $\prod_{i=1}^{j-1} b_i$, we get

$$b_{j-1}x_{j-1}\prod_{i=1}^{j-1} b_i + (a_j - \lambda)x_j\prod_{i=1}^{j-1} b_i + b_jx_{j+1}\prod_{i=1}^{j-1} b_i = 0$$
$$\Rightarrow b_{j-1}^2v_{j-1} + (a_j - \lambda)v_j + v_{j+1} = 0,$$

which gives

$$v_{j+1} = (\lambda - a_j)v_j - b_{j-1}^2v_{j-1}, j = 2, 3, \ldots, n-1. \tag{4}$$

Now, from Equation (1), we have $v_2 = (\lambda - a_1)x_1 = x_1P_1(\lambda)$. Again from Equation (4),

$$v_3 = (\lambda - a_2)v_2 - b_1^2v_1 = x_1\{(\lambda - a_2)P_1(\lambda) - b_1^2\} = x_1P_2(\lambda).$$

Proceeding this way, we see that $v_{j+1} = x_1P_j(\lambda), j = 1, 2, \ldots, n-1$ which can also be written as $v_j = x_1P_{j-1}(\lambda), j = 2, 3, \ldots, n$. This further implies that

$$x_j = \frac{P_{j-1}(\lambda)}{\prod_{i=1}^{j-1} b_i}x_1, j = 2, 3, \ldots, n. \tag{5}$$

Since X is an eigenvector, $X \neq 0$. If $x_1 = 0$, then from Equation (5), we see that all the other components of X become zero. Thus, $x_1 \neq 0$. □

Theorem 4. *The IEPP has a unique solution if and only if $x_j \neq 0$ for all $j = 1, 2, \ldots, n$. The unique solution is given by*

$$a_1 = \lambda_1, a_j = \lambda_j - \frac{b_{j-1}^2P_{j-2}(\lambda_j)}{P_{j-1}(\lambda_j)}, j = 2, \ldots, n$$

$$b_1 = \frac{x_1}{x_2}P_1(\lambda_n) \text{ and } b_{j-1} = \frac{x_1P_{j-1}(\lambda_n)}{x_j\prod_{i=1}^{j-2} b_i}, j = 3, 4, \ldots, n.$$

Proof. Let $x_j \neq 0$ for all $j = 1, 2, \ldots, n$. As per the conditions of IEPP, λ_j is an eigenvalue of A_j for each $j = 1, 2, \ldots, n$. Thus, $P_1(\lambda_1) = 0 \Rightarrow a_1 = \lambda_1$.

$$\begin{aligned} &P_j(\lambda_j) = 0 \\ \Rightarrow &(\lambda_j - a_j)P_{j-1}(\lambda_j) - b_{j-1}^2 P_{j-2}(\lambda_j) = 0 \\ \Rightarrow &a_j = \lambda_j - \frac{b_{j-1}^2 P_{j-2}(\lambda_j)}{P_{j-1}(\lambda_j)}, \end{aligned} \tag{6}$$

which gives the expression for calculating a_j. The expression is valid as $P_{j-1}(\lambda_j) \neq 0$, because by Lemma 2, $P_{j-1}(\lambda_j)$ and $P_j(\lambda_j)$ cannot be simultaneously zero.

Now, since (λ_n, X) is an eigenpair of A_n, so by Equation (5), $x_j = \dfrac{P_{j-1}(\lambda_n)}{\prod_{i=1}^{j-1} b_i} x_1$, which implies that

$$b_1 = \frac{x_1}{x_2} P_1(\lambda_n) \text{ and } b_{j-1} = \frac{x_1 P_{j-1}(\lambda_n)}{x_j \prod_{i=1}^{j-2} b_i}. \tag{7}$$

Since $x_j \neq 0$ hence it follows that, $P_{j-1}(\lambda_n) \neq 0$. Hence the above expression for b_{j-1} is valid and $b_{j-1} \neq 0$ for all $j = 2, 3, \ldots, n$. Successive use of Equations (6) and (7) will give us the values of a_j and b_{j-1} for $j = 1, 2, \ldots, n$.

Conversely, if there exists a unique solution for IEPP, then since X is an eigenvector of A_n, so by Lemma 3, $x_1 \neq 0$. The existence of a solution implies that $b_{j-1} \neq 0$ for $j = 2, 3, \ldots, n$. It then follows from the expressions in Equation (7) that $x_j \neq 0$ for $j = 2, 3, \ldots, n$. □

5. Solution of IEPB

Lemma 5. *Let $P(\lambda)$ be a monic polynomial of degree n with all real zeros and λ_{min} and λ_{max} be the minimal and maximal zero of P respectively.*

- *If $\mu < \lambda_{min}$, then $(-1)^n P(\mu) > 0$.*
- *If $\mu > \lambda_{max}$, then $P(\mu) > 0$.*

The proof immediately follows after expressing the polynomial as a product of its linear factors.

Lemma 6. *If T is a tree, then the minimal and maximal eigenvalues of any matrix $A \in S(T)$ are simple i.e., of multiplicity one. [Corollary 6 of Theorem 2 in [17]]*

In other words, this Lemma says that the minimal and maximal eigenvalues of an acyclic matrix are simple. Again, since for each j, the leading principal submatrix A_j corresponds to a tree so by Lemma 6 the minimal and maximal eigenvalues of A_j must be simple *i.e.*, in particular $\lambda_1^{(j)} \neq \lambda_j^{(j)}$.

Lemma 7. *The sequence $\{P_j(\lambda) = det(\lambda I_j - A_j)\}_{j=1}^{n+m}$ of characteristic polynomials of A_j satisfies the following recurrence relations :*

1. $P_1(\lambda) = (\lambda - a_1)$.
2. $P_j(\lambda) = (\lambda - a_j)P_{j-1}(\lambda) - b_{j-1}^2 P_{j-2}(\lambda), 2 \leq j \leq n+1$.
3. $P_{n+j}(\lambda) = (\lambda - a_{n+j})P_{n+j-1}(\lambda) - b_{n+j-1}^2 P_{n-1}(\lambda) \prod_{i=1}^{j-1} (\lambda - a_{n+i}), 2 \leq j \leq m$.

Lemma 8. *For any $\lambda \in \mathbb{R}$ and $1 \leq j \leq n$, $P_j(\lambda)$ and $P_{j+1}(\lambda)$ cannot be simultaneously zero.*

Proof. Same as Lemma 2. □

By Cauchy's interlacing theorem ([14,18]), the eigenvalues of a symmetric matrix and those of any of its principal submatrix interlace each other. Thus, $\lambda_1^{(j)}$'s and $\lambda_j^{(j)}$'s must satisfy :

$$\lambda_1^{(n+m)} \leq \lambda_1^{(n+m-1)} \leq \ldots \leq \lambda_1^{(2)} \leq \lambda_1^{(1)} \leq \lambda_2^{(2)} \leq \lambda_3^{(3)} \leq \ldots \leq \lambda_{n+m-1}^{(n+m-1)} \leq \lambda_{n+m}^{(n+m)}.$$

Each diagonal element a_i is also a 1×1 principal submatrix of A. Hence $\lambda_1^j \leq a_i \leq \lambda_j^{(j)}, 1 \leq i \leq j$. Since $\lambda_1^{(j)}$ and $\lambda_j^{(j)}$ are the minimal and maximal eigenvalues of A_j, so $P_j(\lambda_1^{(j)}) = 0$ and $P_j(\lambda_j^{(j)}) = 0$. We need to solve these equations successively using the recurrence relations in Lemma 1. For $j = 1$, $P_1(\lambda_1^{(1)}) = 0 \Rightarrow a_1 = \lambda_1^{(1)}$. For $j = 2$, $P_2(\lambda_1^{(2)}) = 0, P_2(\lambda_2^{(2)}) = 0$ which imply that

$$a_2 = \frac{\lambda_2^{(2)} P_1(\lambda_2^{(2)}) - \lambda_1^{(2)} P_1(\lambda_1^{(2)})}{P_1(\lambda_2^{(2)}) - P_1(\lambda_1^{(2)})}, b_1^2 = \frac{(\lambda_2^{(2)} - \lambda_1^{(2)}) P_1(\lambda_1^{(2)}) P_1(\lambda_2^{(2)})}{P_1(\lambda_1^{(2)}) - P_1(\lambda_2^{(2)})}.$$

a_2 will always exist as the denominator in the above expression for a_2 can never be zero. We have $\lambda_1^{(2)} \neq \lambda_2^{(2)}$ and so if $P_1(\lambda_2^{(2)}) = P_1(\lambda_1^{(2)})$, then by Rolle's theorem $\exists\, c \in (\lambda_1^{(2)}, \lambda_2^{(2)})$ such that $P_1'(c) = 0 \Rightarrow 1 = 0$, which is not possible. Thus, $P_1(\lambda_2^{(2)}) - P_1(\lambda_1^{(2)}) \neq 0$. Also, by Lemma 5, $(-1)^1 P_1(\lambda_1^{(2)}) \geq 0$ and so the expression for b_1^2 is non-negative and so we can get real values of b_1.

Now for $3 \leq j \leq n$, we have

$$P_j(\lambda_1^{(j)}) = 0, P_j(\lambda_j^{(j)}) = 0,$$

which gives

$$a_j P_{j-1}(\lambda_1^{(j)}) + b_{j-1}^2 P_{j-2}(\lambda_1^{(j)}) - \lambda_1^{(j)} P_{j-1}(\lambda_1^{(j)}) = 0,$$

$$a_j P_{j-1}(\lambda_j^{(j)}) + b_{j-1}^2 P_{j-2}(\lambda_j^{(j)}) - \lambda_j^{(j)} P_{j-1}(\lambda_j^{(j)}) = 0.$$

Let D_j denote the determinant of the coefficient matrix of the above system of linear equations in a_j and b_{j-1}^2. Then $D_j = P_{j-1}(\lambda_1^{(j)}) P_{j-2}(\lambda_j^{(j)}) - P_{j-1}(\lambda_j^{(j)}) P_{j-2}(\lambda_1^{(j)})$. If $D_j \neq 0$, then the system will have a unique solution, given by

$$\begin{aligned} a_j &= \frac{\lambda_1^{(j)} P_{j-1}(\lambda_1^{(j)}) P_{j-2}(\lambda_j^{(j)}) - \lambda_j^{(j)} P_{j-1}(\lambda_j^{(j)}) P_{j-2}(\lambda_1^{(j)})}{D_j}, \\ b_{j-1}^2 &= \frac{(\lambda_j^{(j)} - \lambda_1^{(j)}) P_{j-1}(\lambda_1^{(j)}) P_{j-1}(\lambda_j^{(j)})}{D_j}. \end{aligned} \tag{8}$$

We claim that the expression for b_{j-1}^2 in RHS is non negative. This follows from Lemma 5. Since $\lambda_1^{(j)} \leq \lambda_1^{(j-1)}$ and $\lambda_{j-1}^{(j-1)} \leq \lambda_j^{(j)}$, so by Lemma 5,

$$(-1)^{j-1} D_j = (-1)^{(j-1)} P_{j-1}(\lambda_1^{(j)}) P_{j-2}(\lambda_j^{(j)}) + (-1)^{j-2} P_{j-2}(\lambda_1^{(j)}) P_{j-1}(\lambda_j^{(j)}) \geq 0.$$

In addition, by Lemma 2, $P_j(\lambda_1^{(j)})$ and $P_{j-1}(\lambda_1^{(j)})$ cannot be simultaneously zero. Thus, $P_{j-1}(\lambda_1^{(j)}) \neq 0$. Similarly, $P_{j-1}(\lambda_j^{(j)}) \neq 0$. This implies that $\lambda_1^{(j)} \neq \lambda_1^{(j-1)}$ and $\lambda_j^{(j)} \neq \lambda_{j-1}^{j-1}$. Thus, we can get non-zero real values of b_{j-1} from Equation (8) if and only if $\lambda_1^{(j)} < \lambda_1^{(j-1)}$ and $\lambda_{j-1}^{(j-1)} < \lambda_j^{(j)}$ for all $j = 2, 3, \ldots, n$.

Now, if $D_j = 0$, then $(-1)^{j-1} D_j = 0$ *i.e.,* $(-1)^{(j-1)} P_{j-1}(\lambda_1^{(j)}) P_{j-2}(\lambda_j^{(j)}) + (-1)^{j-2} P_{j-2}(\lambda_1^{(j)}) P_{j-1}(\lambda_j^{(j)}) = 0$. Since both the terms in this sum are non negative, we must have $P_{j-1}(\lambda_1^{(j)}) P_{j-2}(\lambda_j^{(j)}) = 0$ and $P_{j-2}(\lambda_1^{(j)}) P_{j-1}(\lambda_j^{(j)}) = 0$. However, from Lemma 2, $P_{j-1}(\lambda)$ and $P_{j-2}(\lambda)$ cannot be simultaneously zero. In addition, $P_j(\lambda)$ and $P_{j-1}(\lambda)$ cannot be simultaneously zero.

Thus, the only possibility is that $P_{j-2}(\lambda_1^{(j)}) = P_{j-2}(\lambda_j^{(j)}) = 0$. However, this will then imply that $a_j = \lambda_1^{(j)} = \lambda_j^{(j)}$, which is not possible as by Lemma 6 $\lambda_1^{(j)} \neq \lambda_j^{(j)}$. Thus, $D_j \neq 0$ for all $j = 2, 3, \ldots, n$.

Again, $\lambda_1^{(n+j)}$ and $\lambda_{n+j}^{(n+j)}$ are the eigenvalues of A_{n+j} and so $P_{n+j}(\lambda_1^{(n+j)}) = 0$ and $P_{n+j}(\lambda_{n+j}^{(n+j)}) = 0$. Hence,

$$\begin{aligned} &(\lambda_1^{(n+j)} - a_{n+j})P_{n+j-1}(\lambda_1^{(n+j)}) - b_{n+j-1}^2 P_{n-1}(\lambda_1^{(n+j)}) \prod_{i=1}^{j-1}(\lambda_1^{(n+j)} - a_{n+i}) = 0, \\ &(\lambda_{n+j}^{(n+j)} - a_{n+j})P_{n+j-1}(\lambda_{n+j}^{(n+j)}) - b_{n+j-1}^2 P_{n-1}(\lambda_{n+j}^{(n+j)}) \prod_{i=1}^{j-1}(\lambda_{n+j}^{(n+j)} - a_{n+i}) = 0. \end{aligned} \tag{9}$$

so we get a system of equations linear in a_{n+j} and b_{n+j-1}^2.

$$\begin{aligned} &a_{n+j}P_{n+j-1}(\lambda_1^{(n+j)}) + b_{n+j-1}^2 P_{n-1}(\lambda_1^{(n+j)}) \prod_{i=1}^{j-1}(\lambda_1^{(n+j)} - a_{n+i}) = \lambda_1^{(n+j)} P_{n+j-1}(\lambda_1^{(n+j)}), \\ &a_{n+j}P_{n+j-1}(\lambda_{n+j}^{(n+j)}) + b_{n+j-1}^2 P_{n-1}(\lambda_{n+j}^{(n+j)}) \prod_{i=1}^{j-1}(\lambda_{n+j}^{(n+j)} - a_{n+i}) = \lambda_{n+j}^{(n+j)} P_{n+j-1}(\lambda_{n+j}^{(n+j)}). \end{aligned} \tag{10}$$

We first investigate the conditions under which the coefficient matrix of the above system is singular. By Cauchy's interlacing property, we have

$$\lambda_1^{(n+j)} \leq \lambda_1^{(n+j-1)} \leq \ldots \leq \lambda_1^{(n+1)} \leq a_{n+i} \leq \lambda_{n+1}^{(n+1)} \leq \ldots \lambda_{n+j-1}^{(n+j-1)} \leq \lambda_{n+j}^{(n+j)}, \text{ for all } i = 1, 2, \ldots, m. \tag{11}$$

Thus, $\prod_{i=1}^{j-1}(\lambda_{n+j}^{(n+j)} - a_{n+i}) \geq 0$ and $(-1)^{j-1}\prod_{i=1}^{j-1}(\lambda_1^{(n+j)} - a_{n+i}) \geq 0$. Let D_{n+j} be the determinant of the coefficient matrix of Equation (10). Then,

$$\begin{aligned} (-1)^{n+j-1}D_{n+j} =& (-1)^{n+j-1}P_{n+j-1}(\lambda_1^{(n+j)})P_{n-1}(\lambda_{n+j}^{(n+j)}) \prod_{i+1}^{j-1}(\lambda_{n+j}^{(n+j)} - a_{n+i}), \\ &+ (-1)^{n+j-2}P_{n+j-1}(\lambda_{n+j}^{(n+j)})P_{n-1}(\lambda_1^{(n+j)}) \prod_{i=1}^{j-1}(\lambda_1^{(n+j)} - a_{n+i}). \end{aligned}$$

As a consequence of Lemma 5, both the products in the LHS are non-negative and so $(-1)^{n+j-1}D_{n+j} \geq 0$ for all $j = 1, 2, \ldots, m$. Thus, D_{n+j} will vanish if and only if $(-1)^{n+j-1}D_{n+j}$ will vanish *i.e.*, if and only if

$$P_{n+j-1}(\lambda_1^{(n+j)})P_{n-1}(\lambda_{n+j}^{(n+j)}) \prod_{i=1}^{j-1}(\lambda_{n+j}^{(n+j)} - a_{n+i}) = 0$$

and

$$P_{n+j-1}(\lambda_{n+j}^{(n+j)})P_{n-1}(\lambda_1^{n+j}) \prod_{i=1}^{j-1}(\lambda_1^{(n+j)} - a_{n+i}) = 0.$$

If $P_{n-1}(\lambda_1^{(n+j)}) = 0$, then since $\lambda_1^{(n+j)} \leq \lambda_1^{(n+j-1)} \leq \ldots \leq \lambda_1^{(n)} \leq \lambda_1^{(n-1)}$ and $\lambda_1^{(n-1)}$ is the minimum possible zero of P_{n-1}, we get $\lambda_1^{n+j} = \lambda_1^{(n+j-1)} = \ldots = \lambda_1^{(n)} = \lambda_1^{(n-1)}$. Consequently, $P_{n-1}(\lambda_1^{(n+j)}) = P_n(\lambda_1^{(n+j)}) = 0$ but this contradicts Lemma 8, according to which $P_{n-1}(\lambda_1^{(n+j)})$ and $P_n(\lambda_1^{(n+j)})$ cannot be simultaneously zero. Hence, $P_{n-1}(\lambda_1^{(n+j)}) \neq 0$. Similarly, it can be shown that $P_{n-1}(\lambda_{n+j}^{(n+j)}) \neq 0$. Thus there are the following possibilities :

i. $P_{n+j-1}(\lambda_1^{(n+j)}) = 0$ and $P_{n+j-1}(\lambda_{n+j}^{(n+j)}) = 0$.

ii. $P_{n+j-1}(\lambda_1^{(n+j)}) = 0$ and $\prod_{i=1}^{j-1}(\lambda_1^{(n+j)} - a_{n+i}) = 0$.

iii. $P_{n+j-1}(\lambda_{n+j}^{(n+j)}) = 0$ and $\prod_{i=1}^{j-1}(\lambda_{n+j}^{(n+j)} - a_{n+i}) = 0$.

iv. $\prod_{i=1}^{j-1}(\lambda_1^{(n+j)} - a_{n+i}) = 0$ and $\prod_{i=1}^{j-1}(\lambda_{n+j}^{(n+j)} - a_{n+i}) = 0$.

If (i) happens, then, since $b_{n+j-1} \neq 0$, so from the equations in Equation (10), $\prod_{i=1}^{j-1}(\lambda_1^{(n+j)} - a_{n+i}) = 0$ and $\prod_{i=1}^{j-1}(\lambda_{n+j}^{(n+j)} - a_{n+i}) = 0$. This implies that $a_{n+i} = \lambda_1^{(n+j)}$ for some $i = 1, 2, \ldots, m$ and $a_{n+i} = \lambda_{n+j}^{(n+j)}$ for some $i = 1, 2, \ldots, m$. However, as per the inequality Equation (11), it then follows that $\lambda_1^{(n+j)} = \lambda_1^{(n+j-1)} = \ldots = \lambda_1^{(n+1)}$ and $\lambda_{n+1}^{(n+1)} = \lambda_{n+2}^{(n+2)} = \ldots = \lambda_{n+j}^{(n+j)}$. Since $P_{n+2}(\lambda_1^{(n+2)}) = 0$ and $P_{n+1}(\lambda_1^{(n+1)}) = 0$, so the above equalities imply that $P_{n+2}(\lambda_1^{(n+j)}) = 0$ and $P_{n+1}(\lambda_1^{(n+j)}) = 0$. Hence from the recurrence relation (3) of Lemma 7, we get

$$(\lambda_1^{(n+j)} - a_{n+1})P_{n+1}(\lambda_1^{(n+j)}) - b_{n+1}^2 P_{n-1}(\lambda_1^{(n+j)})(\lambda_1^{(n+j)} - a_{n+1}) = 0$$

which implies that $\lambda_1^{(n+j)} = a_{n+1}$. Similarly, it will follow that $\lambda_{n+j}^{(n+j)} = a_{n+1}$. However, $\lambda_1^{(n+j)} \leq a_{n+1} \leq \lambda_{n+j}^{(n+j)}$ and so $\lambda_1^{(n+j)} = \lambda_{n+j}^{(n+j)}$, but this is not possible as $\lambda_1^{(n+j)}$ and $\lambda_{n+j}^{(n+j)}$ are the minimal and maximal eigenvalues of the acyclic matrix A_{n+j} and by Lemma 6, the minimal and maximal eigenvalues of an acyclic matrix are simple. Hence (i) cannot hold. From the above arguments, it also follows that (iv) cannot hold.

If (ii) holds, then the augmented matrix of the system of Equation (10) will be of rank one and so the system will have infinite number of solutions. Similarly, if (iii) holds, then the system will have infinite number of solutions. However, if we put the additional constraint that $\lambda_1^{(n+j)} < \lambda_1^{(n+j-1)}$ and $\lambda_{n+j-1}^{(n+j-1)} < \lambda_{n+j}^{(n+j)}$ for all $j = 2, 3, \ldots, m$ then $P_{n+j-1}(\lambda_1^{(n+j)}) \neq 0$ and $P_{n+j-1}(\lambda_{n+j}^{(n+j)}) \neq 0$, so that (ii) and (iii) will not hold.

Thus, we see that $D_{n+j} \neq 0$ if and only if

$$\lambda_1^{(n+j)} < \lambda_1^{(n+j-1)} < \ldots < \lambda_1^{(n+1)} < \lambda_{n+1}^{(n+1)} < \ldots < \lambda_{n+j-1}^{(n+j-1)} < \lambda_{n+j}^{(n+j)}, \text{ for all } i = 1, 2, \ldots, m$$

Under this constraint, the unique solution of the system Equation (10) is given by

$$a_{n+j} = \frac{A_j - B_j}{D_{n+j}}, b_{n+j-1}^2 = \frac{(\lambda_{n+j}^{(n+j)} - \lambda_1^{(n+j)})P_{n+j-1}(\lambda_1^{(n+j)})P_{+j-1}(\lambda_{n+j}^{(n+j)})}{D_{n+j}}, \tag{12}$$

where $A_j = \lambda_1^{(n+j)} P_{n+j-1}(\lambda_1^{(n+j)}) P_{n-1}(\lambda_{n+j}^{(n+j)}) \prod_{i=1}^{j-1}(\lambda_{n+j}^{(n+j)} - a_{n+i})$ and $B_j = \lambda_{n+j}^{(n+j)} P_{n+j-1}(\lambda_{n+j}^{(n+j)}) P_{n-1}(\lambda_1^{(n+j)}) \prod_{i=1}^{j-1}(\lambda_1^{(n+j)} - a_{n+i})$.

$$b_{n+j-1}^2 = \frac{(\lambda_{n+j}^{(n+j)} - \lambda_1^{(n+j)})P_{n+j-1}(\lambda_1^{(n+j)})P_{+j-1}(\lambda_{n+j}^{(n+j)})}{D_{n+j}}. \tag{13}$$

The above analysis of the IEP can be framed as the following theorem :

Theorem 9. *The **IEPB** has a solution if and only if*

$$\lambda_1^{(n+m)} < \lambda_1^{(n+m-1)} < \ldots < \lambda_1^{(2)} < \lambda_1^{(1)} < \lambda_2^{(2)} < \ldots < \lambda_{n+m-1}^{(n+m-1)} < \lambda_{n+m}^{(n+m)}$$

and the solution is given by

$$a_1 = \lambda_1^{(1)}, a_2 = \frac{\lambda_2^{(2)} P_1(\lambda_2^{(2)}) - \lambda_1^{(2)} P_1(\lambda_1^{(2)})}{P_1(\lambda_2^{(2)}) - P_1(\lambda_1^{(2)})}, b_1^2 = \frac{(\lambda_2^{(2)} - \lambda_1^{(2)}) P_1(\lambda_1^{(2)}) P_1(\lambda_2^{(2)})}{P_1(\lambda_1^{(2)}) - P_1(\lambda_1^{(2)})},$$

$$a_j = \frac{\lambda_1^{(j)} P_{j-1}(\lambda_1^{(j)}) P_{j-2}(\lambda_j^{(j)}) - \lambda_j^{(j)} P_{j-1}(\lambda_j^{(j)}) P_{j-2}(\lambda_1^{(j)})}{P_{j-1}(\lambda_1^{(j)}) P_{j-2}(\lambda_j^{(j)}) - P_{j-1}(\lambda_j^{(j)}) P_{j-2}(\lambda_1^{(j)})}, j = 3, 4, \ldots, n+1,$$

$$b_{j-1}^2 = \frac{(\lambda_j^{(j)} - \lambda_1^{(j)}) P_{j-1}(\lambda_1^{(j)}) P_{j-1}(\lambda_j^{(j)})}{P_{j-1}(\lambda_1^{(j)}) P_{j-2}(\lambda_j^{(j)}) - P_{j-1}(\lambda_j^{(j)}) P_{j-2}(\lambda_1^{(j)})}, j = 3, 4, \ldots, n+1,$$

$$a_{n+j} = \frac{A_j - B_j}{D_{n+j}}, j = 1, 2, \ldots, m,$$

$$b_{n+j-1}^2 = \frac{(\lambda_{n+j}^{(n+j)} - \lambda_1^{(n+j)}) P_{n+j-1}(\lambda_1^{(n+j)}) P_{+j-1}(\lambda_{n+j}^{(n+j)})}{D_{n+j}}, j = 2, 3, \ldots, m.$$

The solution is unique except for the signs of the non-zero off-diagonal entries.

6. Numerical Examples

We apply the results obtained in the previous section to solve the following :

Example 1. *Given 7 real numbers $\lambda_1 = 1, \lambda_2 = 5, \lambda_3 = -4, \lambda_4 = 3, \lambda_5 = 9, \lambda_6 = -8, \lambda_7 = -3$ and a real vector $X = (-2, 5, -7, 3, 1, 4, 8)^T$, find a matrix $A_7 \in S(P_7)$ such that λ_j is an eigenvalue of A_j for each $j = 1, 2, \ldots, 7$ and (λ_7, X) is an eigenpair of A_7.*

Solution Using Theorem 1, we obtain the following matrix as the solution :

$$A_7 = \begin{pmatrix} 1 & 1.6 & 0 & 0 & 0 & 0 & 0 \\ 1.6 & 4.36 & 4.8 & 0 & 0 & 0 & 0 \\ 0 & 4.8 & -1.0642 & -3.4832 & 0 & 0 & 0 \\ 0 & 0 & -3.4832 & 2.0515 & -39.5368 & 0 & 0 \\ 0 & 0 & 0 & -39.5368 & -347.5644 & 115.7937 & 0 \\ 0 & 0 & 0 & 0 & 115.7937 & -32.6379 & 0.3448 \\ 0 & 0 & 0 & 0 & 0 & 0.3448 & -3.1724 \end{pmatrix}$$

The eigenvalues of the all the leading principal submatrices are :

$\sigma(A_1) = \{\mathbf{1}\}$

$\sigma(A_2) = \{0.3600, \mathbf{5.0000}\}$

$\sigma(A_3) = \{\mathbf{-4.0000}, 0.8364, 7.4594\}$

$\sigma(A_4) = \{-5.2369, 0.5809, \mathbf{3.0000}, 8.0033\}$

$\sigma(A_5) = \{-351.9801, -4.7789, 0.7321, 5.8098, \mathbf{9.0000}\}$

$\sigma(A_6) = \{-389.1678, \mathbf{-8.0000}, -3.1028, 0.8636, 7.4210, 18.1310\}$

$\sigma(A_7) = \{-389.1678, -8.0084, -3.2694, \mathbf{-3.0000}, 0.8638, 7.4211, 18.1333\}$

Example 2. *Given 13 real numbers $0.5, 1, -1.4, 2, -2.2, 3, -3.8, 4.7, -4.4, 5, -6, 6, 7$, rearrange and label them as $\lambda_1^{(j)}, 1 \leq j \leq 7$ and $\lambda_j^{(j)}, 2 \leq j \leq 7$ and find a matrix $A_{4+3} \in S(B_{4+3})$ such that λ_1^j and $\lambda_j^{(j)}$ are the minimal and maximal eigenvalues of A_j, the jth leading principal sub matrix of A_{4+3}.*

Solution Using Theorem 9, we rearrange the numbers in the following way

$$\lambda_1^{(7)} < \lambda_1^{(6)} < \lambda_1^{(5)} < \lambda_1^{(4)} < \lambda_1^{(3)} < \lambda_1^{(2)} < \lambda_1^{(1)} < \lambda_2^{(2)} < \lambda_3^{(3)} < \lambda_4^{(4)} < \lambda_5^{(5)} < \lambda_6^{(6)} < \lambda_7^{(7)}$$

i.e.,

$$-6 < -4.4 < -3.8 < -2 < -1.4 < 0.5 < 1 < 2 < 3 < 4.7 < 5 < 6 < 7.$$

Then, using the expressions for a_j, b_{j-1}^2, a_{n+j} and b_{n+j-1}^2 we get

$$A_{4+3} = \begin{pmatrix} 1.0000 & 0.7071 & 0 & 0 & 0 & 0 & 0 \\ 0.7071 & 1.5000 & 1.9380 & 0 & 0 & 0 & 0 \\ 0 & 1.9380 & -0.0047 & 2.1580 & 0 & 0 & 0 \\ 0 & 0 & 2.1580 & 3.3615 & 1.8991 & 3.1884 & 3.7247 \\ 0 & 0 & 0 & 1.8991 & -3.1612 & 0 & 0 \\ 0 & 0 & 0 & 3.1884 & 0 & -1.5823 & 0 \\ 0 & 0 & 0 & 3.7247 & 0 & 0 & -3.1925 \end{pmatrix}.$$

Here we have taken all the b_is as positive. We can take some of the b_is as negative also. In fact, we can construct 2^6 such matrices for the above problem, the only difference being in the signs of the non-zero off-diagonal entries.

We compute the spectra of all the leading principal submatrices of A_{4+3} to verify the the conditions of the **IEPB** are satisfied. The minimal and maximal eigenvalues of each principal submatrix are shown in bold.

$\sigma(A_7) = \{\mathbf{-6.0000}, -3.1677, -2.1477, -1.4356, 0.8594, 2.8124, \mathbf{7.0000}\}$
$\sigma(A_6) = \{\mathbf{-4.4000}, -2.6174, -1.4386, 0.8449, 2.7245, \mathbf{6.0000}\}$
$\sigma(A_5) = \{\mathbf{-3.8000}, -1.8025, 0.7966, 2.5016, \mathbf{5.0000}\}$
$\sigma(A_4) = \{\mathbf{-2.0000}, 0.7714, 2.3854, \mathbf{4.7000}\}$
$\sigma(A_3) = \{\mathbf{-1.4000}, 0.8953, \mathbf{3.0000}\}$
$\sigma(A_2) = \{\mathbf{0.5000}, \mathbf{2.0000}\}$
$\sigma(A_1) = \{\mathbf{1}\}$

7. Conclusions

The inverse eigenvalue problems discussed in this paper require the construction of specially structured matrices from mixed eigendata. The results obtained here provide an efficient way to construct such matrices from given set of some of the eigenvalues of leading principal submatrices of the required matrix.

The problems IEPP and IEPB are significant in the sense that they are partially described inverse eigenvalue problems *i.e.*, they require the construction of matrices from partial information of eigenvalues and eigenvectors. Such partially described problems may occur in computations involving a complicated physical system where it is often difficult to obtain the entire spectrum. Many times, only the minimal and maximal eigenvalues are known in advance. Thus, the study of inverse problems having such prescribed eigen structure are significant. It would be interesting to consider such IEPs for other acyclic matrices as well.

Acknowledgments: The authors are grateful to the anonymous reviewers for the valuable comments and suggestions.

Author Contributions: Both the authors have contributed equally to the work.

Conflicts of Interest: The authors declare no conflict of interest.

References

1. Chu, M.T. Inverse eigenvalue problems. *SIAM Rev.* **1998**, *40*, 1–39.
2. Ghanbari, K. m-functions and inverse generalized eigenvalue problem. *Inverse Prob.* **2001**, *17*, 211–217.
3. Zhang, Y. On the general algebraic inverse eigenvalue problems. *J. Comput. Math.* **2004**, *22*, 567–580.
4. Ehlay, S.; Gladwell, G.M.; Golub, G.H.; Ram,Y.M. On some eigenvector-eigenvalue relations. *SIAM J. Matrix Anal. Appl.* **1999**, *20*, 563–574.
5. Peng, J.; Hu, X.-Y.; Zhang, L. Two inverse eigenvalue problems for a special kind of matrices. *Linear Algebra Appl.* **2006**, *416*, 336–347.

6. Ghanbari, K.; Parvizpour, F. Generalized inverse eigenvalue problem with mixed eigen data. *Linear Algebra Appl.* **2012**, *437*, 2056–2063.
7. Pivovarchik, V.; Rozhenko, N.; Tretter, C. Dirichlet-Neumann inverse spectral problem for a star graph of Stieltjes strings. *Linear Algebra Appl.* **2013**, *439*, 2263–2292.
8. Pivovarchik, V.; Tretter, C. Location and multiples of eigenvalues for a star graph of Stieltjes strings. *J. Differ. Equ. Appl.* **2015**, *21*, 383–402.
9. Duarte, A.L. Construction of acyclic matrices from spectral data. *Linear Algebra Appl.* **1989**, *113* , 173–182.
10. Duarte, A.L.; Johnson, C.R. On the minimum number of distinct eigenvalues for a symmetric matrix whose graph is a given tree. *Math. Inequal. Appl.* **2002**, *5*, 175–180.
11. Nair, R.; Shader, B.L. Acyclic matrices with a small number of distinct eigenvalues. *Linear Algebra Appl.* **2013**, *438*, 4075–4089.
12. Monfared, K.H.; Shader, B.L. Construction of matrices with a given graph and prescribed interlaced spectral data. *Linear Algebra Appl.* **2013**, *438*, 4348–4358.
13. Sen, M.; Sharma, D. Generalized inverse eigenvalue problem for matrices whose graph is a path. *Linear Algebra Appl.* **2014**, *446*, 224–236.
14. Hogben, L. Spectral graph theory and the inverse eigenvalue problem of a graph. *Electron. J. Linear Algebra* **2005**, *14*, 12–31.
15. Nylen, P.; Uhlig, F. Inverse Eigenvalue Problems Associated With Spring-Mass Systems. *Linear Algebra Appl.* **1997**, *254*, 409–425.
16. Gladwell, G.M.L. *Inverse Problems in Vibration*; Kluwer Academic Publishers: Dordrecht, The Netherlands, 2004.
17. Johnson, C.R.; Duarte, A.L.; Saiago, C.M. The Parter Wiener theorem: Refinement and generalization. *SIAM J. Matrix Anal. Appl.* **2003**, *25*, 352–361.
18. Horn, R.; Johnson, C.R. *Matrix Analysis*; Cambridge University Press: New York, NY, USA, 1985.

Permissions

All chapters in this book were first published in Mathematics, by MDPI; hereby published with permission under the Creative Commons Attribution License or equivalent. Every chapter published in this book has been scrutinized by our experts. Their significance has been extensively debated. The topics covered herein carry significant findings which will fuel the growth of the discipline. They may even be implemented as practical applications or may be referred to as a beginning point for another development.

The contributors of this book come from diverse backgrounds, making this book a truly international effort. This book will bring forth new frontiers with its revolutionizing research information and detailed analysis of the nascent developments around the world.

We would like to thank all the contributing authors for lending their expertise to make the book truly unique. They have played a crucial role in the development of this book. Without their invaluable contributions this book wouldn't have been possible. They have made vital efforts to compile up to date information on the varied aspects of this subject to make this book a valuable addition to the collection of many professionals and students.

This book was conceptualized with the vision of imparting up-to-date information and advanced data in this field. To ensure the same, a matchless editorial board was set up. Every individual on the board went through rigorous rounds of assessment to prove their worth. After which they invested a large part of their time researching and compiling the most relevant data for our readers.

The editorial board has been involved in producing this book since its inception. They have spent rigorous hours researching and exploring the diverse topics which have resulted in the successful publishing of this book. They have passed on their knowledge of decades through this book. To expedite this challenging task, the publisher supported the team at every step. A small team of assistant editors was also appointed to further simplify the editing procedure and attain best results for the readers.

Apart from the editorial board, the designing team has also invested a significant amount of their time in understanding the subject and creating the most relevant covers. They scrutinized every image to scout for the most suitable representation of the subject and create an appropriate cover for the book.

The publishing team has been an ardent support to the editorial, designing and production team. Their endless efforts to recruit the best for this project, has resulted in the accomplishment of this book. They are a veteran in the field of academics and their pool of knowledge is as vast as their experience in printing. Their expertise and guidance has proved useful at every step. Their uncompromising quality standards have made this book an exceptional effort. Their encouragement from time to time has been an inspiration for everyone.

The publisher and the editorial board hope that this book will prove to be a valuable piece of knowledge for researchers, students, practitioners and scholars across the globe.

List of Contributors

Jean-Pierre Magnot
Lyc'ee Jeanne d'Arc, Avenue de Grande Bretagne, F-63000 Clermont-Ferrand, France

Kyunghyun Baek
Department of Physics, Sogang University, Mapo-gu, Shinsu-dong, Seoul 121-742, Korea

Wonmin Son
Department of Physics, Sogang University, Mapo-gu, Shinsu-dong, Seoul 121-742, Korea
Department of Physics, University of Oxford, Parks Road, Oxford OX1 3PU, UK

Seddik Ouakkas and Djelloul Djebbouri
Laboratory of Geometry, Analysis, Control and Applications, University de Saida, BP138, En-Nasr, 20000 Saida, Algeria

Dimplekumar N. Chalishajar
Department of Applied Mathematics, Virginia Military Institute (VMI), 431 Mallory Hall, Lexington, VA 24450, USA

Kulandhivel Karthikeyan
Department of Mathematics, KSR College of Technology, Tiruchengode 637215, India

Annamalai Anguraj
Department of Mathematics, PSG College of Arts and Science, Coimbatore 641014 , India

Blake C. Stacey
Physics Department, University of Massachusetts Boston, 100 Morrissey Boulevard, Boston, MA 02125, USA

H. M. Abdelhafez
Department of Physics and Engineering Mathematics, Faculty of Electronic Engineering, Menoufia University, Menouf 32952, Egypt

Danhua Wang and Gang Li and Biqing Zhu
College of Mathematics and Statistics, Nanjing University of Information Science and Technology, Nanjing 210044, China

Andronikos Paliathanasis
Instituto de Ciencias Físicas y Matemáticas, Universidad Austral de Chile, Valdivia 5090000, Chile

Richard M. Morris
Department of Mathematics, Institute of Systems Science, Durban University of Technology, PO Box 1334, Durban 4000, South Africa

Peter G. L. Leach
Department of Mathematics, Institute of Systems Science, Durban University of Technology, PO Box 1334, Durban 4000, South Africa
School of Mathematics, Statistics and Computer Science, University of KwaZulu-Natal, Private Bag X54001, Durban 4000, South Africa
Department of Mathematics and Statistics, University of Cyprus, Lefkosia 1678, Cyprus

Meysam Alvan
Department of mathematics, Central Tehran Branch Islamic Azad university, Tehran 13185/768, Iran

Rahmat Darzi
Department of Mathematics, Neka Branch Islamic Azad University, Neka 48411-86114, Iran

Amin Mahmoodi
Department of mathematics, Central Tehran Branch Islamic Azad univesity, Tehran 13185/768, Iran

Tohru Morita
Graduate School of Information Sciences, Tohoku University, Sendai 980-8577, Japan

Ken-ichi Sato
College of Engineering, Nihon University, Koriyama 963-8642, Japan

Hong-Yan Xu and Hua Wang
Department of Informatics and Engineering, Jingdezhen Ceramic Institute, Jingdezhen 333403, Jiangxi, China

Phong Le
Department of Mathematics and Computer Science, Goucher College, Baltimore, MD 21204, USA

Sunil Chetty
Department of Mathematics, College of Saint Benedict and Saint John's University, Collegeville, MN 56321, USA

Kalyanasundaram Madhu and Jayakumar Jayaraman
Department of Mathematics, Pondicherry Engineering College, Pondicherry 605014, India

Debashish Sharma
Department of Mathematics, Gurucharan College, College Road, Silchar 788004, India

Mausumi Sen
Department of Mathematics, National Institute of Technology Silchar, Silchar 788010, India

Index

www.ingramcontent.com/pod-product-compliance
Lightning Source LLC
Chambersburg PA
CBHW061959190326
41458CB00009B/2917
9781632385741